HYDRODYNAMIC MODELS IN URBAN DRAINAGE:

APPLICATION AND CALIBRATION.

This thesis is a personal enterprise throughout.

Colofon:

Cover: "De Stad" by Timo.A. Clemens

Author: François H.L.R. Clemens
Prinses Irenelaan 7
7437 XW Bathmen (The Netherlands)
Tel: +31-(0)570-541081
E-mail: F.H.L.RClemens@freeler.nl or F.H.L.R.Clemens@citg.tudelft.nl

HYDRODYNAMIC MODELS IN URBAN DRAINAGE:

APPLICATION AND CALIBRATION

Proefschrift

ter verkrijging van de graad van doctor
aan de Technische Universiteit van Delft,
op gezag van de Rector Magnificus Prof.ir. K.F. Wakker
voorzitter van het College voor Promoties,
in het openbaar te verdedigen

op dinsdag 13 maart 2001 te 13.30 uur

door

François Henri Léon Raymond CLEMENS

Civiel ingenieur
Geboren te Leiden

Dit proefschrift is goedgekeurd door de promotor:
Prof.Dr.Ir. G.S. Stelling

Samenstelling promotiecommissie:

Rector Magnificus	Voorzitter
Prof. Dr. Ir. G.S. Stelling	Technische Universiteit Delft, promotor
Prof. Dr. M. Donze	Technische Universiteit Delft
Prof. Dr. R. K. Price	IHE, Delft
Prof. Dr. Ir. C.B. Vreugdenhil	Universiteit Twente
Prof. Dr. Ir. A.W. Heemink	Technische Universiteit Delft
Prof. Dr. Ir. J. Berlamont	Katholieke Universiteit Leuven, België

Published and distributed by: DUP Science

DUP Science is an imprint of
Delft University Press
P.O. Box 98
2600 MG Delft
The Netherlands
Telephone: +31 15 2785121
Telefax: + 31 15 2781661
E-mail: DUP@Library.TUDelft.NL

ISBN 90-407-2163-7

Keywords: urban drainage, hydrodynamic modelling, calibration, monitoring

Printed in The Netherlands

'It was said of them in older times
that the child is the father of the man.
And yet in growing
knows not the man, and of man's ways.
He watches the aged one's
old and seemingly wise, and wonders;
could he ever grow to be such as they,
ageless in passing as the wind and the
timeless tide'.

Soliloguy I, C. Simpson

ACKNOWLEDGEMENT

Writing this thesis was induced by a number of people, whose remarks initiated me to get started somewhere in December 1998. First of all Professor Guus Stelling, my promotor, I appreciated every minute of our discussions and hope to get the opportunity to extend our co-operation in the future. Theo Olsthoorn who gave the starting signal for this thesis while we discussed some early work I did on calibration, I feel we demonstrated the benefits co-operation between 'Urban drainage' and the 'Drinking water world'.

Professors Jan Kop and Jan Wiggers, my old masters, apart from introducing me in the craft of civil engineering you also infected me with passion for engineering. Professor Marcel Donze, I enjoy the manner in which you sharpen my views.

Wicher Worst, I anticipate on continuing our discussions ranging from science to religion. Jaap Voorhoeve who had the courage to allow me all the freedom I needed in my work. Adriaan Posthumus who shows a real interest in the people he works with and who doesn't stop warning me on the possible hazards of working to hard.

During the last 14 years I had the pleasure to work with a large number of fellow 'Urban drainagists' getting innumerous projects done.
First of all I mention Harry van Mameren who, apart from being 'partner in crime' in almost every crazy idea I launched, is a good friend and appreciated colleague. Marcel Boomgaard; it is fun to work with you on the algorithmic aspects of your great-great-great-grand-children and their impact on urban drainage. Maryon Reydon; a pleasant colleague and the living proof of a person being able to grow above herself.

Paul van Berkum (municipality of Apeldoorn) and Bert Groosjohan (municipality of Deventer) gave permission to use data and measuring results for the examples and the case study in this thesis.

Furthermore; all the colleagues of the department of Urban Drainage at Witteveen + Bos, it was tedious to guide you, but I wouldn't have missed it. Luckily there was Ingeborg Smids, calmly ignoring the formal rules, being a great little lady and an appreciated colleague. Riet Kleverwal; thanks for saving me from the many falling pits in our little bureaucratic world. Rob Veldkamp, Jeroen Langeveld and Hans Korving; it is fun working together, I thank you for the many laughs, discussions and ideas we had and will have.
I had the pleasure to guide several students through their B.Sc or M.Sc. thesis or an apprenticeship, I still wonder who learnt most during our sessions.

In my personal life there is a number of people who supported me during every phase, joyous as well as during the black period of my illness in 1996. The professional and personal concern displayed by J. van Bemmel, N. van Adrichem, F. Meima, B. Nobel and B. Boverhof are invaluable. My parents who conducted the formidable task of raising a child; I am indebted to you. Not many people have the luxury of having a spare couple of parents; Wim and Joke, the both of you did a great job also. My parents in law, I thank you for your support and for lending me your daughter.

Then there is Eric, my old friend, we shared everything between hilarious drunkenship and real fear of death and lived to see the best of ourselves. My beloved wife Anja, since I cannot express what I really would say to you here, I just thank you for not pulling the electrical fuses when I was writing on this thesis during almost every spare moment; I promise not to start on a second one.

Als laatste, maar zeker niet de minste, mijn zoon Timo; bedankt voor het plezier dat je me geeft door er gewoon te zijn. Jouw bijdrage aan mijn proefschrift is niet goed uit te leggen. Afgezien van het feit dat ik veel te weinig tijd voor je had tijdens het schrijven, heb ik een van je eerste intellectuele uitingen misbruikt. De stelling '*Hoe slimmer de vis, hoe duurder hij is*', is helemaal van jou en staat, zoals ik je beloofde, nu een boek.([1])

François Clemens
Bathmen 26 januari 2001

([1]) For those who can't understand this section written in Dutch I will translate the essence of it, since it may become an important issue in Dutch literature. The proposition between quotation marks can be read as 'The smarter the fish, the more expensive it is'. This important observation was made by my son Timo at the age of 4 years; therefore it is his intellectual property.

TABLE OF CONTENTS

CHAPTER 1: INTRODUCTION

1.1 The function of sewer systems

Sewer systems have two main functions; collecting and transporting wastewater and storm water out of urban areas. The collection and transport of wastewater is initiated by the urge to maintain a high level of public hygiene. Combined with a high quality of drinking water this results in an important reduction of the occurrence of infectious diseases (see Section 1.2). Though other methods exist to remove wastewater and storm water out of urban areas the application of underground systems (sewer systems) is by far the most commonly applied. The diversion of storm water, for instance, can be realised using other technologies like infiltration in groundwater or diversion via ditches and open channels, see e.g. Monster & Leeflang (1996). Main reasons from a historic perspective to use an underground system are:

- The fact that hardly any space was and is available in existing urban areas to create an extensive system of open channels.
- The fact that open channels sometimes are used to dispose solid waste, resulting in unhygienic situations (Figure 1).
- Infiltration in ground water is not always possible in a technical sense.

Figure 1: Open ditch in an urban area in Djakarta (Photo by J. Kraaij).

During the last decade some of the alternative systems have been built in newly developed urban areas. The main argument to do this is to avoid interference with the existing water system. Traditional drainage systems contribute to an accelerated discharge of water out of an area and in this manner contribute to a disturbance of natural groundwater levels. This eventually leads to a change in the ecology of the area. Therefore, it is tried to erect new urban areas while maintaining the natural hydrological conditions as much as possible. Experience with maintenance of these systems and the possible effects on public health are limited though. It is known in general that living in a damp environment may cause a significantly higher risk of chronic diseases, see e.g. Dijckmeester (1988), the effects however, can only be felt after a relatively long period of time (5 to 10 years).
With respect to the diversion of wastewater alternative solutions to the concept of a drainage system are:

- Cesspools.
- System of regular collection (in analogy to the collection of solid wastes from households in urban areas).
- Local treatment in small scale treatment facilities. (Ministry of Housing, Land-use Planning and the Environment of the Netherlands (VROM) (1988)).

The first two possibilities mentioned have been in use, see e.g. Koot (1979), but proved to be inferior in urban areas compared to large-scale underground urban collecting systems. The possibility of local treatment is implemented in the Netherlands to reduce the environmental impact of the last remaining free discharges of wastewater in less densely populated areas.
In most of urban areas, however, the classical concept of urban drainage, a gravity drainage system, is applied. In this thesis the discussion is limited to gravity systems (in which the water is transported by gravitational force), four systems are distinguished:

- The combined sewer system.
- The improved combined sewer system.
- The separate system.
- The improved separate system.

In Figure 2 the essentials of these systems are shown. The combined system is the predominant type applied. In this system both wastewater and storm water are collected and transported in one system of pipes and manholes. The combined system is popular since it is relatively cheap to construct, relatively easy to maintain and is robust in its functioning. Its disadvantage however, is that relatively clean storm water is mixed with wastewater. During storm events this mixture is transported to a Waste Water Treatment Plant (WWTP). In most cases a WWTP suffers from a loss in removal efficiency when diluted wastewater is entering the plant. Moreover, when a pressurised transporting system between drainage system and WWTP is present, during storms the raw wastewater in the transporting system is entering the WWTP at maximum discharge. This also will lead to a loss in removal efficiency of the plant. A second drawback of the combined system is caused by sewer overflows. During heavy rainfall occasionally a mixture of storm water and wastewater is diverted to surface water bodies via sewer overflows. (In literature the acronym CSO, Combined Sewer Overflows is used). This may result in a water quality problem. A reduction of this problem can be achieved using improved combined sewer systems. In these systems an extra storage capacity is realised in series with an overflow construction. This facility reduces the frequency of spills to surface water. Furthermore, their ecological impact is also reduced by separation and removal of gross solids (floatable as well as settleable fractions). Facilities like Storage Settling Tanks (see e.g. NWRW (1989)), Storage Settling sewers (see i.e. Province of Utrecht (1990)), Swirl separators (see e.g. Field (1974)), Parallel Storage Settling Sewers (e.g. Kollen & Clemens, 1991) are applied on a large scale in order to upgrade existing combined systems. A recent summary of possible constructions is found in Berlamont (1998).

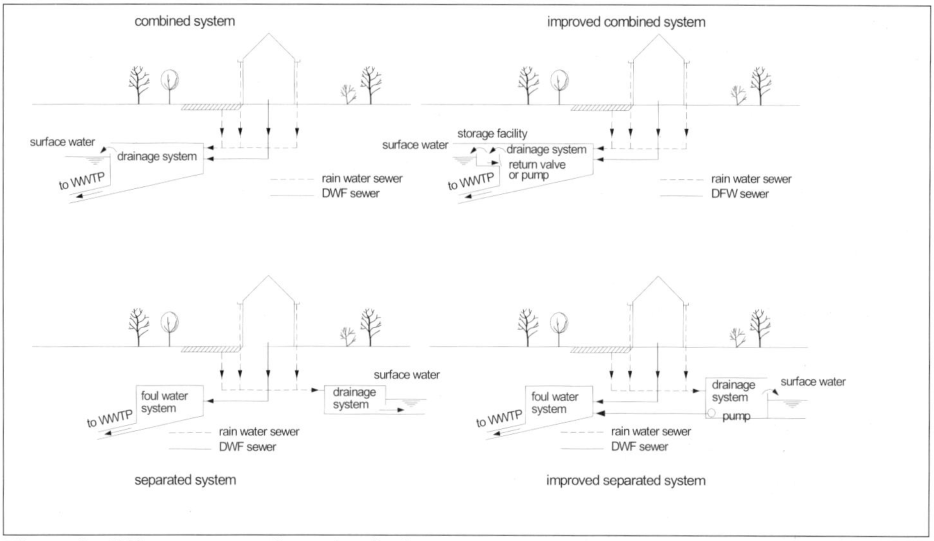

Figure 2: different types of gravity drainage systems.

The essential drawbacks of the combined system are circumvented when using a separate sewer system. In this system the wastewater and the storm water are collected by, and transported in, two separate systems. These systems were widely introduced in the early 1970's; the wastewater is transported to a WWTP while the storm water is directly discharged into surface waters. In practice though, faulty connections occur; wastewater discharges into the storm water system and vice versa. It is known that about 5% of the individual connections into a separated system are faulty. This number increases in time due to changes made in indoor sewer systems by house owners. For this reason, and because of the fact that storm water is contaminated with wastes from paved areas before entering the sewer system, the improved separated system was introduced. In this system the first amount of storm water running of from streets is transported to the WWTP with the wastewater. The pumping capacity of improved separated systems is about half that of a combined system, thus the problem of temporary reduction of the efficiency of WWTP's is reduced significantly. So the improved sewer system is the preferred type by water authorities since the mid- 1980's, though it causes an increase in investments of urban drainage by about a factor of 1.6 when compared with combined systems. In Table 1 the percentage in operation of the different types of system in the Netherlands is shown (Figures obtained from the Foundation RIONED, 2000). The combined system is still the predominant type.

system	2000
combined	72 %
(improved) separated	28 %

Table 1: Gravity systems in the Netherlands in 2000.

1.2 Brief history

The use of underground collecting systems for wastewater as well as for storm water is known to exist at least since the Roman Empire. One of the main sewers in Rome, the world famous

Cloaca Maxima, functioned until the end of the 1980's while constructed more than 2000 years ago, see e.g. Koot (1979). Archaeological research in Pompeii has revealed much of the ingenuity of the Romans. Dicker (1993) is recommended for further reading on this subject. The reason why the Romans already built large scale, expensive, underground infrastructure may only be guessed at. It may well be they already knew of the possible hazards that accumulation of wastes in urban areas may pose. During the Middle Ages the situation in urban areas in Europe with respect to hygiene was in a deplorable state; as is commonly known frequent pandemics of the plague evolved. During this period the construction of drainage systems is started in some cities, though not on a large scale. In the 19th century, when medical science made swift progress, this situation changed. English medical doctors suspected that hygiene was linked to the occurrence of epidemics. One of the first scientific demonstrations of the link between the absence of adequate drinking water, a proper disposal of waste water and the occurrence of epidemic diseases was given by Snow (1854). He showed that a public water well in the city of London was contaminated with bacteria due to leaking cesspools resulting in a minor epidemic of cholera. This relation was commonly recognised in the second half of the 19th century and large sewer systems were constructed in about every major city in Europe.So sewage was collected and transported out of urban areas. It was commonly dumped in some nearby stream, preferably downstream of the city. In the Netherlands early sewer constructions have existed since the middle ages (e.g. Deventer, Arnhem, Maastricht). Large-scale construction of modern systems in urban areas started by the end of the 19th century until the 1930's. During the years after the Second World War, especially during the period 1950-1970, urbanisation accelerated, resulting in large scale construction of combined sewer systems (see Figure 3).

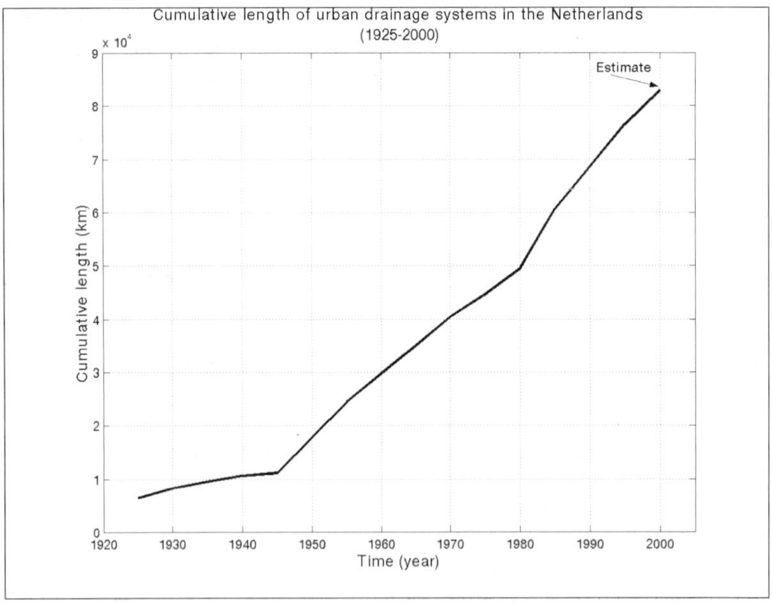

Figure 3: Development of the cumulative length of urban drainage systems in the Netherlands in the period 1925-2000), Foundation RIONED (1999).

In the 19th century it was considered sufficient to divert the wastewater from the urban area into surface waters. This was acceptable because the self-cleansing capacity of these waters

was sufficient to avoid hindrance to the majority of inhabitants. This changed rapidly when population grew and the water quality deteriorated on a large scale causing these waters to be unfit for use as fishing water or process water for industries. In this sense it was realised that water quality has also an economic value.
By the end of the 19th century it was seen and realised that free discharge of sewage can have serious adverse effect on water quality, leading to anaerobic conditions and undesirable odours. An increasing amount of biological degradable waste dumped into surface water led to studies focused on the main processes taking place in this surface water. It was soon realised that for biological digestion of wastes aerobic processes were important. The oxygen balance of surface waters was mathematically described in a study by Streeter & Phelps (1925). The development of WWTP's started in the 19th century (see e.g. Imhoff (1932) and van Zon (1986)) and aimed at a reduction of the loads of biodegradable matters in the wastewater discharged. Throughout Europe the sewer systems of the major cities were connected to these facilities. In the Netherlands the large scale construction of WWTP's took place during the late 1960's and 1970's. Free discharges of wastewater hardly exist any more in the Netherlands; 97% of the households discharges into a sewer system connected to a WWTP (foundation RIONED, 2000). In some countries however, (like former east-block countries and Belgium, see Berlamont (1998)) this percentage is significantly less, though extensive building programs are planned in these countries.

1.3 Rules and regulations in the Netherlands (brief overview)

The CEN (Comité Europêen de Normalisation), technical committee TC 165 has issued the European standard covering the principles to be applied with respect to the hydraulic design and the considerations to be taken into account with respect to environmental protection. As van Luijtelaar (1999) reported, especially the criteria with respect to operation under storm conditions set in the European standard cannot be met in flat areas, whereas the environmental protection criteria may be overruled by local (national) norms (as is the case in the Netherlands). In the Netherlands collection, transport and treatment of wastewater are public tasks for which the responsibility is held by municipalities (collection and transport) and water boards (treatment). These water boards also are responsible for the quality of the surface waters, and put constraints on the discharges of wastewater. The presence of wastewater systems([2]) does not reduce the effects of wastewater on surface water quality to zero.
This is due to two reasons;

- WWTP's do not have a 100 % removing efficiency for all possible substances in the sewage.
- A large number of overflow constructions do exist in the sewer systems. In the Netherlands an estimated 15.000-16.000 of such constructions are present in combined sewer systems.

The overflow constructions are a necessity because it is economically not feasible to built a drainage system having a hydraulic capacity large enough to avoid flooding under all possible storm loads, the overflows function as a safety valve. As a result frequent discharge of (diluted) sewage to open water bodies occurs (this occurs about 5 to 10 times per year).
The combination of location, frequency of occurrence, quality and quantity of the discharged sewage and characteristics of the receiving water determines whether such a situation is to be accepted. Quality requirements of effluents discharged by WWTP's are based on a national

([2]) A wastewater system is the combination of a collecting, transporting system and a Waste Water Treatment Plant.

law dating from 1970 (the WVO, Wet verontreiniging oppervlaktewater (Law on contamination of surface waters)). This law is also the basis for regulations with regard to sewer overflows. A major problem with respect to regulations related to sewer overflows is the fact that a direct quantification of the effects of these events on water quality is hard to give. Moreover, water boards are not able to formulate acceptable levels of quantity and quality of the discharges of sewage overflow on a given surface water. Therefore, water boards apply rather 'practical' regulations with respect to sewer overflows. At present the aims with respect to the reduction of pollutional load are based upon international agreements made in 1985. In the North Sea Action Plan and the Rhine Action Programme it was agreed to reduce the pollutant discharges in the Rhine catchment area and the North Sea by 50% relative to 1985 for all sources. In the Netherlands there are 26 water boards having their own particular set of regulations. There is however a certain uniform basis found in the so-called 'reference system' which is postulated to give the agreed 50% reduction of pollution due to CSO's. This system was defined by a national committee (CUWVO (1991)):

> *Any combined sewer system should emit a pollution load less than or equal to a fictitious system having 7 mm in sewer storage, 2 mm storage in a settling facility and a pumping capacity equal to the DWF production plus 0,7 mm/h.*

This formulation proves hard to apply in practise, because no uniform methods have been defined for quantifying the pollution load; the reference system is ill defined. This leads to a situation that can be characterised as a lack of legal security, as was demonstrated by van Mameren & Clemens (1993). The introduction of a standardised method for hydrodynamic calculations (VROM, see section 1.4) and some playing rules regarding mean concentration used to quantify the pollutional load leads to a more or less clear situation (e.g. WRW, 1999). This implies that regulations are based upon calculations of the quantity of sewer overflows in terms of permissible volumes rather than on the effects these spills have on receiving water bodies. To render it possible to base regulations on these effects the following working knowledge is to be obtained:

- Quantification of the permissible pollutional load on a given water body.
- Quantification of the pollutional load emitted from a sewer overflow.

Both questions are far from being answered; this is due to the complexity of processes involved. This is best illustrated by the fact that the topic 'urban drainage and water quality' is an ongoing important topic at international conferences held on urban drainage (see e.g. Gujer & Krejci (1987) and Harremoës & Madsen (1998)). A first prerequisite however, is that a good understanding of the transporting medium, water, is obtained. A further observation is that in practice measures aiming at the reduction of the pollutional load are based upon calculations rather than on observations. This implies that there seems to be an almost unlimited faith in the correctness of the calculations made. With respect to flooding, no formal regulations exists in the Netherlands, though an accepted limit for flooding is a frequency of 0.5 year^{-1}. Contrary to European regulations no differentiation is made with respect to allowable flooding frequencies in relation to the specific use of the areas involved. Basically every municipality in the Netherlands can set its own acceptable limit.

1.4 models in urban drainage

1.4.1 Developments in models

Models play an important role in urban drainage; Yen (1987) presented the development of urban drainage from art to science (Figure 4). Parallel to this scheme the development of models is shown. Initially system design was based on craftsmanship only, science developed and in particular hydraulics was introduced design became more sophisticated. The relations between discharge and water head found in hydraulics were the first engineering models applied in urban drainage. The development of hydrology and further refinement of hydraulics increased the scientific basis of designs. However, craftsmanship, 'common sense' and practical experience of the designing engineer were (and still are) needed to a large extent. The development of computers and numerical methods resulted in what is now called Hydroinformatics. By the end of the 1960's the first computer models for use in urban drainage were developed. Initially only steady state calculations were made on simplified geometric descriptions of drainage systems. The increase in speed, storage capacities and the sharp decrease in investments necessary to obtain calculational power resulted in a quick development of hydrodynamic calculational tools. The introduction of the Personal Computer in the 1980's had a major effect in this respect. These developments enabled practising engineers to study the time dependent behaviour of water flows in a sewer system. A problem however, was to define 'design storms', that is, the statistical characteristics of the rain had to be caught in a limited number of events. Therefore it was foreseen that time series calculation would become a demand, still limited however by calculational possibilities.

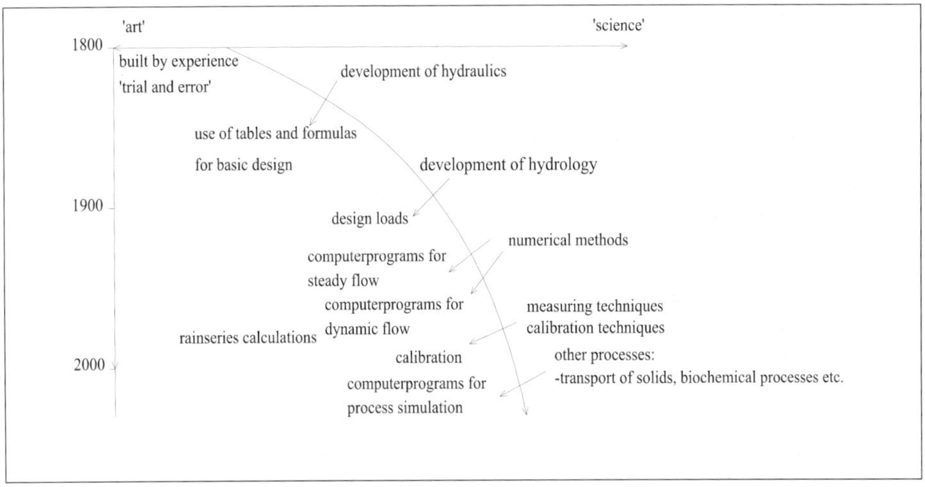

Figure 4: From art to science (after Yen (1987)).

At present time series calculations are a practical possibility, except for large systems. In Figure 5 the decrease of processing time to perform hydrodynamic calculations on a large catchment (circa 3500 nodes) is shown. Parallel to the development of models describing water quantity, models were developed describing water quality in the sewer system incorporating processes like sedimentation, biodegradation, and erosion. These models however, are still not at the stage of practical application.

It may even be questioned if they ever will, considering the many processes, parameters and initial conditions needed to feed these models (see e.g. Ahyerre et al (1998)). At this stage the hydraulic models that describe water flow in a drainage network are the most refined and best scientifically embedded compared to other (sub) models applied in urban drainage. With respect to hydrology, pure deterministic models seem to be unfit for practical application, so grey-box models are widely adopted (see Chapter 2 of this thesis).

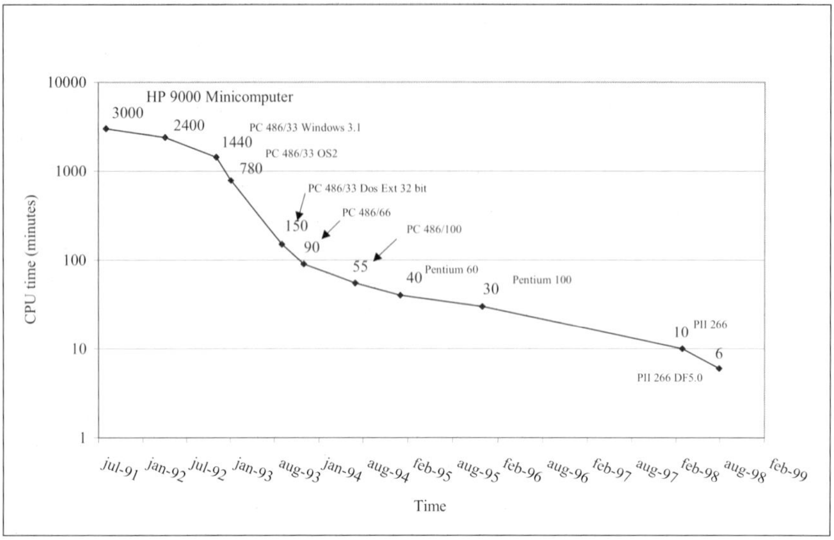

Figure 5: CPU time for a large catchment from 1991 to 1998.

Deterministic models describing in-sewer processes, as the ones mentioned before, are still in a stage of scientific development and offer only a limited value for the practising engineer as discussed by Sieker & Verworn (1997) pp.71-195. Apart from models based on a deterministic approach, application of grey-box or black-box models based on input-output analysis have been developed. Some of the methods applied are neural networks, fuzzy logic, genetic algorithms and system identification. These methods circumvent the need for a detailed insight in the processes taking place, but they call for a large set of data of observed system behaviour. It is stated that models and their application in urban drainage are still not fully matured. The enthusiasm for computer models displayed by practising engineers during the 1980's and the 1990's diverted the attention from the fact that field data are needed to understand and quantify the processes involved in e.g. transport of solids and other processes in the sewer systems. Therefore the following logical path to follow is proposed:

- Calibration and verification of hydrodynamic models on a routine basis.
- Monitoring of urban drainage systems with respect to water quantity.
- Scientific research into the processes in sewer systems aimed at refining existing models describing water quality in sewer systems.
- Introducing process models.

The main subject reported in this thesis is the verification and the calibration of hydrodynamic models.

1.4.2 The role of models in urban drainage

Restricting the discussion to models describing the water quantity aspects of urban drainage the main applications can be summarised as follows:

- Design of new systems.
- Redesign of existing systems.
- Operational application (as in Real Time Control applications).

Each field of application puts its own demands on the models applied. When designing a new system the behaviour of a not yet existing design has to be judged, in many cases without exact information with respect to hydraulic loads being available. In these cases the model is used by the designing engineer to evaluate the safety of his design in terms of flooding and environmental impact. If an existing system no longer meets the demands put on it (due to changes in the system, changing demands or an increase of loads) the system needs to be redesigned. In this case the designing engineer has to judge whether the alterations he will propose will indeed be a sufficient remedy for the problem encountered in practice.
This implies he/she has to be sure that the model applied describes the actual situation to a fair extent and a verification of the model with reality is needed. In an operational application (e.g. Real time Control), the model is used to give a prediction of the system's behaviour when taking certain on-line measures (like opening or closing sluices, starting or stopping pumping stations). This prediction is based on the actual situation of the system obtained by on-line measurements and in some cases on a prediction of storm intensities. It is clear that the constraints put on the model applied will differ in the three cases mentioned. When designing it is impossible to get an exact model of the reality, simply because this reality does not exist yet. The quality of the model and its input data must be such that a safe design is obtained; this is the main responsibility of the designing engineer. When redesigning a system the actual situation has to be described properly by the model used. This asks for a high quality of the input data, especially with respect to the geometry of the system under study (see also Chapter 3 of this thesis.). For operational applications calculational speed is essential; the model should run significantly faster than reality. Furthermore the input data are a mixture of geometry (database) and actual measuring data at a few points in the system, so both data sources must have a certain minimal degree of accuracy. The models used in these applications should have a high degree of accuracy at least with respect to the effects of control actions taken. Especially in this field of application neural networks can be used to advantage; a well-trained neural network is fast and does not produce information on points beyond the direct sphere of interest, see e.g. Loke et al (1999).

1.5 Guidelines for hydrodynamic models in urban drainage in the Netherlands[3]

1.5.1 General

In the late 1980's the department of environment of the Netherlands initiated a project to standardise the use of models in urban drainage (Faber (1995,1999). The initial goal was to standardise quality and quantity models together. By the end of the project however, the conclusion was drawn that the models for water quality were not at a sufficient level of development to allow standardisation. It was also recognised that the developments with respect to quantity models had not ended yet. Therefore a system was drawn up in which enough possibilities were left open not to hamper development while still ensuring a minimum degree of standardisation. One basic decision was not to prescribe a software package or a detailed guideline for the calculational process itself, but merely to make guidelines which:

- Give uniformity of starting points.
- Describe reality as closely as feasible.

(³) This section is based on two articles: van Mameren & Clemens (1997) and van Luijtelaar & Rebergen (1997), more details can be found in these articles.

- Gain wide acceptance for practical use.
- Assure exchangeability of data sets and calculation results.
- Assure repeatability of results.
- Give enough flexibility to avoid hindrance of scientific progress in the field of urban drainage.

This was achieved by designing a system more or less in the same way as a quality-system (ISO-9000) might look: procedures with a prescribed format for the results including prescribed verification- and control-routines.

1.5.2 Philosophy of the Dutch guidelines

In the guidelines a distinction is made between calculations made with respect to flooding and calculations made with respect to overflow events. This distinction is based on the way in which the statistical properties of storm events are handled. In principle two approaches are possible:
- Using the statistical properties of storm events as model input (statistics a priori).
- Applying statistical methods on the results of a rain series calculation (statistics a posteriori).

Mathematically, statistics a priori and using the results to describe the functioning of a drainage system in a statistical sense, is tantamount to assuming that the relation between storm events and overflow-discharges are known exactly in an analytical manner. Since numerical techniques are still needed to calculate this relation, it is clear enough that applying statistics a priori is not the right way. So, ideally one would like to make rain series calculations using a complete model. This would allow for application for statistics a posteriori to eliminate the risks involved in applying statistics to the storm data (statistics a priori). With regard to the development of hard- and software, it is to be expected that within some years this method will be in practice on a large scale. It is already effectively in use for medium sized systems (up to 5000 branches using a 10-year rain series). At this moment (2000) it is not feasible to apply this method on a large scale and for every urban drainage system, due to the excessive calculational effort needed for extensive systems.
So, a more practical approach was suggested maintaining the advantages of the rain series and avoiding its drawbacks of it as well as possible. This approach is schematically shown in Figure 6 and differentiates between calculation aimed at flooding (hydraulic capacity) and calculations aimed at quantifying the operation of CSO's.

Hydraulic capacity: single event calculations

When studying flooding or when designing a system with respect to hydraulic capacity the drainage system must be modelled in detail. So for these cases a set of standard storm events in combination with a complete model is used. In the Dutch guidelines 10 artificial standard storm events are defined, with return times ranging from 0.25 years to 10 years. These events were derived from the standard rain series applied in the guidelines. The choice of the return period applied while studying flooding is not prescribed in the guidelines. It is advised in the guidelines to use observations when available. For example: when a well-documented storm event including observations of a flooding is available, the use of this specific event to verify the (complete) model is recommended.

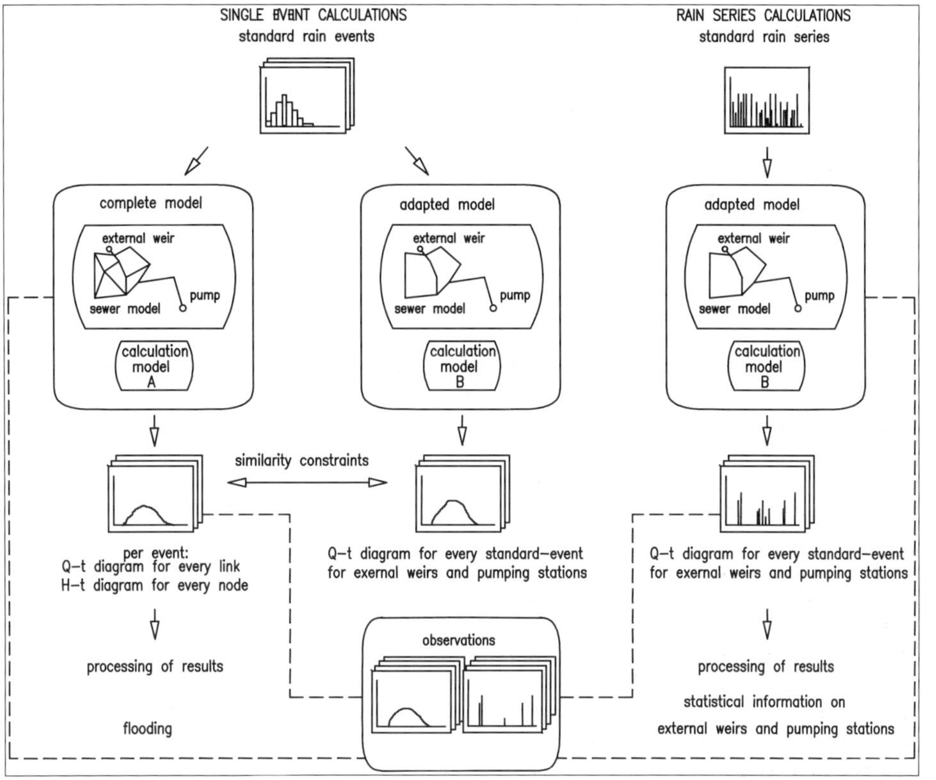

Figure 6: General layout of the Dutch guidelines.

Rain series calculations

When studying the storm water outlets or CSO's in relation to, for example, the quality of the receiving surface water, one is only interested in the processes taking place at these locations with respect to return periods of overflow volumes, peak-discharges and average number of overflow-events. This means that the model applied may differ from the complete model: the adapted model. In this model there is no need for all nodes and conduits to be incorporated as individual elements, but there is a need for a simulation time spanning over a long period, say 10-25 years.

The rain series are prescribed in the guidelines: the continuous rain series as observed in De Bilt over the period of 1955-1979 with a time resolution of 15 minutes (KNMI (1993)). The use of the rain series makes it possible to apply statistics a posteriori on the calculation results. So it can be stated that when studying the hydraulic capacity of a sewer system a high resolution in the space dimension is needed, and the storm is simplified. Studying the behaviour of outlets, a high resolution and a high amount of calculational points in time are needed, while the resolution in space can be limited to the points of interest.

The guidelines put only one demand on the adapted model: the results (Q-t relation for every outlet) must match with the results of the complete model for a set of 5 or 6 standard storm events.

1.6 Scope of the research

In the Netherlands, hydrodynamic models in urban drainage are in practice seldom verified against field observations although large investments are based on the results of these models. In the Dutch guidelines the urge for calibration against field data is highlighted, however no methods to this end are described. Furthermore, a correct description of the hydraulics is a prerequisite for ever being able to correctly model other processes taking place in drainage systems. The attention given in scientific literature to verification and calibration in urban drainage is limited (see Chapter 5) and the results are rarely applied in day to day engineering practice. The subject of this research is therefore, to formulate methods to calibrate hydrodynamic models that can be used by practising engineers enabling them to quantify the quality of the parameters gained from this calibration. Furthermore sources of errors are studied; this comprises modelling errors, data errors and measuring errors. In this manner the uncertainties in the model-results can be quantified enabling for a probabilistic design approach. Since in practice mostly commercially available models are used, in this thesis also such a model is applied: HydroWorks® 3.3. NT supplied by Wallingford Software Ltd. Although not investigated in detail, the results and main conclusions in this thesis also apply to other well-known models like MOUSE®, Hystem/extran, SWMM or SOBEK-URBAN®.
Main difficulties in using hydrodynamic models are:

- Acquiring a correct database containing all geometrical information from the system.
- Acquiring the amount, types of run-off area and the corresponding inflow points in the drainage system.
- Introducing the correct parameters describing run-off, hydraulics.
- Obtaining information on the hydrodynamic and run-off processes by means of a measuring project.
- Obtaining well-calibrated models with a quantification of the quality of the results of the calibration process.

In practice it is believed that these difficulties are hard, if ever, to overcome. This has resulted in a tendency to look at 'alternative modelling methods' like neural networks, intelligent agents, see e.g. Verwey et al (1994) pp.179-239. These methods are interesting from a scientific point of view, though in the opinion of the author are not as informative as the 'classic' deterministic models for designing engineers.
In the author's opinion it is to early yet to turn to alternative models for practical use, therefore in this thesis the following topics are studied:

- The effects of data-errors on the calculational results of hydrodynamic models.
- Methods to design monitoring networks.
- The use of automated parameter optimisation in calibration processes.
- Measuring methods and monitoring network design.

The main goal is to obtain, describe and demonstrate methods to enhance the day-to-day practical application of hydrodynamic models by means of combining models and measuring results. To this end the methods employed have to meet the following conditions:

- Directly applicable in practice.
- Accessible for modelling specialists.
- No urge for huge calculational power (i.e. the methods must be capable of being applied using a modern top-end Personal Computer system).

1.7 short readers guide

In Figure 7 the relations between the chapters in this thesis are shown.

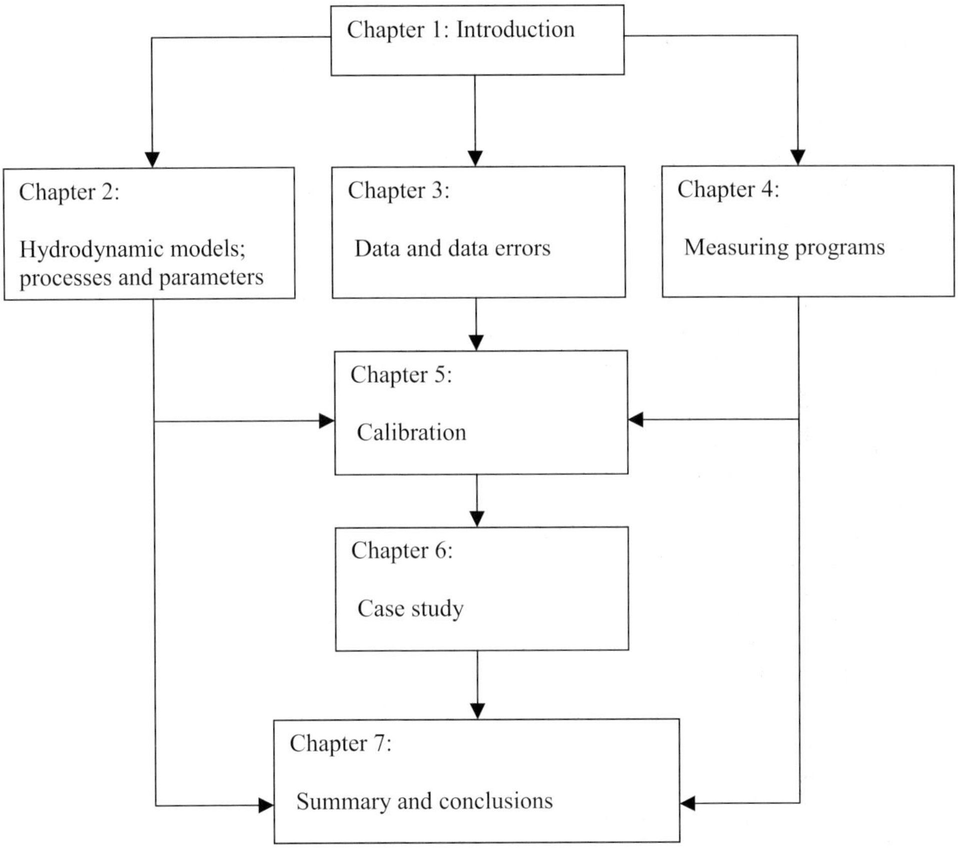

Figure 7: general lay-out of this thesis.

Chapter 2 deals with processes and models, a distinction is made between run-off models and hydrodynamic models describing the fluid flows in sewer systems. In Chapter 3 the geometrical description of sewer systems and methods to recognise and eliminate errors from databases is dealt with. Also the effects of errors and uncertainties in model input are discussed and quantified using Monte-Carlo simulations. A prerequisite in judging models is the availability of information on the actual hydrodynamic behaviour of urban drainage systems. Chapter 4 describes some measuring methods used in measuring programs in urban drainage and deals with methods for defining a sampling interval depending on system characteristics and measuring accuracy. The main subject of this thesis, the process of calibration of models in urban drainage, is reported in Chapter 5. Several parameter optimisation methods (classic gradient based methods as well as a genetic algorithm) are discussed and are demonstrated using artificial examples. In these examples the influence of systematic errors in models as well in measuring data on the achievable quality of a model calibration are demonstrated. Chapter 6 reports the results of a field measuring program and the actual calibration of a commercially available hydrodynamic model. Chapter 7 comprises

the summary in English and Dutch as well as the main conclusions and recommendations as a result of this study.
The chapters 2, 3, 4, 5 and 7 can be regarded as more or less individual texts, while chapter 6 is to be read in conjunction with the other chapters.

CHAPTER 2: HYDRODYNAMIC MODELS; PROCESSES AND PARAMETERS

2.1 Introduction

2.1.1 General

A model is a description of reality in some form. Here the discussion is restricted to mathematical models of physical processes used by engineers to solve technical problems in practice. In engineering sciences, simulation using models is widely applied to gain understanding of physical systems. As H.A. Simon states (Simon (1996), page 13), simulation is possible only *'because distinct physical systems can be organised to exhibit nearly identical behaviour'*.

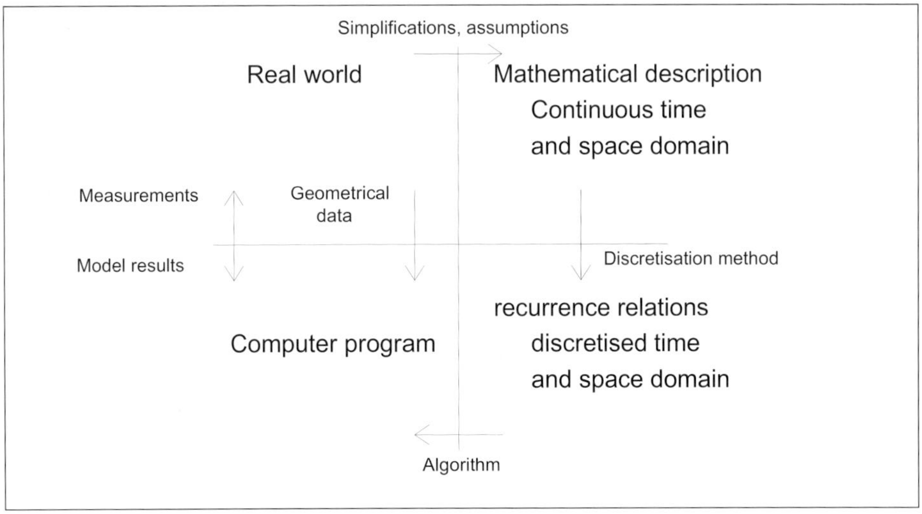

Figure 8: stages between reality and computer program (after Stelling & Booij (1999)).

In Figure 8 the stages between reality and a computer program describing this reality are shown. Stelling & Booij (1999) describe these stages in terms of successive isomorphical mappings, eventually resulting is an electrical circuit (the computer) simulating the behaviour of some physical system. In each step assumptions, simplifications and adaptations are, often unconsciously made. Therefore using a computer program to model physical reality always results in an approximation of this reality, having limitations in accuracy and validity.
When turning to hydrodynamic models, the observable reality is that of water movement as a function of time and space. This observable reality is mapped into some form of mathematical description([4]) based upon fundamental principles like mass-conservation and momentum

([4]) In this respect some philosophical doubt may be introduced: 'In as far mathematical formulas describe reality they are incorrect, in as far they are correct they do not describe reality'. This observation is believed to be stated by Albert Einstein. A further, more pictorial , discussion on models versus reality may be found in René Magritte's paintings, e.g. 'Le soir qui tombe' and 'Ceci n'est pas une pipe'.

conservation. In classical physics these principles are widely accepted and prove to be correct by observation. The form in which the mathematical image appears is that of a (set of) differential equation(s) along with a set of initial and boundary conditions. For practical engineering problems analytical solutions seldom exist. This is certainly so in the field of hydrodynamics in which the Navier-Stokes equations still resist every attempt to be solved analytically for the general case([5]). Therefore, numerical techniques are applied to obtain an approximate solution in terms of difference equations. The resulting 'map' of the original mathematical image deviates with respect to the solution space. The original differential equations are valid in a continuous space domain whereas the recurrence relations obtained using numerical techniques necessarily are only valid in a discretised space. Taking the process of mapping reality into a computer program one step further, an algorithm is used to implement the recurrence relations in a computer program. In such an algorithm, in many cases, some stopping criterion for solving zero's in equations has to be chosen, or foreseeable zero divisions have to be avoided, resulting in (slight) deviations between the actual outcome of the algorithm and the original recurrence relations. Finally the algorithm is implemented in a computer code; this, even in the unlikely case in which programming errors (or bugs) are absent, introduces a further deviation. This is because that, when using a computer, the full domain of real numbers is not available (implicitly supposed throughout the preceding traject). A computer is limited by the fact that there are certain maximum and minimum numbers and there exists a finite possible difference between two successive numbers. Following the steps as described, a computer image of the processes regarded as essential for the problems to be modelled is obtained. In order to obtain meaningful results for a given case the geometrical properties are introduced in the computer (the geometrical database). As will be discussed in Chapter 3, the geometrical database is also a map (suffering from imperfections) of the geometry of the real world, possibly introducing errors in the computer image. When using the results of a computer program, an isomorphism with the real world is implicitly assumed. This assumption can only be verified when measuring data are available (Chapter 4) based on which calibration/verification of the model (process model as well as geometrical model) is possible. In the remainder of this chapter, it is shown what kind of assumptions, simplifications and adaptations are normally made when modelling the hydraulic processes in an urban drainage system using a computer programme.

2.1.2 Models in urban drainage

2.1.2.1 Processes

As discussed in the previous section, a model is a description of reality; in this thesis the focus is on mathematical models used in hydrodynamic calculations in urban drainage. This implies that the mathematical description of processes involved in the water movements in urban drainage systems is studied. When discussing these models a distinction has to be made in components constituting a model. The definitions as formulated by van Mameren & Clemens (1997) are used; a model is built up from three basic components (see Figure 9):

- A description of the hydraulic processes, mostly referred to as the process model.
- A geometrical description of the system under study, the database or geometrical model.
- Hydraulic loads, making a distinction between dry weather conditions and storm conditions.

([5]) In fact the Navier-Stokes equations are one of the seven main unsolved mathematical problems recently (may 2000) identified as such by the Clay Mathematical Institute. (see www.claymath.org).

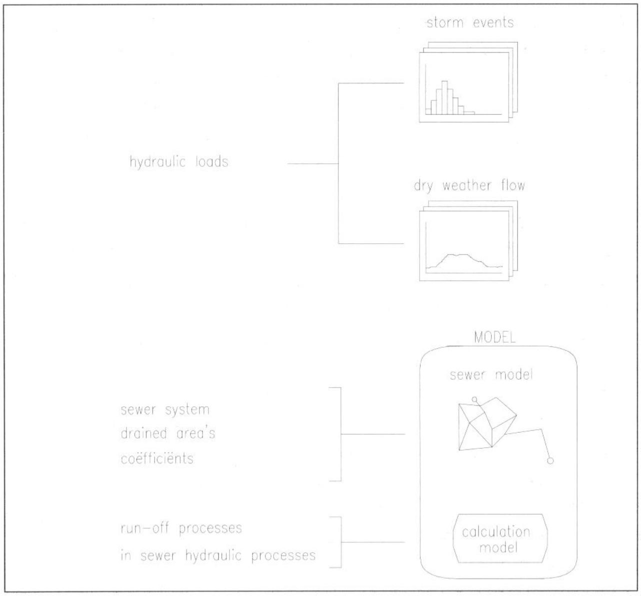

Figure 9: Components of a model in urban drainage.

The hydraulic processes involved are usually distinguished into two main groups:

- The processes in the urban drainage system (the hydraulic model).
- The processes involved in the transformation from rain intensities to run-off (the hydrological model).

In fact, calculation models in urban drainage contain therefore two more or less separate process descriptions:

- The hydrological model (describing run-off processes).
- The hydraulic model.

This has implications for the calibration, since in most practical cases only water levels and/or discharges in the sewer system are measured, implying that runoff is not quantified as a separate quantity. The output of the hydrological model is the input for the hydraulic model and it creates a time variable boundary condition. The implications of this fact with respect to calibration are discussed in chapter 5 and 6. Starting in the early 1970's software tools were developed to be able to handle the massive calculation effort involved in major hydrodynamic simulations on urban drainage networks.

2.1.2.2 Loads

When studying the processes taking place in an urban drainage system two modes are distinguished:

- Dry Weather Flow (DWF).
- Storm conditions.

In purely DWF systems, the DWF is the main contributing flow of water. Although in practical cases some storm water will enter the system due to faulty connections and storm water from roofs connected to the system to create some regular flushing of the system, see Meijer (1998). On the other hand, in pure storm water systems only storm water will enter the system; in this case, also some DWF may enter the system due to mis-connections. In combined systems and improved separated systems, both sources of water are present. The DWF is a result of different sources of water:

- Domestic wastewater.
- Industrial waste water.
- Drain water due to leakage.

Domestic wastewater consists for the major part of discharged drinking water after use for cooking, toilet flushing, washing etc. Therefore, the figures for water consumption are usually used in order to estimate the quantity of the DWF. Drinking water consumption in households shows a periodic pattern in time, as does the actual DWF in a sewer system. However there is a phase shift and a difference in amplitude between them. This is because drinking water is temporarily stored in the households before it is discharged and because a certain portion of the drinking water does not enter the drainage system at all (due to e.g. evaporation and garden sprinkling). Therefore, the time patterns present in drinking water consumption should not be used for making accurate estimates of the DWF pattern. Especially in older systems, drainage water can mount up to 50% of the DWF flow. This is due to the combination of two factors:

- Leaking joints.
- High groundwater levels.

Where leaking joints are present and the groundwater levels are beneath (part) of the system, DWF exfiltrates. This latter process is in practice hard to detect. Before entering the drainage system, storm water flows over the receiving areas like streets and roofs. The processes transforming rain intensity patters into inflow patterns in the drainage system are (see Figure 10):

- Wetting of dry surface.
- Infiltration.
- Storage in local surface depressions.
- Evaporation.
- Flow over the receiving area.

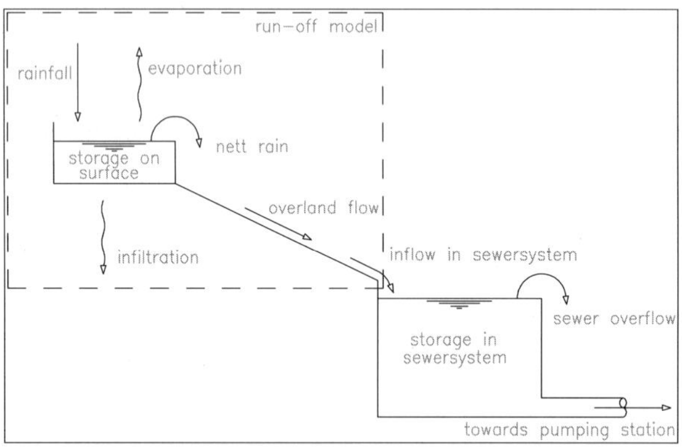

Figure 10: Processes involved in modelling urban drainage systems.

This implies that the inflow intensity pattern deviates from the run-off intensity pattern; therefore, when modelling an urban drainage system the pattern in the run-off intensities and the locations at which this storm water enters the systems must be known. To this end hydrological models are used. Run-off and its practical modelling is discussed are Section 2.2.

2.2 Hydrological models

2.2.1 General

The transformation of rain falling on a surface into the actual amount of storm water discharged into a drainage system is the subject of extensive research over the last decades. Theoretically, every process involved can be accurately described using a deterministic model. For practical use this approach is inhibited due to the many model parameters involved and the large number of initial and atmospheric conditions that must be known. This will become clear in the next sections on the individual processes involved in the run-off process.

2.2.2 Storage in surface depressions and initial losses

When rain falls on a surface this will not necessary result in an immediate discharge into the drainage system. When the surface is dry at the outset of the rain, initial losses will occur. These initial losses depend on the type of surface and the humidity and temperature of the surface at the start. A certain amount of the rain is caught in small local depressions in the pavement.

Since it is virtually impossible to describe the geometry of the pavement for a whole catchment area in detail in the modelling practice, an average constant value for the depth of the depression is used varying with the type of pavement.

Exact figures for initial losses and storage losses are scarce. In Table 2 results of various field data obtained from literature (see e.g. Pecher (1969), NWRW 4.3 (1989) and van de Ven (1989)) are summarised. Apart from the type of pavement, its state of maintenance is an important factor influencing the parameters for depression storage. Therefore, the magnitude of depression storage changes over time in a given catchment. Furthermore, it must be mentioned that water stored in depressions vanishes over time between successive storms due to evaporation and infiltration. The available amount of surface storage at the start of an individual storm therefore depends on the history as well.

	Flat roofs	Tilted roofs	Impervious road areas	Semi-impervious roads
Initial losses	0-0.5 mm	0.1 mm	0.07-0.7 mm	0-1.5 mm
surface storage	2-2.5 mm	0.1 mm	0.3-1.7 mm	0.8-6.0 mm

Table 2: Initial losses and surface storage.

A poor state of maintenance of paved areas generally increases the surface storage capacity. In this sense, neglecting maintenance is advantageous since it decreases the total hydraulic load on the receiving urban drainage system. Kidd (1978) related the surface storage to the terrain slope by:

$$b = 0.77 i_t^{-0.49} . \qquad \text{(eq.2.1)}$$

In which:

b	surface storage	(mm)
i_t	terrain slope	(%)

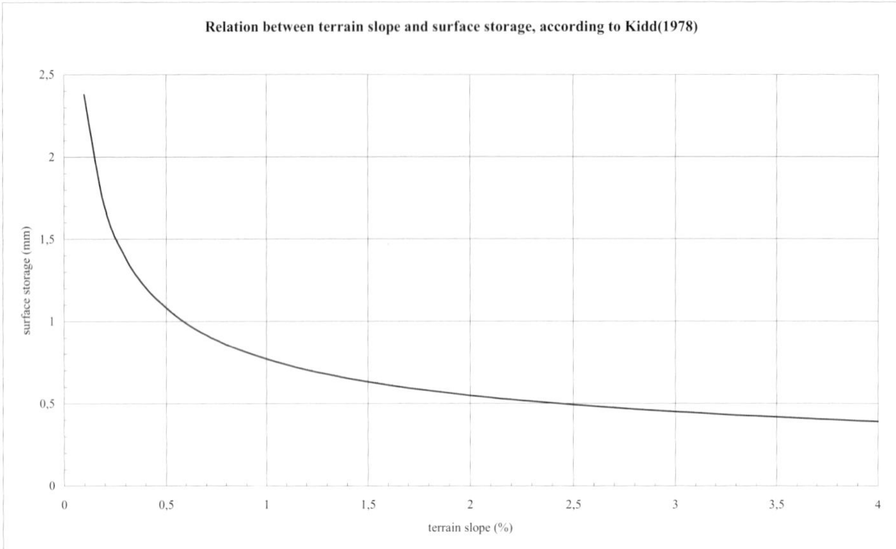

Figure 11: Relation between terrain slope and surface storage (Kidd (1978)).

2.2.3 Evaporation

A relevant process in relation to run-off is evaporation. Surface storage is made available due to infiltration and evaporation. The evaporation rate depends on several variables:

- Temperature.
- Wind speed.
- Atmospheric humidity.
- Rate of heat influx.
- Intensity of sunshine.
- Colour of the surface.

The variables mentioned are in general not known in any detail when modelling an urban drainage system; therefore monthly average figures are normally applied, the so-called Penman evaporation (Raudkivi (1979)). In his research into run-off models, Van de Ven (1989) concluded that it is impossible to quantify the evaporation term for individual storms. Therefore, a practical approach is usually adopted. This implies that the evaporation either is neglected during a storm or is set to a constant value equal to that of an open water surface. In Figure 12 the monthly-averaged evaporation-rate values for De Bilt in the Netherlands are shown.

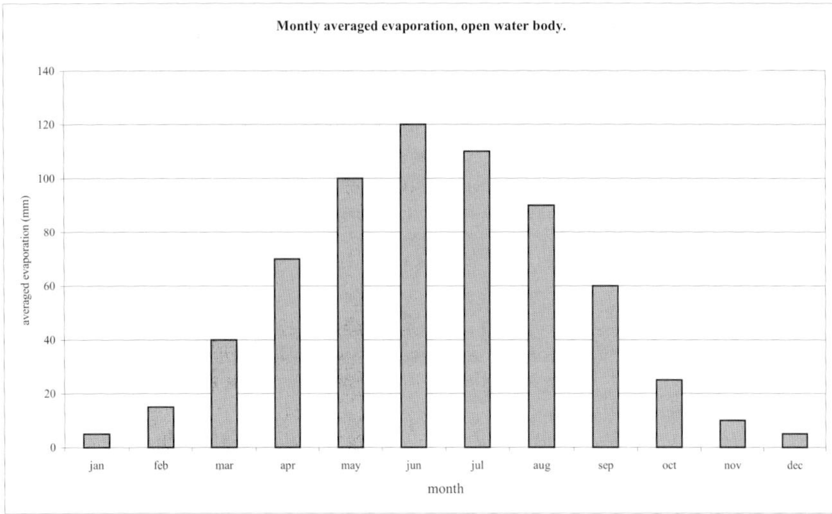

Figure 12: Monthly averaged evaporation values, de Bilt, period 1955-1979.

2.2.4 Infiltration

Rain falling on a pervious or semi-pervious area (like grass or clinkers) will partly infiltrate into the groundwater. The process of infiltration is complicated; depending of the initial conditions the infiltration rate will decrease with time as the unsaturated zone becomes filled (Mein & Larson (1973)). As soon as this zone has become saturated, the minimum infiltration rate is reached. The infiltration capacity of the soil increases only after the precipitation has stopped and the storage in the unsaturated zone has been reduced or emptied. In literature several values for infiltration rates are reported (see e.g. Ando (1984), Bebelaar & Bakker (1981), de Roo (1982), van Dam & Schotkamp (1983) and van de Ven (1989)), in Table 3 the ranges found are shown.

	infiltration values (mm/h)
Concrete clinkers	7-353
Tiles	1-254
Grass	10-500
soil without vegetation	10-100

Table 3: Ranges for the infiltration capacity obtained from literature.

Several models are applied in practice to describe the infiltration process:
Hillel&Gardner (1970):

$$I_{cum} = \sqrt{at+b} - c \qquad \text{(eq.2.2)}$$

In which:

I_{cum} cumulative infiltration since t=0 in mm

a,b,c parameters depending on transmissivity, humidity and crust resistance.

Philip :

$$I_{cum} = a\sqrt{t} + bt \qquad \text{(eq.2.3)}$$

In which:

I_{cum} cumulative infiltration since t=0 in mm

a,b,c parameters depending on transmissivity, humidity and crust resistance.

A widely accepted model is the model suggested by Horton (1940). In this model, it is assumed that maximum and minimum values limit the infiltration rate. When infiltration starts at a given rate, it will decrease with time as the soil becomes saturated. Eventually it will reach a minimum value when the storage capacity of the pores is filled. As soon as the surface area is dried up by evaporation and infiltration, this storage capacity tends to become available again. This implies the infiltration rate is increasing again with time. These processes are described by the following formulas:

Decrease: $$f(t) = f_e + (f_b - f_e)e^{-k_a t} \quad \text{(eq.2.4)}$$

Increase: $$f(t) = f_b - (f_b - f_e)e^{-k_d t} \quad \text{(eq.2.5)}$$

This model contains four parameters, the maximum infiltration capacity f_b (assuming the unsaturated zone is fully available), a minimum infiltration capacity f_e (the storage in the unsaturated zone is filled) and the recession factors k_a and k_d . These values depend on the type of soil and the momentary groundwater level. A simple model is the constant infiltration rate model where the infiltration rate is set to a constant depending only on the characteristics of the particular surface. Van de Ven (1989) made a comparison of several infiltration models based on in-situ measurements. He concluded that the models as defined by Hillel & Gardner (1970), Horton (1940) and Philip did not show significant differences. Furthermore, he concluded that the constant infiltration rate model was less accurate than the other models. However, when studying the reported experimental results, the constant rate model could be used for practical purposes according to van de Ven. This is mainly of importance since this simple model calls for only one parameter to be estimated, making calibration in practice simpler and enhancing the reliability of the parameter values obtained in a calibration. In Table 4, the value for R^2, model efficiency in a comparison with field measurements, is tabulated for the infiltration models mentioned. As can be seen the Hillel&Gardner model gives the best results, the differences however are small.

	Hillel&Gardner	Philip	Horton	Constant
Copper slug	0.993	0.988	0.977	0.966
Concrete clinker	0.993	0.992	0.997	0.954
Concrete tiles	0.995	0.991	0.989	0.97

Table 4: R^2 values for several infiltration models reported by van de Ven (1989).

Furthermore, it is argued that since exact values for evaporation cannot be given, a refined model for infiltration is of academic value only. In relation to the hydraulic load on an urban drainage system, it is only of importance to have an estimate for the net-rain and an estimate for the available storage capacity on surface areas.

2.2.5 Run-off

2.2.5.1 General

Once the amount of rain that has fallen becomes larger than the sum of the initial losses, the losses in local depressions and the evaporation, run-off occurs. So, the moment at which after the start of the storm water starts to run-off (t_s) is calculated from the integral equation:

$$\int_{t=0}^{t=t_s} (r(t) - i(t) - e(t))dt = S + W \quad \text{(eq.2.6)}$$

In which:

S	the surface storage	mm
W	the initial loss	mm
r(t)	the rain intensity as a function of time	mm/h
i(t)	the infiltration rate as a function of time	mm/h
e(t)	the evaporation as a function of time	mm/h

So, for $t>t_s$ run-off to the drainage system occurs if r(t)>i(t)+e(t). The amount of rain resulting in run-off is defined as the net-rain-intensity $p_n(t)=r(t)-i(t)-e(t)$. Several models are developed to describe the transformation from $p_n(t)$ into the actual discharge entering a drainage system q(t). Some of the most widely used models will be discussed briefly.

2.2.5.2 The Nash-model

The Nash model or Nash-cascade (Nash & Sutchiffe (1970)) is a cascade of n identical linear reservoirs (Figure 13). The momentary value for q(t) is calculated using the following convolution between net-rain intensity and a transfer function h(t):

$$q(t) = \int_0^t h(t-\tau)p_n(\tau)d\tau \qquad \text{(eq.2.7)}$$

The transfer function h(t) is defined as:

$$h(t) = \frac{1}{k\Gamma(n)}\left[\frac{t}{k}\right]^{n-1} e^{\frac{-t}{k}}, \text{ for } t \geq 0 \qquad \text{(eq.2.8)}$$

In which:

k	reservoir constant	$(mm/h)^{1/n}$
n	number of reservoirs	-
t	time	h
p(t)	net-rain intensity as function of time	mm/h
q(t)	run-off discharge as function of time	mm/h
Γ(n)	the gamma function of n (if n is a integer then Γ(n)=n!)	-

This model contains two parameters. Due to the definition it is possible to define a cascade with a non-integer number of reservoirs.

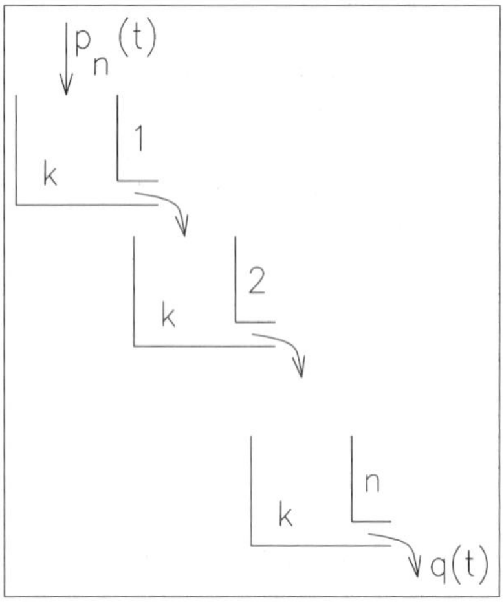

Figure 13: The Nash cascade.

2.2.5.3 Non-linear reservoir model (NLR-model)

The non-linear reservoir model is defined by:

$$q(t) = Kp_n(t)^b \qquad \text{(eq.2.9)}$$

In which

q(t)	Run-off
K	Reservoir constant
$p_n(t)$	Netto-rain
b	Power

If the parameter b is equal to unity then this model is equal to a Nash-model with n=1. The non-linear model is often applied with a value for b of 2/3. This is based on the application of the stationary, uniform equation of water motion in one dimension. This parameter value however, is only of theoretical importance and does most of the time not apply to practical cases. The values for the constants in the family of reservoir models have been subject to a wide spectrum of researches. As an example the Desbordes model is briefly discussed, it is linear due to that fact that b=1. The value of K however is defined as:

$$K = K_{Desb}\, A_r^{\,0.18}\, P_{nt}^{\,-0.36}\, (1+C)^{-1.9}\, T_3^{\,0.21}\, L^{0.15}\, H_{pe}^{\,-0.07} \qquad \text{(eq.2.10)}$$

In which:

K_{Desb}	proportionality constant depending on the type of surface	(-)
A_r	sub-catchment area	(ha)
P_{nt}	sub-catchment slope	(%)
C	the proportion of sub-catchment area that is impermeable (between 0 and 1)	(-)
T_3	the duration of the rainfall sub event	(s)
L	sub-catchment length	(m)
H_{pe}	total accumulated effective rainfall for the rainfall sub-event(m)	

The value for K depends not only on characteristics of the area under consideration but also on the characteristics of the storm event in terms of duration and accumulated rain.

Apart from these generally applied models, the Volterra model and the Laguerre model are also well known. A detailed description of these models is found in van der Kloet & van de Ven (1981). These models however, pose the problem for practical applications that a large number of model parameters have to be estimated (in some cases up to 10). As will be seen in the chapter on calibration, an increase in the number of the model parameters has a negative effect on the process of calibration as well as on the reliability of the calibrated model parameters. Therefore the Volterra and Laguerre models are not discussed in this thesis. Intensive research done by van de Ven (van de Ven, 1989)) has shown that the differences between run-off models are only marginal in practice. In Table 5 some of his results obtained from field measurement are shown.

	Nash 'n'	Nash 'k'	NLR b	NLR k
Municipal area	0.65	415	1.02	5.6
Parking lot	1.05	225	1.07	4.0

Table 5: Some parameter values for the Nash model and the non-linear reservoir model (NLR). b in mm, k for the NLR in $mm^{1-b}min^{-1}$

The power b in the non-linear reservoir model is close to unity, implying a linear model can be used. In the Netherlands, a standard model (NWRW 4.3 (1989)) has been chosen for practical implementation. This model is in its essence built up from:

- A surface depression storage and initial losses model.
- An infiltration model according to Horton.
- A single linear reservoir model.
- A simple evaporation model using an average evaporation value varying per month (the so called Penmann evaporation).

2.2.5.4 Comparison

A practical comparison of some of the models discussed is summarised in Figure 14
From these results the following practical conclusions are draw:

- All models show a more or less equal characteristic response to the storm imposed.
- Simple models (like the NWRW 4.3 model) perform reasonably well, while posing the advantage of being simple to implement.

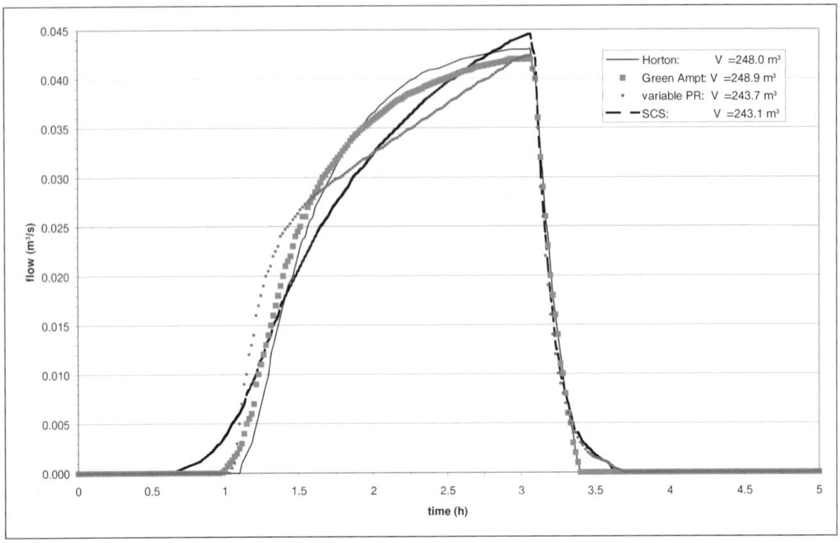

Figure 14: Some results for different run-off models imposing a block storm.

When comparing several complete run-off models (i.e. surface storage model + infiltration model + Routing model) with respect to the number of parameters to be specified per area the following figures are obtained (see Table 6):

	Model parameters	Numerical parameters	Empirical parameters
Horton	4	1	
Nash	2	1	
Kidd	1		2
Hillel&Gardner	3	1	
Philip	2	1	
Desbordes	9	2	2

Table 6: Number of parameters in several models.

From a practical point of view, a model with the smallest number of parameters is to be preferred. This is even more so when it is realised that some parameters (like 'n' and 'k' for the Nash cascade) cannot be quantified from simple measurements since they have no appropriate physical meaning. In this sense, run-off models are to be regarded as black or grey box models. The notion to use measurable parameters and models with a minimum number of parameters is also supported when calibrating models. As will become clear in Chapter 5 of this thesis, reduction of the number of parameters in a calibration process increases the quality of the calibrated result. Furthermore, when using models with physically meaningful parameters it is possible to crosscheck the parameter values obtained from a calibration with independent separate measurements.

2.3 Hydrodynamics

2.3.1 General

The run-off from surfaces during storms and the DWF enters the drainage system in which it is transported. The resulting flow in the drainage system is described by equations developed in the field of fluid dynamics. In this section the processes, basic equations and assumptions are described.

2.3.2 Flow equations

2.3.2.1 3-Dimensional fluid flow

Conservation of momentum and conservation of mass describe the movement of fluids. The momentum conservation equation reads in its general form (3 space dimensions and the time dimension) as:

$$\frac{\partial u_j}{\partial t} + \nabla_i(u_i u_j) + \frac{1}{\rho}\nabla_j p - \nu \nabla_i^2 u_j = \begin{bmatrix} 0 \\ 0 \\ g \end{bmatrix} \qquad \text{(eq. 2.11)}$$

In which:

u	velocity vector in three dimensions	(m/s)
t	time	(s)
p	pressure	(N/m^2)
ρ	density	(kg/m^3)
ν	viscosity	(m^2/s)
g	gravitational acceleration	(m/s^2)

Each term in this equation represents a process involved in the flow. Generally speaking, potential energy (due to the position in the earth's gravity field) is transformed into kinetic energy held by the movement of the fluid. Due to internal friction in the fluid and friction forces induced by walls this kinetic energy in turn is transformed into heat, noise or deformation of objects. The conservation of mass is described by([6]):

$$\nabla_i u_i = 0 \qquad \text{(eq. 2.12)}$$

This set of partial differential equations is known as the Navier-Stokes equations.
At low Reynolds numbers([7]) the system is stable, at large Reynolds numbers the non-linear convective term becomes large relative to the viscosity term, resulting in an unstable system in which infinitesimal differences in initial conditions result in major differences in the solution. This behaviour is commonly known as turbulence. Turbulence results in an increase in internal friction. Energy from the main stream is dissipated via a cascade of motions on smaller and smaller scales until eventually on the molecular scale the energy results in a raise

([6]) The fluid is assumed to be incompressible implying the rate of change of the density of the fluid, following the motion is zero: $\frac{D\rho}{Dt} = 0$, in that case the mass-conservation equations take the form $\nabla_i u_i = 0$ (see Batchelor (1983)).
([7]) The Reynolds number is a measure for the forces induced by momentum relative to viscous forces, Re=uR/υ, in which R is a characteristic measure of space dimension related to the geometry of the problem studied. E.g. for a free surface flow the hydraulic radius may be applied.

of temperature. For fully turbulent situations the Reynolds equations are applied. This set of equations is derived directly from the Navier-Stokes equations by introducing the assumption that velocity, pressure and tension consist of a mean part and a fluctuating part superimposed on it, i.e. $u_i = \overline{u_i} + u_i'$ (eq. 2.13). Implementation results in six extra terms in the Navier-Stokes equation, the so-called Reynolds stresses q_{ij}. A full derivation can be found Rodi (1980). In this manner the problem of turbulence is reduced to finding proper expressions for the Reynolds tensions. The occurrence of the six extra terms in the equations is known as the closure problem; in practice several model concepts are used to circumvent this. One of the most widely applied models is the k-ε model (Rodi, 1980), which adds two equations to the momentum- and mass balance equations: one equation describing local energy dissipation (ε) and one equation describing the local energy held by the fluid (k). In order to find solutions for practical applications numerical methods are necessary. Modern turbulence models are successful in their application in scientific as well as in engineering practice. This is partly due to the fact that, apart from a description of the geometry of the problem, only few model parameters are necessary, some of which are directly related to the fluid involved and these can be obtained from literature. The storage and calculation capacities of computers however limit the practical application. Application in urban drainage of 2 and 3-D flow models is limited (in 1999) to design of special constructions like storage settling tanks (e.g. Clemens et al (1991), Kluck (1997), De Cock, Vaes & Berlamont (1998)). Generally it may be stated that 2 and 3-D flow simulations are applied on small scale problems whereas, in urban drainage, on the scale of whole systems 1-D flow simulations are applied. As outlined in Chapter 3, data collection for relatively simple one-dimensional models already poses a serious problem. The 3-D geometrical description of an urban drainage system of even a small catchment is yet a practical impossibility. Therefore, in the opinion of the author, a full 3-D process model of fluid flow in urban drainage systems is not a realistic option in the near future.

2.3.2.2 Reduction to 1 space dimension

As outlined, a fundamental process description of hydrodynamic processes is not a feasible alternative in urban drainage yet. Therefore, a 1-dimensional approach is usually adopted. This means that the velocity vector in the momentum and mass-balance equation are integrated into a one dimensional form. The dependent variables are integrated quantities. In order to get a closed system of equations simplifying assumptions are needed. Local effects such as caused by weirs, are taken into account by estimating their effect on the integrated equations. The reduction into one space dimension is defensible; when studying the geometry of a conduit in an urban drainage system the flow is to a large extent 1-dimensional, except for the velocity field at the direct entrance. In manholes the velocity field is 3-dimensional, but this can be circumvented in a 1-d model as will be explained later.

The equations applied in a 1-dimensional approach are the well-known De Saint-Venant equations (De Saint-Venant (1871)):

Momentum balance:

$$\underbrace{\frac{\partial Q}{\partial t}}_{I} + \underbrace{\frac{\partial}{\partial x}\left[\beta\frac{Q^2}{A}\right]}_{II} + \underbrace{gA\frac{\partial h}{\partial x}}_{III} + \underbrace{c_f\frac{Q|Q|}{R_h A}}_{IV} = 0 \qquad \text{(eq. 2.14)}$$

Mass balance:

$$\frac{\partial Q}{\partial x} + \frac{\partial A(h)}{\partial t} = \frac{\partial Q}{\partial x} + B(h)\frac{\partial h}{\partial t} = 0 \qquad \text{(eq. 2.15)}$$

In which:

Q	discharge	(m^3/s)
A	cross-sectional area	(m^2)
B	width of the free water surface	(m)
g	gravitational acceleration	($\approx 9.813\ m/s^2$)
R_h	hydraulic radius	(m)
c_f	resistance constant	(-)
h	water level	(m)
x	location along x-axis	(m)
t	time	(s)
β	Boussinesq's number	(-)

The assumptions applied are:

- hydrostatic pressure.
- velocity components in y and z direction are negligible compared to the velocity component in x direction ($u_y=u_z<<u_x$).

The individual terms in the momentum balance equations (eq. 2.14) are:

I	acceleration term
II	convective term
III	gravitational term
IV	friction term

The third term may also be written as:

$$gA\frac{\partial h}{\partial x} = gA\left[\frac{\partial a}{\partial x} + \frac{\partial z_b}{\partial x}\right] = gA\left[\underbrace{\frac{\partial a}{\partial x}}_{IIIa} - \underbrace{i_b}_{IIIb}\right] \qquad \text{(eq. 2.16)}$$

In which:

a	water depth	(m)
z_b	bottom level	(m)
i_b	bottom slope	(-)

Term IIIa is the pressure term and term IIIb is the gravity term.

The De Saint-Venant equations (using the dynamic wave approach) form a hyperbolic system of partial differential equations. This implies that in order to obtain a well-posed problem the initial condition (Q and h at t=0) and two boundary conditions have to be defined.
Several possible simplifications of eq. 2.14 and 2.15 are applied depending on the possibility of neglecting terms in the momentum balance equation. For instance when term I and II (the inertia term) can be neglected, which is the case when the flow varies only slowly with time and space, either the kinematic or diffusion wave equation is obtained. In Table 7 the possible simplifications for the momentum equations are summarised along with the identification normally used.

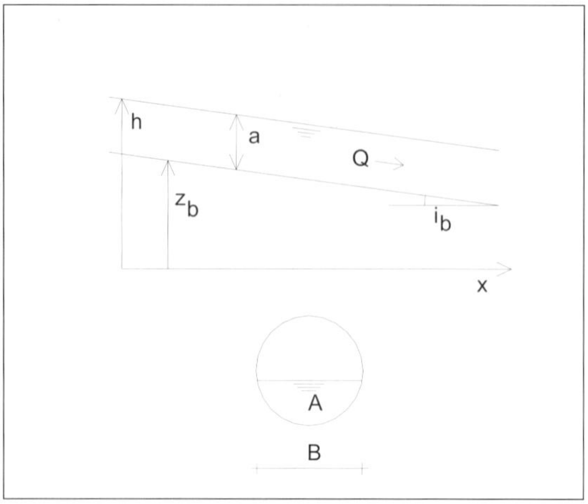

Figure 15: Definition sketch.

Terms taken into account	Identification of the simplified equation	main assumptions
I+II+III+IV	Dynamic wave	no additional assumptions
I+II+III	Gravity wave	friction is small compared to gravity and inertia terms
III+IV	Diffusion wave	inertia terms are small compared to gravity and friction
IIIb+IV	Kinematic wave	inertia terms are small compared to gravity and friction and $\frac{\partial a}{\partial x} << i_b$

Table 7 : Possible simplifications for the momentum balance equation.

The simplified versions of the momentum equation are mostly used to drive analytical solutions for special (academic) cases, or to simplify the numerical calculation for practical cases. Due to the integration of two space dimensions, several new parameters enter the equations. For instance, the geometry of the problem like hydraulic radius, wetted area and hydraulic depth are such parameters. The parameter β([8]) in the convective term (term II) is the Boussinesq number (Boussinesq, (1897)), defined by:

$$\beta = \frac{\int_0^A u^2 dA}{\bar{u}^2 A} \qquad \text{(eq. 2.17)}$$

The value for β is >1 and can eventually reach values >1.2 in extreme cases, e.g. when a sediment bed is present in a channel with a circular cross section at shallow water depths, see Kleywegt (1992). These equations basically describe the flow of a fluid as a function of place

([8]) The Boussinesq number accounts for the fact that the velocity is not uniformly distributed over the cross section of the flow, in fact it is a correction factor due to simplifying the 3-dimensional flow equations.

in one dimension and time in a conduit; the derivation of this famous set of equations can be found in several textbooks (e.g. Chow, (1959)).
In the 3-D description, no geometrical parameters enter the equations; the geometry of a given problem is completely defined by the locations in the 3-D space of the boundary conditions. In the next sections, some of the most important consequences of simplification into a 1-D system are discussed. In this discussion the main focus is on hydraulic parameters and structures of importance for urban drainage problems.

2.3.2.3 Frictional losses

The energy dissipation by internal friction in the fluid and due to wall friction is accounted for in the friction term. Due to the fact that a 1-dimensional simplification is applied, the friction term consists of components regarding material properties (the wall-roughness), the geometry (profile shape), which is essentially a function of water depth and the mean velocity in the channel. The friction term in the momentum balance equations reads:

$$-c_f \frac{Q|Q|}{AR_h} \qquad \text{(eq. 2.18)}$$

Strictly speaking this approach is only valid in the uniform, stationary case. For short lengths relative to the hydraulic radius (i.e. $L<40\text{-}50\ R_h$), as mostly occurs in urban drainage systems, measurable deviations may occur. This is due to the fact the development of a uniform velocity profile depends on the velocity. It must be stated however, that these effects become only manifest at relatively high flow velocities ($u>2\text{-}3$ m/s). In urban drainage these velocities may easily occur in steep areas. The friction coefficient can be calculated using several empirical formula's most of which use the Nikuradse's equivalent sand roughness (see Nikuradse (1932, 1933)) to introduce the influence of the wall roughness on the friction term. Some well known empirical formula's are:

$$c_f = \frac{g}{C^2}, \text{ with } C = 18\log\left(\frac{12R_h}{k_n}\right) \text{ (Colebrook-White)} \qquad \text{(eq. 2.19)}$$

With k_n equivalent roughness (m)

or

$$c_f = \frac{g}{C^2}, \text{ with } C = \frac{R_h^{\frac{1}{6}}}{n} \text{ (Manning)} \qquad \text{(eq. 2.20)}$$

in which n is 'Mannings n'. The value of n depends on the type of material, its value can be found in several textbooks (see e.g. Chow (1959)).

2.3.2.4 Local losses

Other sources of energy dissipation are present at locations where the basic assumption (1 dimensional flow) is not valid (so-called local losses). For example, the flow in manholes, junctions and bends has significant velocity components in three dimensions (resulting in recirculation zones as depicted in Figure 16).

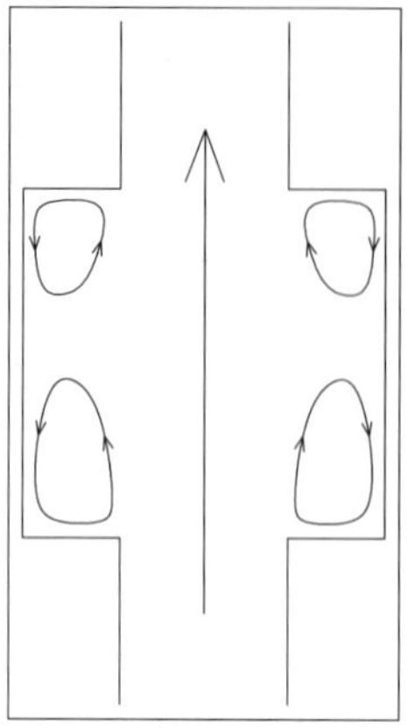

Figure 16: Flow pattern in a manhole.

These can be included in the 1-D concept by introducing a local loss coefficient relating the head losses to the main component of the velocity vector:

$$\Delta H = \xi \frac{u^2}{2g} \qquad \text{(eq. 2.21)}$$

in which:

ξ	head loss coefficient	(-)
u	streaming velocity	(m/s)
g	gravitational acceleration	(m/s^2)
ΔH	head loss	(m)

The actual value for ξ depends on:

- Geometry.
- Flow velocity and direction.
- Water levels.

The effect of the local losses (and friction in general) is relatively small when flow velocities are relatively small (e.g. < 0.5 m/s). In rather flat systems friction is not a dominant factor while in steep systems this aspect is important and should therefore be modelled correctly. Many laboratory studies have been carried out to obtain values or relations between geometry and local loss coefficient. Some results discussed in literature will be presented here. Lindvall (1984) studied the loss coefficient of two manhole geometry's shown in Figure 17.

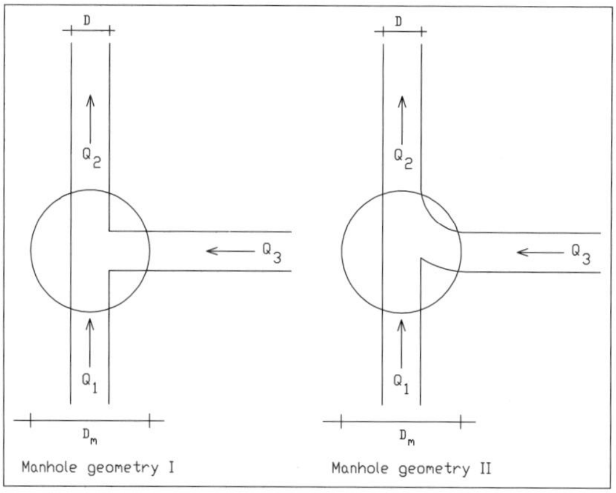

Figure 17: Manhole geometries.

For the flow situation 'straight forward' (i.e. $Q_3=0$) the following relationships were derived from experimental data:

$$\left.\begin{array}{l}\xi_I = 0.07 + 0.083\left[\frac{D_m}{D} - 1.7\right]\tanh\left[0.5\left[\frac{a}{D} - 1.0\right]\right] + \\ 0.40\left[1.1 + \tanh\left[\frac{D_m}{D} - 3.0\right]\right]e^{\left[-15\left[\ln\left[\frac{a}{1.5D}\right]\right]^2\right]}\end{array}\right\} \quad \text{(eq. 2.22)}$$

And

$$\left.\begin{array}{l}\xi_{II} = 0.04 + 0.025\left[\frac{D_m}{D} - 1.7\right]\tanh\left[0.5\left[\frac{a}{D} - 1.0\right]\right] + \\ 0.10\left[1.35 + \tanh\left[\frac{D_m}{D} - 2.8\right]\right]e^{\left[-7\left[\ln\left[\frac{a}{1.5D}\right]\right]^2\right]}\end{array}\right\} \quad \text{(eq. 2 23)}$$

In which:

a	water depth above invert level	(m)
D	diameter of the conduits	(m)
D_m	diameter of the manhole	(m)
ξ	head loss coefficient	(-)

These relations were reported to be valid for $1.7 < \frac{D_m}{D} < 4.1$ and $\frac{a}{D} > 1$ (full pipe flow).

These relations are shown in Figure 18 and Figure 19.

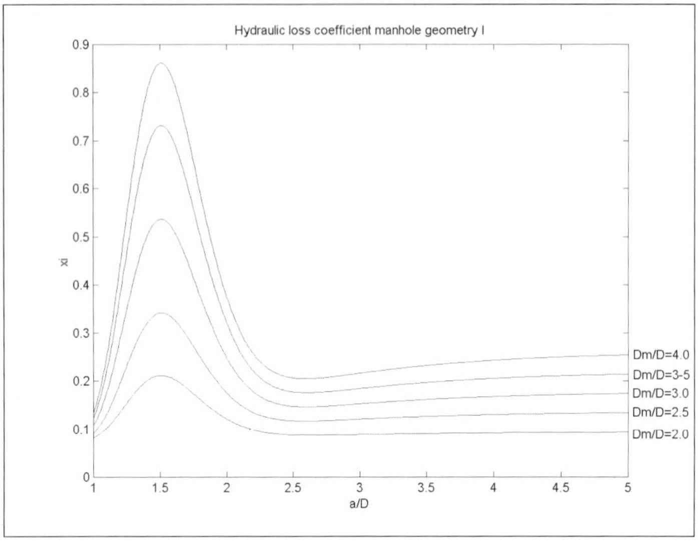

Figure 18: Hydraulic loss coefficient manhole geometry I.

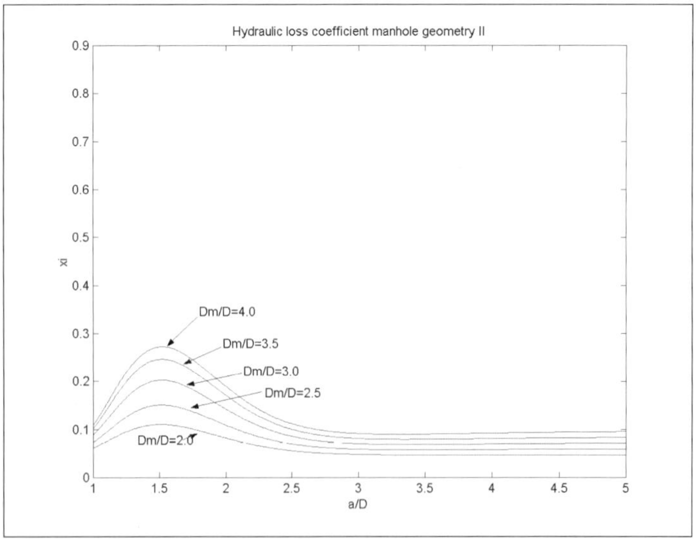

Figure 19: Hydraulic loss coefficient manhole geometry II

As can be seen the value for the loss coefficient strongly depends on the geometry of the manhole and in the domain $1 < \frac{a}{D} < 2.5$ also on the ratio between water depth and pipe diameter. Similar observations are reported for other manhole geometries by Howarth & Saul (1984), Pedersen & Mark (1990) and Sakakibara et al (1996). The applicability of values reported in literature is limited to the geometries tested in the laboratory. In practice, an almost infinite number of different types of geometry for manholes and junctions is found making general statements with respect to values for ξ impossible. Possibly a certain standardisation of geometry may change this in the future.

In the early days of hydraulic calculation, local losses were ignored; that is to say, they were not introduced in the model separately. In order to account for their influence the hydraulic roughness of the conduits was raised. In the Netherlands it was accepted practice to use a value of k_n=1.5 mm (value according to Nikuradse (1933)) in stationary calculations. The effect of one manhole can be translated into an equivalent pipe length, defined by

$$L_{eq} = \frac{\xi C^2 R_h}{2g}. \qquad \text{(eq.2.24)}$$

For instance, when introducing a conduit with a diameter of 1 m and an equivalent roughness of 1.5 mm this expression becomes $L_{eq} = \xi * 48.44 \quad m$. The value for ξ lies roughly between 0.05 and 0.9, implying that the equivalent length is between 2.4 and 43.6 m (relative to a mean conduit length of 20-30 m this may form a considerable increase in hydraulic resistance). From these estimates it is clear that the manholes may play an important role in the hydraulic calculations, especially when poor hydraulic designs are implemented (high values for ξ) or/and when high velocities may occur (> 0.5 m/s).

2.3.2.5 Structures

In most drainage systems several types of structures are incorporated. Weirs, sluices, orifices, local contractions are of particular importance.
The flow-head relationship generally reads as (for sub-critical flow):

$$Q(\Delta h) = mA\sqrt{2g\Delta H} \qquad \text{(eq. 2.25)}$$

in which:

A	cross section	(m^2)
g	gravitational acceleration	(m/s^2)
ΔH	difference in head over the structure	(m)
m	empirical coefficient	

This Q-h relations follows directly from applying the momentum- and mass balance equation (for stationary conditions) locally assuming friction losses due to wall friction to be zero. The coefficient m originates from the application of the1-D simplification, especially when neglecting the two components in the velocity vector causing energy loss due to local turbulence. The value of m has to be derived either from measurement, detailed 3-D calculations or tables (see e.g. Idel'cik (1963)).

The discharge over a weir is generally described in 1-D modelling by the Kindsvater-Carter equation (Kindsvater & Carter (1957)):

$$Q(h) = Lm\frac{2}{3}\sqrt{\left(\frac{2}{3}g\right)}h^{\frac{3}{2}} \qquad \text{(eq. 2.26)}$$

This formula is straightforward in its application and calls for only one parameter to be given (apart from the weir length L); namely the overflow coefficient m.
The value for this parameter is extremely well known for specified geometries with respect to weir construction and initial flow conditions, because of the use of weirs as laboratory

measuring equipment. These conditions however, are usually not met (unless by coincidence) in the practical situations in urban drainage systems.
In practice, non-of the conditions as described before is met in urban drainage:

- The physical properties are not known in sufficient detail.
- The weir geometry hardly ever meets the prescribed conditions.
- The flow pattern in the direct vicinity of the weir is mostly irregular.

So, textbook figures for the overflow coefficient may not be applied safely, indeed applying them may even lead to unaccounted for flooding, since the actual coefficients are mostly significantly lower that the textbook figures. Field calibration of weir constructions applied in urban drainage generally result in Q-h relations of the form:

$$Q(h)=\alpha h^{\beta} \qquad \text{(eq. 2.27)}$$

in which the value of β may deviate substantially from the theoretical value of 1.5; see also Section 4.4.3.1. of this thesis and NWRW theme 8.1 (1987) or Foundation RIONED et al (1999)

2.3.2.6 Boundary conditions, initial conditions, loads

The De Saint-Venant equations form a hyperbolic system of partial differential equations. This can be seen when analysing the characteristics of the equations. The primary characteristics are obtained as follows:

$$\left(\frac{dx}{dt}\right)_{1,2}=\frac{Q}{A}\pm\sqrt{g\frac{A}{B}} \qquad \text{(eq. 2.28)}$$

and the relationship in the Q-h plane is:

$$dQ=\left[g\frac{Q|Q|}{C^{2}R_{h}A}\right]dt-\left(\frac{dt}{dx}\right)_{1,2}\left[gA-\frac{Q^{2}}{A^{2}}B\right]dh \qquad \text{(eq. 2.29)}$$

As can be seen the primary characteristics are (for physically meaningful values of A, Q and B) real and distinct, implying that the system of equations is hyperbolic (see Figure 20). The primary characteristics represent the velocities of infinitesimal disturbances in the water; the first characteristic is the maximum velocity with which such a disturbance (a wave) can propagate([9]).
A practical problem is only well posed when a set of initial and boundary conditions is given. This implies that for any given problem the initial values for Q and h at t=0 should be known (initial condition). The type of boundary conditions is less obvious and may change in the time domain. For instance, if for given initial conditions the two characteristic directions have an opposite sign this implies that information travelling from the upstream as well as the downstream boundary influences the solutions in the x-t plane. Therefore, at both the upstream and downstream boundary a value for either h or Q should be prescribed as a function of time. In this case, the flow is called subcritical. When both characteristic directions are positive, information from the downstream boundary cannot influence the solution in the x-t domain (supercritical flow). In this case, two values (Q and h) should be prescribed as a function of time at the upstream boundary.

([9])Based on the characteristics the system of differential equations can be solved for any given well-posed set of initial and boundary conditions. This method however, is relatively cumbersome to implement in a computer code for an urban drainage system and will therefore not be considered any further. An elaborate description of this particular method is given in e.g. Abbott & Basco (1989).

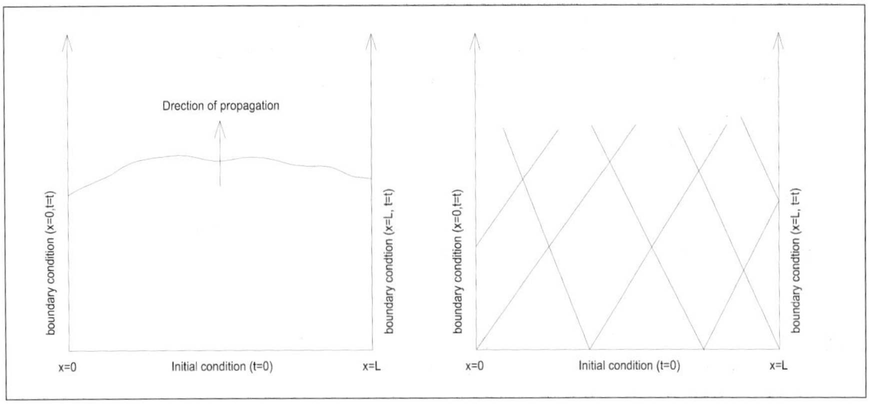

Figure 20: Hyperbolic problem.

The Froude number is normally used to make this distinction. When assuming the Boussinesq number to be unity this is easily derived from the characteristic relations. If $\frac{Q}{A} < \sqrt{g\frac{A}{B}}$ the two characteristic directions have an opposite sign, if $\frac{Q}{A} = \sqrt{g\frac{A}{B}}$ the values for the characteristic relations become 0 and $2\frac{Q}{A}$, while when $\frac{Q}{A} > \sqrt{g\frac{A}{B}}$ both directions are >0.
The Froude number can be defined by:

$$Fr = \frac{Q}{A\sqrt{g\frac{A}{B}}} \qquad \text{(eq.2.30)}$$

and the presented conditions become $Fr < 1, Fr = 1, Fr > 1$, respectively.
Because the discharge and some geometrical properties that are functions of the water depth are needed to define the Froude number, the Froude number may have different values in every point in the x-t plane. This immediately implies that the necessary type of boundary conditions may change as a function of time.
This is certainly so in an urban drainage system, where the type of boundary conditions applies for individual conduits. An example is given in Figure 21. Initially (at t=0) the water level in manhole 2 is below the invert level of the conduit, implying a free overflow at the downstream end. In this case the boundary condition is given by a Q-h relation based on critical flow and the geometry of the conduit (see annexe II). After some time (t=t) the water level in node 2 has risen above the invert level, and the boundary condition changes into a 'prescribed' water level as a function of time.

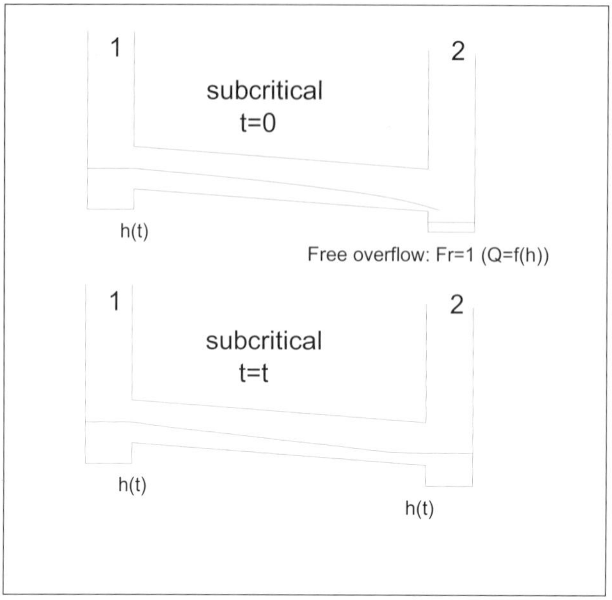

Figure 21 : Two different types of boundary conditions.

Essentially the boundary conditions for conduits are internal boundary conditions; of course some external boundary conditions may have to be given also. For instance, if at some location an outflow occurs the external water level (h(t)) should be supplied as the boundary condition at that location.
Boundary conditions must also be supplied for overflows, pumps, and free outflows. With respect to pumping-stations the switch-on and switch-off levels are supplied along with the pumping characteristics. For weirs, the discharge is zero as long as the upstream water level is below the weir level. When the water level is at or above the weir level the discharge is defined by the Q-h relation (assuming modular flow) for that particular construction. For individual conduits several situations can occur, either the flow is sub-critical or supercritical, in the former situation a boundary condition is supplied at the upstream end as well as at the downstream end (either h(t) or Q(t)). When the flow is supercritical, two boundary conditions should be supplied at the upstream end. The following types of boundary conditions are distinguished:

- Closed boundary: u(t)=0.
- Prescribed water level: h=h(t).
- Prescribed discharge: Q=Q(t).
- Internal boundary (e.g. a pump, internal weir or other structure).
- A Q-h relation (e.g. a free outflow).

The initial values for water levels discharges and state of operation of e.g. pumps, has to be known for each element present in the system to be modelled. Since there is no objective norm by which to judge which type of initial condition is to be used, a large number of possibilities are applied in practice. Either all discharges and water depths are set to a (small) fixed value, or some equilibrium state is adopted as the initial condition. The latter approach has the favourable property that the processes modelled at t>0 are not interfered with by initialising processes.

The hydraulic load on the system is the inflow from external sources into the system, which can be supposed to occur mainly at the manholes. The actual magnitude of these external discharges at manholes either is supplied by the user or is the result of the run-off model running in conjunction with the hydrodynamic model.

2.4 Discretisation

2.4.1 General

As stated before, no general analytical solution is known for the De Saint-Venant equations. In practice this implies numerical methods have to be applied in order to achieve a solution. For some special academic cases analytical solutions exist, these may be used to verify the software code based on numerical methods. A large body of literature exists with respect to numerical methods to solve partial differential equations: e.g. finite difference methods, finite elements methods, boundary element method, method of characteristics (see e.g. Vreugdenhil (1989) or Abbot & Basco (1989)). In hydrodynamic models, difference methods are the most popular for practical use, since they are more convenient in use when compared to the other possibilities. Finite difference methods transform partial differential equations into finite difference equations by approximating functions and their derivatives using estimates based on discrete parameter values, for example the differential equation:

$$\frac{da}{dt} = f(t,a) \qquad \text{(eq. 2.30)}$$

is approximated by:

$$\frac{a^{t+\Delta t} - a^{t}}{\Delta t} = f(t,a^{t},...) \qquad \text{(eq. 2.31)}$$

In this manner a discretisation of the differential equations is obtained enabling a computer code to be written for it. Mostly implicit difference schemes are applied. By definition a numerical scheme is called implicit if the function at the right side in eq. 2.31 depends on $a^{t+\Delta t}$.

The general form of such a numerical scheme reads as:

$$\underline{\underline{A}}\,\underline{\psi}^{n+1} = \underline{\underline{B}}\,\underline{\psi}^{n} + \underline{R} \qquad \text{(eq. 2.32)}$$

Or:

$$\underline{\psi}^{n+1} = \underline{\underline{D}}\,\underline{\psi}^{n} + \underline{C} \qquad \text{(eq. 2.33)}$$

In which:

$$\underline{\underline{D}} = \underline{\underline{A}}^{-1}\underline{\underline{B}} \qquad \text{(eq. 2.34)}$$

Here $\underline{\underline{A}}$, $\underline{\underline{B}}$ and $\underline{\underline{D}}$ are a matrix operators defined by the applied difference method and $\underline{\psi}^{n}$ is a vector containing the unknown variables at time level n for all calculation points in the model.

2.4.2 Translation of the De Saint-Venant equations into finite difference equations

Since the model applied in this thesis is based on the well-known Preissmann (also known as Box-, Keller- or Wendroff scheme); see Preissmann (1961). This scheme will be used as an example of how the set of partial differential equations translate into a set of finite difference equations.

In the Preissmann scheme the trapezoidal interpolation in the space dimension is used together with the implicit midpoint interpolation for the time dimension:

$$\left.\begin{aligned} f(t)_x &\approx \theta f_x^{n+1} + (1-\theta) f_x^n \\ f(x)_t &\approx \frac{1}{2}\left(f_{j+1}^t + f_j^t\right) \end{aligned}\right\} \qquad \text{(eq. 2.35)}$$

$$\left.\begin{aligned} \frac{\partial f(t)_x}{\partial t} &\approx \frac{f_x^{n+1} - f_x^n}{\Delta t} \\ \frac{\partial f(x)_t}{\partial x} &\approx \frac{f_{j+1}^t - f_j^t}{\Delta x} \end{aligned}\right\} \qquad \text{(eq. 2.36)}$$

Combining these semi-discretisations yields:

$$\left.\begin{aligned} f(x,t) &\approx \frac{\theta}{2}(f_{j+1}^{n+1} + f_j^{n+1}) + \frac{1-\theta}{2}(f_{j+1}^n + f_j^n) \\ \frac{\partial f}{\partial x} &\approx \theta \frac{f_{j+1}^{n+1} - f_j^{n+1}}{\Delta x} + (1-\theta)\frac{f_{j+1}^n - f_j^n}{\Delta x} \\ \frac{\partial f}{\partial t} &\approx \frac{f_{j+1}^{n+1} - f_{j+1}^n + f_j^{n+1} - f_j^n}{2\Delta t} \end{aligned}\right\} \qquad \text{(eq. 2.37)}$$

in which θ is a weighting parameter, defined relative to the central point of the scheme (see Figure 22). The De Saint-Venant equations (eq. 2.14 and 2.15) are a system of non-linear, first order, partial differential equations resulting in a system of non-linear difference equations after imposing the discetisation (e.g. according to the Preissmann scheme). Since no general method is available for directly solving large system of non-linear equations either some form of linearisation is to be applied or an iterative solution procedure has to be adopted.

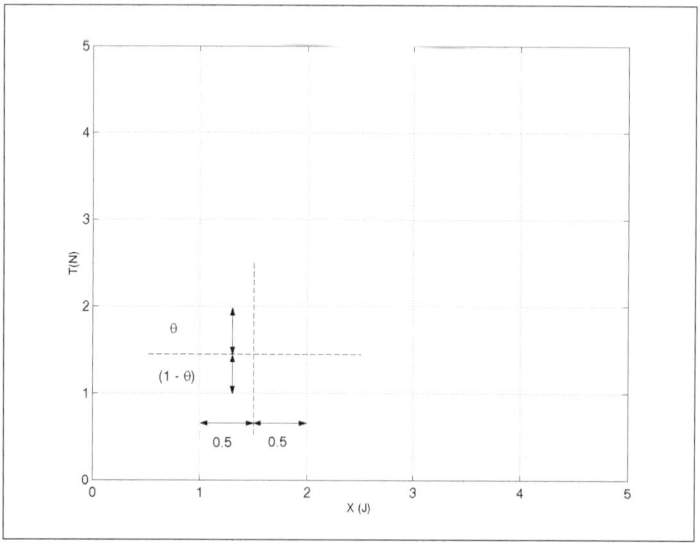

Figure 22: The Preissmann scheme in the x-t plane.

In fact, several methods to linearise the difference equations are possible; it is difficult to say which method of linearisation is best since no specific condition to this end can be formulated in general. For some terms, however it is possible to judge a linearisation as 'preferable'. Schuurmans (1991) has shown that the resistance term in the momentum equation is best linearised by:

$$\left(\frac{c_f}{2AR^2}\right)_j^{n+1/2} \left|Q_j^n\right| Q_j^{n+1} + \left(\frac{c_f}{2AR^2}\right)_{j+1}^{n+1/2} \left|Q_{j+1}^n\right| Q_{j+1}^{n+1} \qquad \text{(eq 2.38)}$$

For the convective term in the momentum balance equation, Stelling (1984) made plausible that a linearisation in the form:

$$\frac{\partial}{\partial x}\left[\frac{Q^2}{A}\right] \approx \frac{\left[\frac{Q^{n+1}Q^n}{A^{n+1/2}}\right]_{j+1} - \left[\frac{Q^{n+1}Q^n}{A^{n+1/2}}\right]_j}{\Delta x} \qquad \text{(eq. 2.39)}$$

is to be preferred in order to minimise the possible adverse effects of this term concerning the stability properties of the scheme.

Since there are many different possibilities for linearisation, many different implementations of the Preissmann scheme for solving the De Saint Venant equations have been applied.

As an example, a possible linearisation technique is shown for the mass-balance equation; a similar approach can be used for the momentum balance.

The mass-balance equation reads as:

$$\frac{\partial h}{\partial t} + \frac{1}{B(h)}\frac{\partial Q}{\partial x} = 0 \qquad \text{(eq. 2.40)}$$

in which B(h) is the width of the free water surface as a function of depth. For brevity the following notations are applied:

$$\left.\begin{aligned} Q^{n+1} &= Q^n + \Delta u \\ h^{n+1} &= h^n + \Delta h \\ B^{n+1} &= B^n + \Delta B \end{aligned}\right\} \qquad \text{(eq. 2.41)}$$

Applying a Taylor expansion, B is approximated by:

$$B^{n+1} = B^n + \left.\frac{dB}{dh}\right|_{h^n} \Delta h + \left.\frac{d^2B}{dh^2}\right|_{h^n} \frac{\Delta h^2}{2} + \ldots.. \qquad \text{(eq. 2.42)}$$

Applying the Preissmann scheme to eq. 2.40, using eq. 2.42 while dropping higher order terms and assuming $\Delta h \Delta Q \approx \Delta Q \Delta Q \approx \Delta h \Delta h \approx 0$, results in:

$$\left.\begin{aligned} &\frac{\Delta h_{j+1} + \Delta h_j}{2\Delta t} + \frac{2}{B_{j+1}^n + B_j^n}\left[\theta \frac{\Delta Q_{j+1} - \Delta Q_j}{\Delta x} + \frac{Q_{j+1}^n - Q_j^n}{\Delta x}\right] - \\ &\frac{2\theta}{\left(B_{j+1}^n + B_j^n\right)^2}\left(\frac{dB_j}{dh}\Delta h_j + \frac{dB_{j+1}}{dh}\Delta h_{j+1}\right)\left(\frac{Q_{j+1}^n - Q_j^n}{\Delta x}\right) = 0 \end{aligned}\right\} \qquad \text{(eq.2.43)}$$

In this manner, a linearised difference equation is obtained in terms of $\Delta u_j, \Delta u_{j+1}, \Delta h_j, \Delta h_{j+1}$:

$$A_j \Delta h_j + B_j \Delta h_{j+1} + C_j \Delta Q_j + D_j \Delta Q_{j+1} = RR_j \qquad \text{(eq. 2.44)}$$

A similar form is obtained for the momentum equation:

$$E_j \Delta h_j + F_j \Delta h_{j+1} + G_j \Delta Q_j + H_j \Delta Q_{j+1} = RM_j \qquad \text{(eq. 2.45)}$$

For a system of 4 calculation points in the space direction, this results in a system of equations like:

$$
\underbrace{\begin{bmatrix}
R_1 & 0 & 0 & 0 & 0 & 0 & 0 & 0 \\
A_1 & C_1 & B_1 & D_1 & 0 & 0 & 0 & 0 \\
E_1 & G_1 & F_1 & H_1 & 0 & 0 & 0 & 0 \\
0 & 0 & A_2 & C_2 & B_2 & D_2 & 0 & 0 \\
0 & 0 & E_2 & G_2 & F_2 & H_3 & 0 & 0 \\
0 & 0 & 0 & 0 & A_3 & C_3 & B_3 & D_3 \\
0 & 0 & 0 & 0 & E_3 & G_3 & F_3 & D_3 \\
0 & 0 & 0 & 0 & 0 & 0 & 0 & R_2
\end{bmatrix}}_{\underline{D}}
\underbrace{\begin{bmatrix}
\Delta h_1 \\ \Delta Q_1 \\ \Delta h_2 \\ \Delta Q_2 \\ \Delta h_3 \\ \Delta Q_3 \\ \Delta h_4 \\ \Delta Q_4
\end{bmatrix}}_{\Delta v}
=
\underbrace{\begin{bmatrix}
RB_1 \\ RR_1 \\ RM_1 \\ RR_2 \\ RM_2 \\ RR_3 \\ RM_3 \\ RB_3
\end{bmatrix}}_{\underline{f}_n}
\qquad \text{(eq. 2.46)}
$$

The first and last row of the matrix represent the boundary conditions for a sub-critical case. The situation at t+Δt is then calculated by $v^{n+1} = v^n + \Delta v^n$. Here $\underline{\underline{D}}$ is the differential operator for one conduit banded with a width of 5. The rows in this matrix contain, alternately, the approximation of the derivative of the mass-balance and momentum equation.
This is in fact a first iteration in a multi-dimensional Newton-Rhapson process for one time step.

$$\Delta v = \underline{D}^{-1} \underline{f}_n \qquad \text{(eq. 2.47)}$$

This method for linearising was followed in the CAREDAS model and has the advantage that per time step only one matrix inversion has to be performed. The price to be paid however is that the principle of (mass) conservation is violated. As long as ΔQ, Δh and the increments for coefficients and geometrical properties (e.g. B) are small relative to their absolute values, the violation of the conservation principle is small. An important argument to apply just one Newton 'iteration' is that of saving calculation time in practical application. This argument however has lost its validity in recent years due to increasing computing power. Therefore, it is preferred to take the principles of mass and momentum conservation as a criterion, since it is in line with the fundamental physical principles of the real world processes. This urges to make some iterations per time step to realise the conservation principles to within a (user) prescribed accuracy. In practice, 2-5 iterations per time step are needed to this end. In annexe II, a calculation method based on these principles is described in some detail. The method was implemented in a computer code and used for the examples used in section 2.4.3 on stability problems.
When the number of space steps is large, the matrix $\underline{\underline{D}}$ (in eq.2.47) contains mostly zero elements claiming a large amount of storage capacity. Therefore, it is sound practice only to store the non-zero elements and keep track of their position in the matrix. When inverting a banded matrix the storage space needed is equal to the bandwidth times the number of rows present.
In the simple example used so far, a rather efficient system is obtained. This may change however, when a network of manholes and conduits is modelled. In this case the order in which the equations are stored (i.e. the internal numbering of mesh points) defines the bandwidth of the matrix obtained. Therefore, it has a large influence on storage facilities and CPU time needed for the matrix inversion.
So far only one conduit was considered. When a system of conduits and nodes is modelled, the following equation for the nodes is inserted:

$$A_m \frac{dh}{dt} + \sum Q_i = Q_{node,external} \qquad \text{(eq. 2.48)}$$

(The sign of the discharges must be defined; e.g., a discharge is positive when directed toward the node). The normal equations apply for each conduit in the system. In some computer programs, the space step Δx is set equal to the conduit length([10]), implying that the equation of motion for a conduit is discretised into one calculation cell. The mass balance equation is valid for the nodes.

The influence of the numbering order on the matrix is illustrated by two alternative numbering orders for the variables([11]) in the system of Figure 23. A naive numbering order would be the sequence([12])

A: $[H_1,H_2,H_3,H_4,H_5,Q_{11},Q_{21},Q_{12},Q_{42},Q_{43},Q_{23},Q_{24},Q_{34},Q_{45},Q_{35},Q_{36},Q_{56}]$.

An optimised sequence B is:

B: $[H_1,Q_{11},Q_{21},Q_{42},Q_{43},Q_{12},Q_{23},H_2,H_4,Q_{24},Q_{45},Q_{24},Q_{35},H_3,Q_{36},Q_{56},H_5]$.

In Figure 24 the effect of using sequences A and B on the scatter of non-zero elements in the matrix. Several optimisation routines are available to obtain a numbering order resulting in a nearly minimal bandwidth of the matrix; see e.g. King (1970). Using techniques like these, the optimal bandwidth is defined by the maximum number of conduits connected to a node.

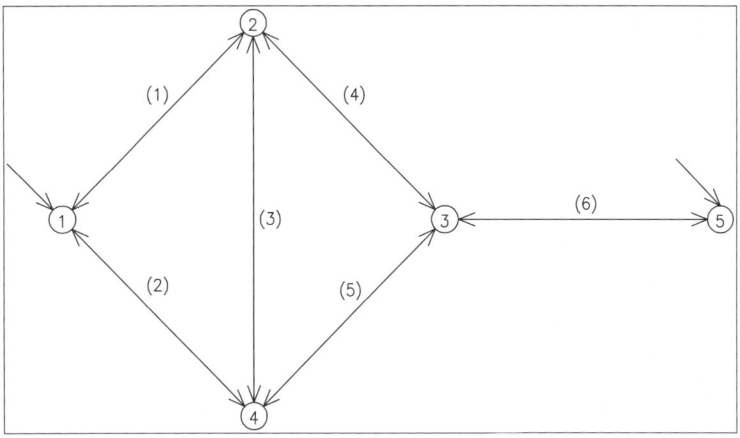

Figure 23: Simple network containing 5 nodes, 6 conduits (numbers between brackets) and two external inflows at nodes 1 and 5.

Apart from band optimisation techniques, other advanced methods to effectively invert large sparse matrices are widely applied. Especially for large matrices iterative methods like the Gauss-Seidel method of successive over-relaxation or the Conjugate Gradient method are mostly preferred. A discussion on this subject is beyond the scope of this thesis. The interested reader is referred to literature (see e.g. Golub & Van Loan (1996)).

([10]) This has effects on the modelling of certain flow modes, like neglecting hydraulic jumps. Furthermore, it inhibits the user to vary the discretisation applied in the space domain making the calculation vulnerable for numerical instabilities (see Section 2.4.4).

([11]) For brevity, we assume that the water level in a manhole is equal to the water level at the invert of the conduits connected to it. In practice this is not always the case especially when differences in invert levels exist. In software tools applied in practice this is of course taken into account.

([12]) For Q_{ij} the subscript i denotes the number of the manhole and the subscript j denotes the conduit number.

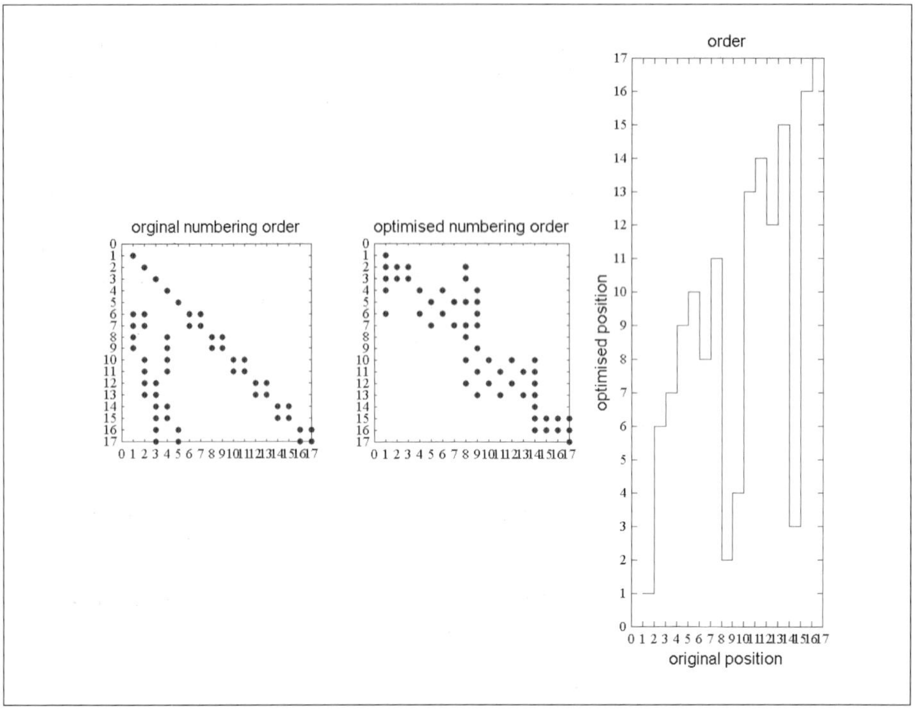

Figure 24: Profile of the system matrix for different numbering sequences.

2.4.3 Stability, consistency, convergence

2.4.3.1 General

A full stability analysis of the numerical schemes applied in hydrodynamic calculation in urban drainage is beyond the scope of this thesis. However, some aspects will be highlighted since stability related problems and their (software engineering) solutions put constraints on their applicability in practice.

The difference equations applied in numerical calculations are not identical to the continuous differential equations in the original process description. In fact, the solution of a difference equation is only an approximation of the real solution. In order to achieve an acceptable (in terms of physical meaningful results) approximation a numerical scheme should be (Stelling & Booij (1999)):

- Consistent.
- Convergent.
- Zero Stable.

Furthermore, for practical application a numerical scheme should posess the following properties

- Absolute stability
- A-stability

A-stability will not be discussed here. Without going into details it is noticed that for a numerical scheme being A-stable no (stability-related) limit for the step size exists.

The consistency of a numerical scheme is defined as the condition that the difference (or truncation error) between difference equation and differential equation tends to zero when the step sizes applied tend to zero (i.e. $\Delta t \to 0, \Delta x \to 0$). Consistency is most cases easily proven by applying Taylor series (see e.g. Richtmyer & Morton (1967)).

A numerical scheme is convergent when the solution obtained for a problem (using a given set of difference equations and a set of boundary and initial conditions) converges to the solution obtained by the corresponding continuous equations to within an arbitrary, but finite, limit. Convergence is studied using the behaviour of successive solutions with different fixed meshes (i.e. $\Delta t = C_t, \Delta x = C_x$) and studying the sequences of the solution in the time-space domain. The concepts of stability, consistency and convergence are linked together by the Lax theorem stating:

> '*For a consistent difference approximation to a well posed linear evolutionary problem, the(zero) stability of the scheme is necessary and sufficient for convergence.*'

The formal proof of the Lax theorem is found in Morton & Mayers (1994). This theorem implies that when a given numerical scheme is consistent and zero stable, convergence is guaranteed.
There are several formal definitions of stability, for brevity the results reported by Godunov & Ryaben'kii (1964) and van der Houwen (1968) are mentioned only.

Zero stability according to Godunov& Ryaben'kii implies that:

$$\left\| D \right\|_{dx=\Delta}^{k} \leq C_k, k = 1, \ldots\ldots, \frac{t}{\Delta t}, \ldots\ldots, \frac{T}{\Delta t}. \qquad \text{(eq. 2.49)}$$

In which C_k is a positive constant independent of t.
This means that when operator D is bounded if the grid distances $\Delta x \to 0$ and $\Delta t \to 0$ while keeping the time-space domain constant, the corresponding numerical scheme is zero stable.

Van der Houwen defines absolute stability by:

$$Lim_{T\to\infty} \left\| D \right\|_{dx=\Delta}^{k} \leq C_0, \qquad \text{(eq. 2.50)}$$

in which C_0 is a positive constant independent of t. This implies that a numerical scheme has the property of absolute stability when the operator D is bounded on a mesh with fixed Δx and Δt while extending the time domain to infinity. A more rigorous treatment of formal definitions on stability is found in van der Houwen (1968) and Godunov & Ryaben'kii (1964).
Proving a numerical scheme to be stable for a given problem by using the definitions is hard, and in many cases a practical impossibility. For instance, a numerical formulation can be referred to as a finite difference equation of which the left side is the output (values for the variables at $t=t+\Delta t$) and the right side the input (values for the variable at $t=t$)([13]):

$$\sum_{i=1}^{i=N_x} \beta_i u_i^{j+1} = \Delta t \sum_{i=1}^{i=N_x} \alpha_i u_i^{j} \qquad \text{(eq. 2.51)}$$

Using Z-transformations and contour integration to obtain the inverse transformation of the obtained transfer function, an analytical solution for $t=t+\Delta t$ can be found. This may be used

([13])The polynomial at the left side is called the first characteristic polynomial, if the roots of this polynomial smaller than or equal to unity the numerical method is zero stable.

for the next time level t=t+2Δt and solved again. By studying the successive solutions to stay bounded, stability may be proved. An advantage of this method is that initial and boundary conditions can be incorporated exactly. A drawback however, is that in doing so serious algebraic problems are encountered while no general view for all possible cases is obtained. Stelling (1983) stated that any method for investigating the stability of numerical schemes should meet two conditions:

- *'The method should be fairly simple to apply.*
- *It should not grossly overestimate the 'true' stability conditions. This might lead to very small time steps, which in practical applications imply costly and thereby not very competitive computing codes.'*

In practice, numerous methods for stability analysis, more or less coping with these requirements, are reported (see e.g. Ganzha & Vorozhtsov (1996)). The discussion here is restricted to the spectral method or normal mode analysis as described by Godunov & Ryaben'kii (1964).

To this end the behaviour of the homogeneous solution of the difference equation is studied, distinguishing three cases with respect to the spatial domain:

- A fully infinite space domain (i.e. $-\infty < x < \infty$)
- Positive half open space domain (i.e. $0 < x < \infty$)
- Negative half open space domain (i.e. $-\infty < x < 0$)

The general solution of a linear difference equation having linear coefficients is (See e.g. Oppenheim & Schafer (1989)):

$$u_k^j = r^j \hat{u}_i \qquad \text{(eq 2.52)}$$

Substituting this solution in the difference equation results in the so-called resolvent equation, having solutions in its turn of the form

$$\hat{u}_k = az^k \qquad \text{(eq 2.53)}$$

in which z is the root (or roots in the case of a system of equations) of the characteristic equation of the resolvent equation. Substituting (eq. 2.53) in (eq. 2.52) the normal modes are obtained:

$$\tilde{u}_k^j = ar^j z^k . \qquad \text{(eq. 2.54)}$$

Considering the three space sub-domains defined above, three conditions on the value of z (which is a complex number) have to be considered:

- $|z| = 1, i.e. z = e^{i\xi}, 0 \le \xi < 2\pi$, the Cauchy problem
- $|z| < 1$ (eq. 2.55)
- $|z| > 1$

When applying the first condition for |z| the well-known von Neumann stability analysis is obtained (Richtmyer & Morton (1967)). Based on this particular analysis many numerical schemes (including the Preissmann scheme) are studied and found to be unconditionally stable, though show instability-like behaviour in some practical applications.

The von Neumann stability analysis is based on the study of the Fourier transform of the operator $\underline{D}$. As was explained, this method is restricted to problems having only initial conditions while the space boundaries are located at infinity. In this respect it is just a partial study of stability analysis as defined by Gudonuv & Ryaben'kii (1964).

Since the main computer code applied in this thesis is based upon the Preissmann scheme (Preissmann (1961)) this particular scheme is discussed in some detail. Other numerical schemes have been applied in urban drainage also (e.g. the exponential Muskingum scheme, Koussis & Bowen (1984)) but will not be discussed here.

The Preissmann scheme is an implicit central difference scheme. A well-known problem of this family of schemes is the fact that at shallow water depths (e.g. DWF) and pipes at almost surcharged conditions, spurious oscillations may occur. This may result in negative values for the water depth initiating emergency stops of the computer program applied. This can be circumvented by applying 'tricks' in software engineering. These tricks however put constraints on the use of a given computer code; the user should be aware of these limitations. Furthermore, one should be able to distinguish numerically induced instabilities from physical instabilities as may occur due to the non-linearity of the physical problem studied.
In literature the problem with the Preissmann-scheme in relation to application in urban drainage are well known. Cunge & Mazoudou (1984), for example, state in their discussion on the software package CAREDAS that the problem with shallow water depth related instability is due to non-linearity of the geometries applied. They suggest to increase the hydraulic resistance with decreasing water depth as a remedy. As will be shown in the next sections, the stability problems discussed are merely problems with oscillations that are inherent to the Preissmann scheme and as such are not entirely induced by the application of non-linear geometries.

2.4.3.2 The stability of the Preissmann scheme.

In order to make the De Saint-Venant equations (eq.2.14 and 2.15) accessible for stability analysis, a linearised version is applied:

$$\frac{\partial a}{\partial t} + \bar{a}\frac{\partial u}{\partial x} = 0 \qquad \text{(eq.2.56)}$$

$$\frac{\partial u}{\partial t} + g\frac{\partial a}{\partial x} + wu = 0 \qquad \text{(eq. 2.57)}$$

in which:

a	local water depth	(m)
u	local flow velocity	(m/s)
t	time	(s)
w	resistance factor	(s^{-1})

with boundary conditions of the form: $a(0,t)=e^{st}$, $u(1,t)=0$, $x\ \varepsilon\ [0,1]$, here s is a complex number determining the periodicity of the boundary condition for the water depth a at x=0, $\bar{a}$ is the constant, mean water depth and w is the constant hydraulic resistance parameter and the convective term is neglected.

As a first, illustrative example, it is assumed that the balance between friction and free surface slope in eq, 2.57 is dominant, reducing eq. 2.57 to:

$$g\frac{\partial a}{\partial x} + wu = 0 \qquad \text{(eq. 2.58)}$$

Applying the Preissmann scheme on the system of differential equations eq. 2.56 and eq. 2.58, taking θ=1, results in:

$$\left.\begin{array}{l} \dfrac{a_{j+1}^{n+1} - a_{j+1}^{n} + a_{j}^{n+1} - a_{j}^{n}}{2\Delta t} + \bar{a}\dfrac{u_{j+1}^{n+1} - u_{j}^{n+1}}{\Delta x} = 0 \\ g\dfrac{a_{j+1}^{n+1} - a_{j}^{n+1}}{\Delta x} + \dfrac{w}{2}\left(u_{j+1}^{n+1} + u_{j}^{n+1}\right) = 0 \end{array}\right\} \qquad \text{(eq. 2.59)}$$

First the following eigen solutions are subsituted into eq. 2.59: $a_j^n = \tilde{a}_j r^n, u_j = \tilde{u}_j r^n$, resulting in the resolvent equations:

$$\left.\begin{aligned} &\frac{\tilde{a}_{j+1}r - \tilde{a}_{j+1} + \tilde{a}_j r - \tilde{a}_j}{2\Delta t} + \bar{a}\frac{\tilde{u}_{j+1}r - \tilde{u}_j r}{\Delta x} = 0 \\ &\frac{g(\tilde{a}_{j+1}r - \tilde{a}_j r)}{\Delta x} + \frac{w}{2}\left(\tilde{u}_{j+1}r + \tilde{u}_j r\right) = 0 \end{aligned}\right\} \qquad \text{(eq. 2.60)}$$

The general solution for the resolvent equation is:

$$\left.\begin{aligned} &\tilde{a}_j = \alpha_1 z_1^j + \alpha_2 z_2^j \\ &\tilde{u}_j = C\left[\alpha_1 z_1^j + \alpha_2 z_2^j\right] \end{aligned}\right\} \qquad \text{(eq. 2.61)}$$

In which z_1 and z_2 are the roots of the characteristic equations of the resolvent equations given by:

$$\begin{vmatrix} \dfrac{zr - z + r - 1}{2\Delta t} & \dfrac{\bar{a}}{\Delta x}(zr - r) \\ \dfrac{g}{\Delta x}(z-1) & \dfrac{w}{2}(z+1) \end{vmatrix} = \frac{w}{2}(r-1)(z+1)^2 - \frac{g\bar{a}\Delta t}{\Delta x^2} r(z-1)^2 = 0 \qquad \text{(eq.2.62)}$$

Suppose r=-1, in that case the roots for eq. 2.62 are:

$$z_{1,2}\frac{w + \dfrac{g\bar{a}\Delta t}{\Delta x^2} \pm 4\sqrt{w\dfrac{g\bar{a}\Delta t}{\Delta x^2}}}{\dfrac{g\bar{a}\Delta t}{\Delta x^2} - w} = \frac{w + \phi \pm 4\sqrt{w\phi}}{\phi - w} \qquad \text{(eq.2.63)}$$

in which $\phi = \dfrac{g\bar{a}\Delta t}{\Delta x^2}$ (eq. 2.64). This implies that if $\phi \to 0$ (caused by either $\Delta t \to 0$, $\Delta x \to \infty$ or $\bar{a} \to 0$) then;

$$\lim_{\phi \to 0} \frac{w + \phi \pm \sqrt{w\phi}}{\phi - w} = \frac{w}{-w} = -1 \qquad \text{(eq 2.65)}$$

This implies that the values for $z_{1,2}$ can become negative, depending on water depth, time and space step, hydraulic friction, resulting in a 2Δx wave in the space domain. This is easily seen when inserting a negative value for $z_{1,2}$ in the general solution of the resolvent equation (eq. 2.61); for odd values of j a negative part, and for even values of j a positive part evolves. It is this kind of wiggle (formally known as 2Δx waves) that often occurs in calculations([14]), depending on the formula describing the roots $z_{1,2}$. From eq. 2.63, it is seen that the norm of the negative roots increases with decreasing water depth and increasing resistance. The amplitude depends on the boundary condition (determining the values for α_1 and α_2). Whether or not these 2Δx waves persist in a practical situation, depends on the amplification in time. To this end, the same procedure as outlined for the simplified case, now not assuming

([14]) The occurrence of 2Δx waves is not restricted to finite difference methods or hydrodynamics, Vermeer & Verruijt (1981) describe this phenomenon in the application of finite element methods to solve consolidation processes. Zienkiewicz & Taylor (1991) describe the occurrence of 'wiggles' in finite element schemes for solving convection dominated problems. The general accepted method for suppressing them is adding numerical diffusion (the so-called Petrov-Galerkin method).

θ=1 and not assuming $\frac{\partial u}{\partial t}$ is negligible, leads to a system of equations in $\tilde{a},\tilde{u}$ which has only solutions if the determinant of the resulting system matrix equals zero, this implies:

$$\left.\begin{aligned}
&\begin{vmatrix}\alpha & \beta\\ \gamma & \delta\end{vmatrix} = \alpha\delta - \beta\gamma = 0, \textit{with}\\
&\alpha = (r-1)(z+1)\\
&\beta = \frac{2\Delta t a}{\Delta x}(z-1)(\theta r + 1 - \theta)\\
&\gamma = \frac{2g\Delta t}{\Delta x}(z-1)(\theta r + 1 - \theta)\\
&\delta = (z+1)(r - 1 + \theta w \Delta t r + w\Delta t(1-\theta))
\end{aligned}\right\} \qquad \text{(eq. 2.66)}$$

Together with $z = e^{i\xi}$ the corresponding expression for the von Neumann analysis is obtained:

$$\left.\begin{aligned}
&\alpha = (r-1)(e^{i\xi}+1)\\
&\beta = \frac{2\Delta t a}{\Delta x}(e^{i\xi}-1)(\theta r + 1 - \theta)\\
&\gamma = \frac{2g\Delta t}{\Delta x}(e^{i\xi}-1)(\theta r + 1 - \theta)\\
&\delta = (e^{i\xi}+1)(r - 1 + \theta w \Delta t r + w\Delta t(1-\theta))
\end{aligned}\right\} \qquad \text{(eq. 2.67)}$$

This leads to a characteristic equation of the form:

$$Ar^2 + Br + C = 0 \qquad \text{(eq.2.68)}$$

In which:

$$\left.\begin{aligned}
&A = (z+1)^2(1 + w\theta\Delta t) - (z-1)^2 4\theta^2 g a \frac{\Delta t^2}{\Delta x^2}\\
&B = (z+1)^2(-2 + w(1-2\theta)\Delta t) - (z-1)^2 8 g a \frac{\Delta t^2}{\Delta x^2}(\theta - \theta^2)\\
&C = (z+1)^2(1 + w(\theta-1)\Delta t) - (z-1)^2 4 g a \frac{\Delta t^2}{\Delta x^2}(1-\theta)^2
\end{aligned}\right\} \qquad \text{(eq. 2.69)}$$

For the von Neumann case, imposing $z = e^{i\xi}$, this is([15]):

$$\left.\begin{aligned}
&A = 1 + w\theta\Delta t + 4\theta^2 g a \frac{\Delta t^2}{\Delta x^2}\tan^2\left[\frac{1}{2}\xi\right]\\
&B = -2 + w(1-2\theta)\Delta t + 8 g a \frac{\Delta t^2}{\Delta x^2}(\theta - \theta^2)\tan^2\left[\frac{1}{2}\xi\right]\\
&C = 1 + w(\theta-1)\Delta t + 4 g a \frac{\Delta t^2}{\Delta x^2}(1-\theta)^2\tan^2\left[\frac{1}{2}\xi\right]
\end{aligned}\right\} \qquad \text{(eq. 2.70)}$$

Miller (1971) has shown that any polynomial of the form $Ar^2 + Br + C = 0$ with complex coefficients A, B and C, satisfies the condition $|r| \leq 1$ only if one of the following conditions is met([16]):

([15]) The formula $\frac{e^{i\xi}-1}{e^{i\xi}+1} = -\tan^2\left[\frac{1}{2}\xi\right]$ is used

([16]) When a symbol is over-lined this implies it is the complex conjugate of the original symbol.

$$|C| < |A| \wedge 2|\overline{A}B - \overline{B}C| \leq |A|^2 - |C|^2 \text{ or: } |C| = |A| \wedge \overline{A}B = \overline{B}C \wedge |B| < |A| \qquad \text{(eq.2.71)}$$

It is easily verified that for the von Neumann case with w=0 these conditions are met when $\theta \geq 0.5$ regardless of wave number or Courant number.
Since A, B and C depend on a number of parameters, the results for a limited number of parameter combinations tabulated in Table 8 are discussed.

	I	II	III	IV	V	VI	VII	VIII
θ	0.55	0.55	0.55	0.55	0.60	0.60	0.60	0.60
a	0.01	1.00	0.01	1.00	0.01	1.00	0.01	1.00
w	0	0	10	10	0	0	10	10

Table 8: Parameter combinations studied.

The other factors in the analysis are: Δx=0.1 m, g=9.813 m/s^2; for Δt two values; Δt=1.0 s and Δt=0.1 s are studied.

Case 1: $|z| = 1$, von Neumann case.

Implementing z=e$^{i\xi}$ results in:

$$r_{1,2} = \frac{1 + (1-\theta)\phi_{1,2}}{1 - \theta\phi_{1,2}} \qquad \text{(eq.2 .72)}$$

with

$$\phi_{1,2} = -0.5w\Delta t \pm \sqrt{0.25w^2\Delta t^2 - 4ga\frac{\Delta t^2}{\Delta x^2}\tan^2[0.5\xi]} \qquad \text{(eq.2.73)}$$

For the case in which the resistance is zero this is:

$$r_{1,2} = \frac{1 \pm (1-\theta)2i\sqrt{ga}\frac{\Delta t}{\Delta x}\tan[0.5\xi]}{1 \mp \theta 2i\sqrt{ga}\frac{\Delta t}{\Delta x}\tan[0.5\xi]} \qquad \text{(eq.2.74)}$$

For which the scheme is stable, regardless of the wave number ξ or the courant number ($Cr = \sqrt{ga}\frac{\Delta t}{\Delta x}$), while 0.5<θ<1.0. When, however, the hydraulic resistance is not ignored, this is also the case but is less easily verified. In Figure 25 some stability conditions are shown in the complex plane as a function of Δt. Generally, increasing hydraulic resistance increases the stability of the numerical scheme.

Case 2: $|z| > 1$

In this case no simple analytical solutions are obtainable, therefore the roots of the characteristic equation are solved numerically for z in the domain $z = \chi e^{i\xi}, \chi > 1,\ 0 \leq \xi < 2\pi$ for parameter combinations I through VIII.

Case 3: $|z| < 1$

In addition, in this case a numerical approach is applied in the domain $z = \chi e^{i\xi}, \chi < 1,\ 0 \leq \xi < 2\pi$ for parameter combinations I through VIII.

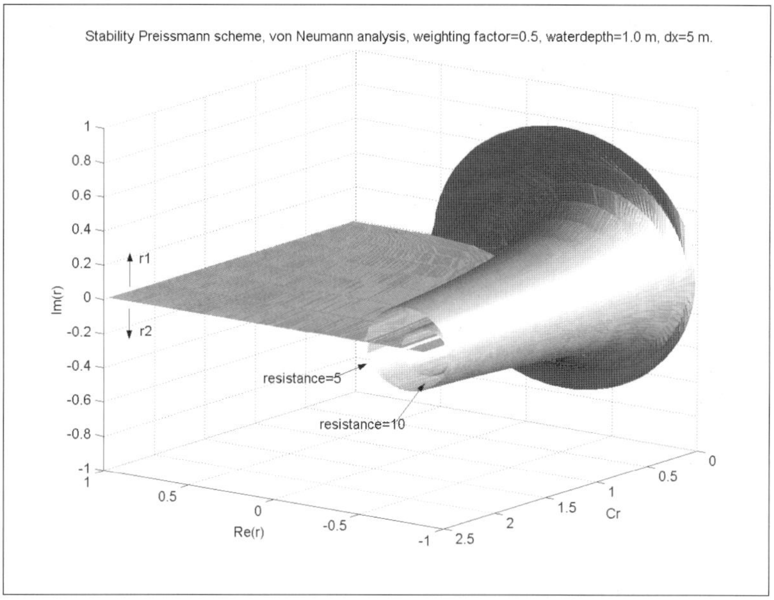

Figure 25: Stability of the Preissmann-scheme for the linearised De Saint-Venant equations.

The results are shown graphically in annexe III. The main conclusions drawn from this analysis are:

- Applying the Von Neumann stability analysis, the Preissmann scheme is found to be unconditionally stable for $\theta \geq 0.50$ (a well-known result).
- An increase in the value of θ results in all cases in a increase of the stability region in terms of z-domain where the amplification factor is < 1.
- In the domains $|z| > 1$ and $|z| < 1$ there are locations at which instability may occur, depending on the initial and boundary conditions.
- An increase in resistance results in an increased stability for all cases for z. In this respect the resistance plays a double role; it increases the risk of 2Δx waves, while on the other hand it gives rise to extra damping of these waves in the time domain. Therefore, the remedy Cunge & Mazaudou (1984) suggest to avoid problems with shallow water depths, namely to increase the hydraulic resistance with decreasing water depth, is no fool-proof method to avoid the occurrence of 2Δx waves.

No general conclusions on the specific type of initial and boundary conditions resulting in instability can be drawn. Based on practical experience these may be categorised as:

- Very shallow water depths.
- Flows at high Froude numbers.
- Imposing large Courant numbers (i.e. C_r>3-4).
- Very steep gradients in water level or discharge either in time or in space.

When faced with instabilities or instability like phenomena, users tend to decrease the time step, but this proves not always to be an adequate solution to avoid them. This is demonstrated by applying the stability analysis for an extra case in which the parameter values in combination VIII are applied but now using a value for Δt of 0.01 s instead of 0.1 s.

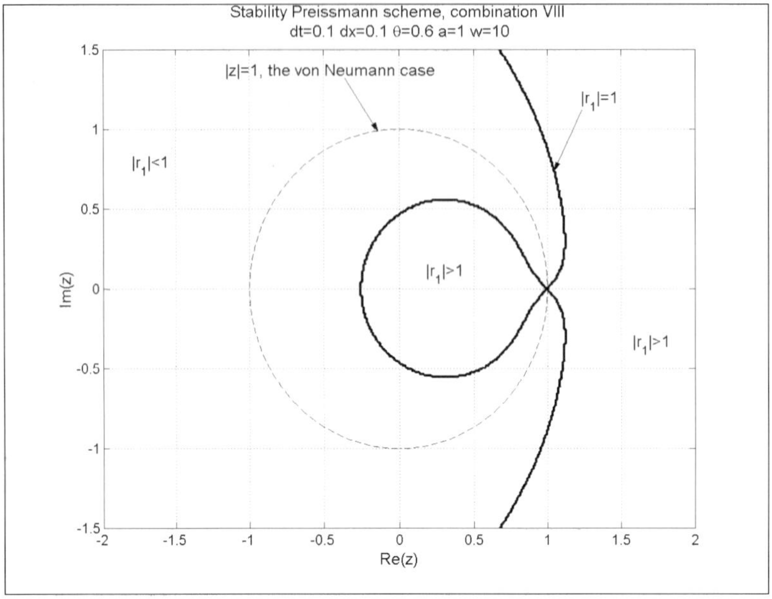

Figure 26: Parameter combination VIII, Δt=0.1 s.

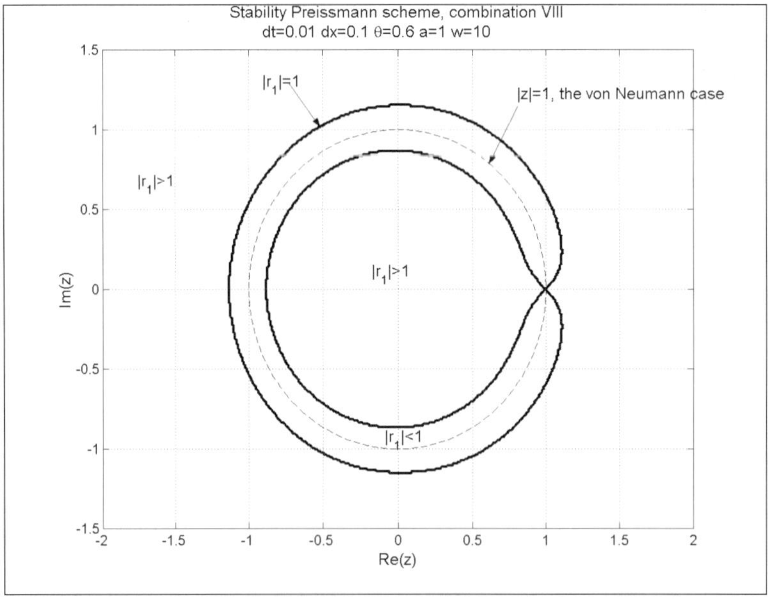

Figure 27: Parameter combination VIII, Δt=0.01 s. As can be seen the stability region is significantly smaller than for the case in which Δt=0.1 s (see Figure 26).

As can be seen from Figure 26 and Figure 27 a smaller time step may result in an increase of the amplification factor and in an increase of the domain for z in which the amplification factor is larger than unity. In most practical cases, it is more helpful to reduce the stability problems encountered by decreasing the value for Δx or reconsidering the methods applied to model the problem under study. This is especially so with respect to the definition of initial and boundary conditions and the numerical scheme applied.

Example:

As an example of the effect of decreasing Δt when faced with instabilities the following case is described:

Consider a one-pipe system, using the Preissmann scheme and the non-linear solution technique described in annexe II. $L=10$ m, $\Delta x=1$ m, $k_n=1.0$ mm, $i_b=0.001$, $D=1.00$ m. The initial condition is the result of the outcome the kinematic wave approach (resulting in an initial state in which only gravity and resistance contribute to the momentum balance) for $0<x<10$ at $t=0.00$ s. The boundary conditions are: $Q=0.001 m^3/s$ at $x=0$ for $t>0$, at $x=10$ m a free overflow condition is imposed, implying the Froude number is unity : $Q(t,h_{x=10}) = A(t,h_{x=10})\sqrt{g\dfrac{A(t,h_{x=10})}{B(t,h_{x=10})}}$ (see annexe II also).

	1	2	3	4	5	6	7	8	9	10
Δt (s)	0.01	0.10	1.00	10.00	20.00	0.01	0.10	1.00	10.00	20.00
K_n (mm)	1.00	1.00	1.00	1.00	1.00	5.00	5.00	5.00	5.00	5.00

In Figure 28 and Figure 29 the results are shown after 1 time step. As can be seen for the case in which $k_n=1$ mm the amplification of initial errors increases with decreasing time step.. When $k_n=5$ mm the amplitude of the $2\Delta x$ waves is increased when compared with the case for $k_n=1$ mm.

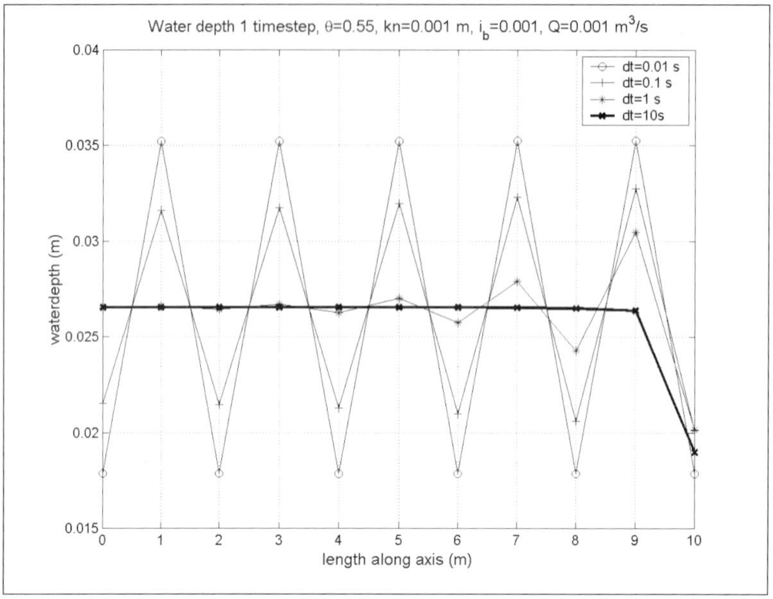

Figure 28: Oscillation in the space domain in a single conduit at shallow water depth using the Preissmann scheme. k_n =1 mm.

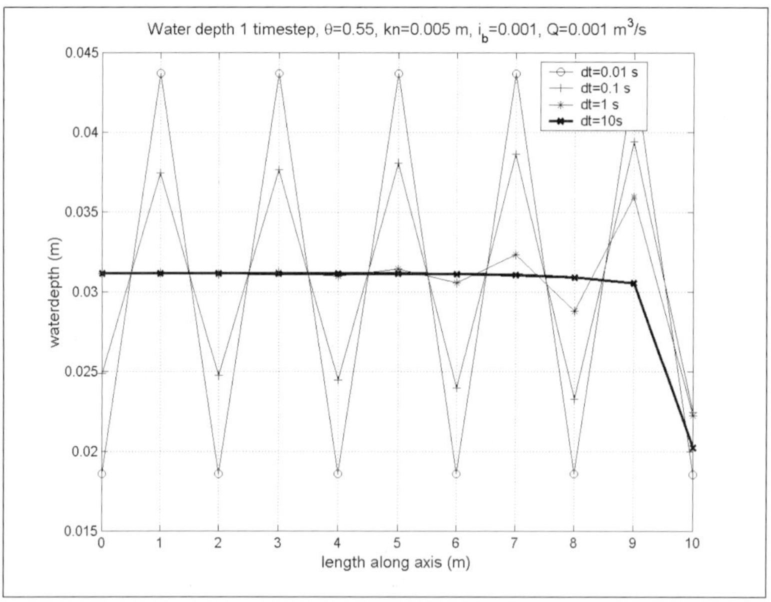

Figure 29: Oscillations in the space domain in a single conduit at shallow water depth using the Preissmann scheme. k_n=5 mm.

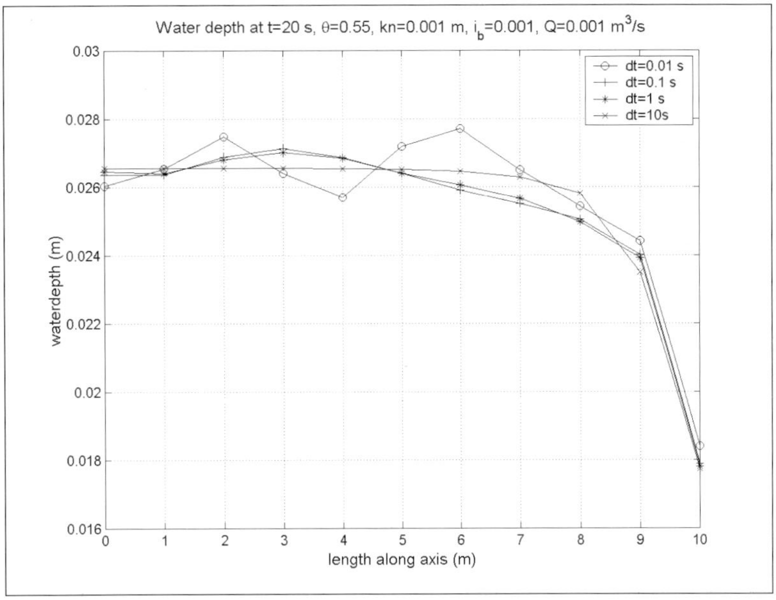

Figure 30: Oscillations in the space domain at t= 20 s, k_n=1 mm.

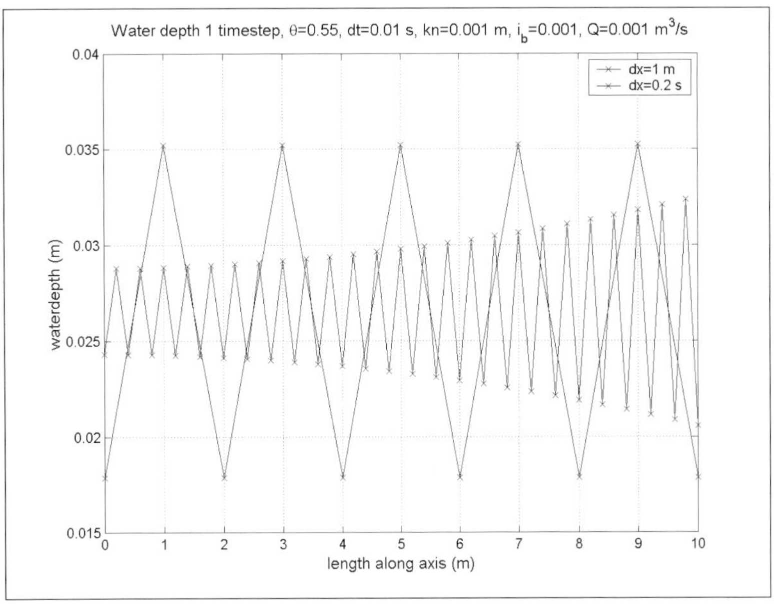

Figure 31 : The effect of Δx on the amplitude of the 2Δx waves.

Figure 30 shows the situation at t=20 s; as can be seen for the smaller time steps the oscillations are still present. At even smaller discharges, negative water depths occur at small time steps, forcing the computer program to an emergency stop due to the occurrence of physically meaningless results. The effect of decreasing the space step Δx is illustrated in Figure 31. *As can be seen the amplitude is decreased significantly when taking Δx=0.2 m instead of Δx=1m.*

In practice, when sticking to the use of numerical schemes suffering from this type of stability problems, the problem of 2Δx-waves is circumvented either by limiting the minimal discharge or by limiting the minimal allowable water depth. Both software-engineering tricks introduce a modelling error. This modelling error seems of minor importance when referring to all-day engineering practice in which the attention is mainly focussed on simulations under (extreme) storm conditions. However, when setting out to calibrate a model this model error may introduce a serious problem. This is illustrated in the following manner: in Figure 32 the development of a measured water level is shown. In order to get information on the parameters of the run-off the simulation should start at the outset of the storm. At this moment the sewer system is in DWF mode, the model will only show variation in the water level as soon as the discharge raises above a given threshold value. This implies that during the period between start and model reaction, a period occurs in which a systematic error is present.

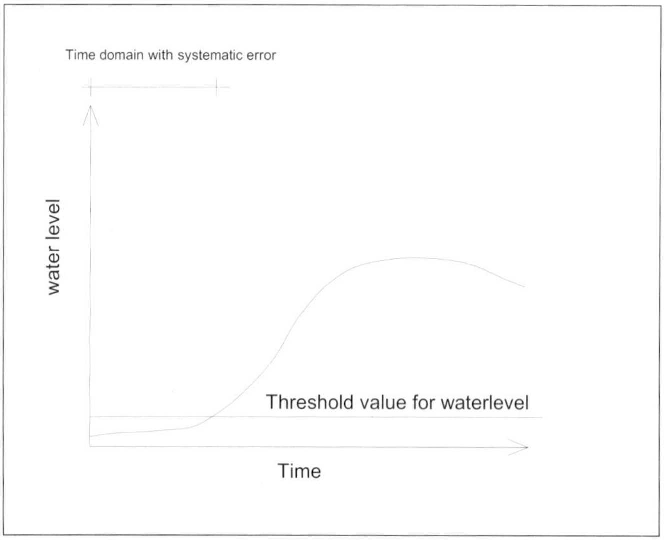

Figure 32: Effect of a threshold value in a model.

In the software package HydroWorks®, used as main tool in this thesis, the following method to circumvent oscillations, as described in this section is applied. The program starts to initialise a given situation based on the hydraulic load at t=0, this mostly implies DWF conditions. When this load leads to a water depth in a certain conduit less than a (user defined) threshold value, this value is adopted regardless of the actual hydraulic load. If the supplied threshold in combination with other parameters like Δt and Δx fails to result in a stationary solution the initialisation process fails. In fact, this is due to oscillations as described. This implies that the user should be well aware of the influence the several parameters have. Since numerical schemes not suffering the inherent problem of 2Δx waves are available, it may be stated that the Preissmann scheme is not to be used in applications in urban drainage. Stelling (1983) has shown that application of staggered grid schemes avoids the existence of spurious solutions associated with 2Δx waves completely.

2.4.3.3 Non-linear instability

Up to now the stability of the linearised equations has been investigated. In actual fact however, the equations used are non-linear. In particular the convective term may cause stability problems in numerical schemes that are stable according to the results of a stability analysis as described in Section 2.4.3.2. This is especially the case when large variations of flow in space occur; then the convective term may become a dominant factor in the equations. This is also the main reason to set the Boussinesq number to unity, though it is known that β>1. Avoiding the adverse effects of a value larger than unity is regarded as more important than using the theoretically correct value. The nature of the instabilities due to the convective term may be illustrated using the simple non-linear wave equation:

$$\frac{\partial a}{\partial t} + c\frac{\partial a}{\partial x} = 0 \qquad \text{(eq. 2.75)}$$

in which c is the wave celerity defined by $c = u + \sqrt{ga}$. The value of c is the slope of the primary characteristic. Due to local variation of the water depth (a), the value of c varies. This may cause the characteristics to intersect in the x-t plane, meaning that 'information' overtakes 'information' at its downstream end. Physically such a point is the location at which a wave breaks, implying three water levels exist at one location (see Figure 33). In a numerical grid only one value for the water depth can be represented; this gives rise to oscillations between the possible water depths in the numerical solution.

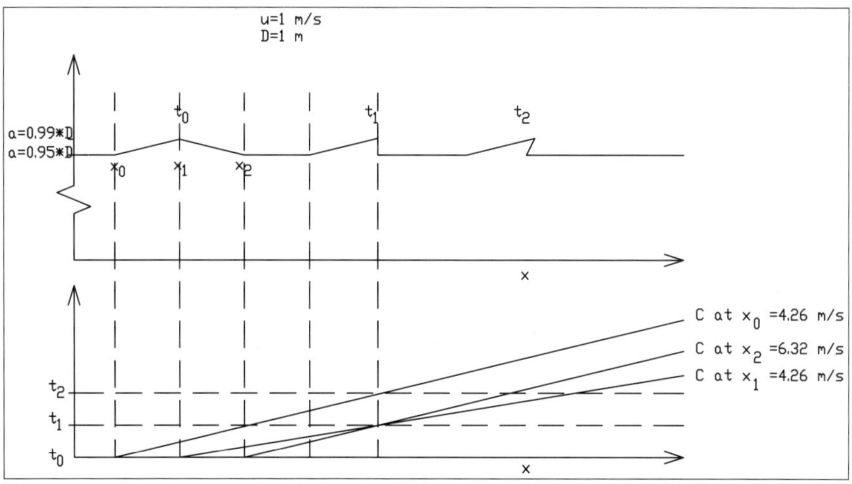

Figure 33: Non-linear instability.

2.4.3.4 Stability at nearly full flow

Another problem encountered is when pipes start to flow full, instability-like phenomena evolve. It should be stated however, that in reality when pipes tend to flow full a kind of chaotic behaviour may occur (see Figure 34). This is because air escaping between the water level and the pipe wall influences the water flow.

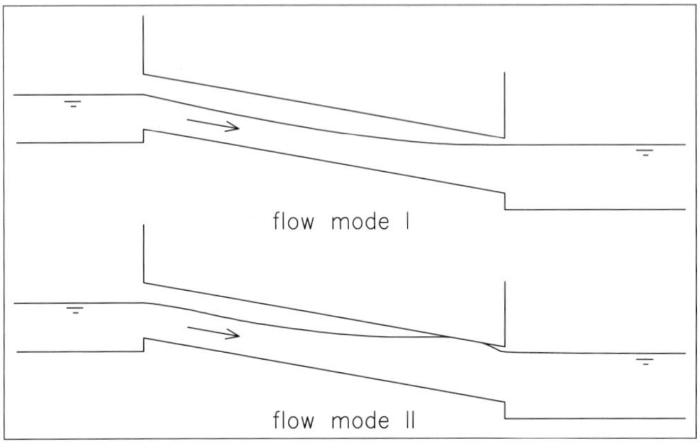

Figure 34: Physical instability. Due to air entrainment the flow may change between modes I and II in an unpredictable pattern in time.

It was observed in laboratory experiments, see e.g. de Somer (1984), that even at steady state flow under certain geometrical conditions unpredictable

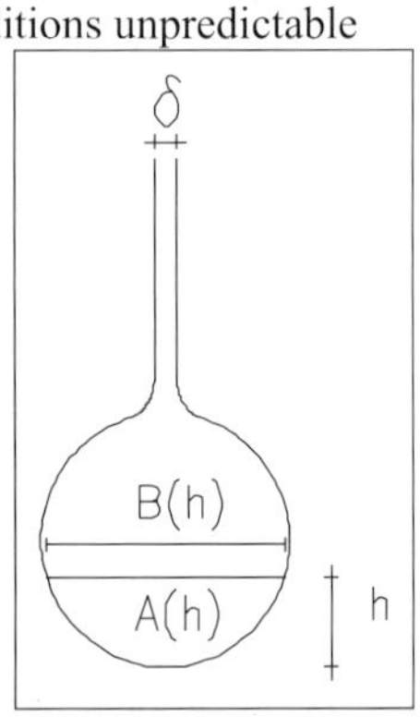

Figure 35 : The Preissmann slot.

transitions in flow mode can occur. The origin of numerical instabilities at nearly full flow can be characterised as a boundary condition problem. Essentially, it is such that nowhere in the system the water level is fixed. Under pressurised conditions this implies that the flow is defined by a difference in pressure only. Therefore, the absolute water level, or equivalently, the local pressure, is undefined due to a lack of a boundary condition in this respect. It is accepted practice to circumvent this by applying the so-called Preissmann slot([17]). In fact the cross-sectional geometry of, e.g., a circular pipe is slightly changed in order to avoid the occurrence of a transition between free surface flow and pressurised flow by adding a slot

([17]) The idea of this piëzometric slot originates from A. Preissmann and was further developed by Cunge & Wagner (1964) for practical application.

with a small width (δ) on top of the pipe. In this manner the water level directly follows from the mass-balance equation. In Figure 35 this slot is depicted. The width of the slot δ is defined theoretically by setting the celerity of a disturbance at the free surface in the slot equal to the celerity of a disturbance in the full-pressurised case. The celerity of a surface wave is defined by: $c = u \pm \sqrt{gH_h}$, in which H_h is the hydraulic depth: $H_h = \frac{A(h)}{B(h)}$. Taking a circular cross-section as an example, A en B are functions of the water depth (see annexe I); when the conduit is almost completely filled it is easily seen that the value of c becomes unbounded([18]):

$$\lim_{h\to 2R} B(h) = 0 \Rightarrow \lim_{h\to 2R} h_h(h) \to \infty \Rightarrow \lim_{h\to 2R} c(h) \to \infty \qquad \text{(eq. 2.76)}$$

Applying the Preissmann slot the width of this slot (δ) is defined by stating: $c_{\text{free surface}} = c_{\text{pressurised}}$

$$u + \sqrt{g\frac{\pi R^2}{\delta}} = u + \sqrt{\frac{1}{\frac{\rho}{K} + \frac{\rho D}{Ed}}} \Rightarrow \delta = \frac{\rho\left[\frac{1}{K} + \frac{D}{Ed}\right]}{g\pi R^2} \qquad \text{(eq2.77)}$$

The resulting values for the slot width δ though, may become small resulting in high values for the Courant number (i.e. >3 to 4). Therefore, in practice, the slot width is set to a fixed value of e.g. 0.1-5% of the diameter of the conduit. This does introduce a discrepancy with reality. As long as no water hammer problems are studied these, discrepancies however, are insignificant.

2.4.4 Accuracy

The accuracy of a numerical scheme is defined by its deviation from the analytical solution. Due to the lack of analytical solutions for the system of equations used, estimates for this deviation can be obtained by studying the influence of the principal parameters in the numerical scheme (Δx, Δt and θ) on the calculation results. To this end, the simplified linearised equations are used to obtain an insight in the accuracy of the numerical methods applied.

$$\left.\begin{aligned} &\frac{\partial a}{\partial t} + \bar{a}\frac{\partial u}{\partial x} = 0 \\ &\frac{\partial u}{\partial t} + g\frac{\partial a}{\partial x} + wu = 0 \end{aligned}\right\} \qquad \text{(eq. 2.78)}$$

In which $\bar{a}$ is the, constant, mean water depth and w is the constant hydraulic resistance term. In order to judge the accuracy of a numerical scheme the amplitude and wave celerity error for the reproduction of solutions for periodic initial conditions in a Cauchy problem (i.e. equivalent to the von Neumann stability analysis) are studied. The relative amplitude (relative to the solution when applying the differential equation) is the norm of the amplification factor

---

([18]) In practice, this celerity is limited to $c = \sqrt{\frac{1}{\frac{\rho}{K} + \frac{\rho D}{Ed}}}$ representing the velocity of a pressure wave travelling through the system. In which ρ is the density of water in kg/m^3, d is the wall thickness of the pipe in m, D is the diameter of the pipe in m, E is elasticity modulus of the pipe material in N/m^2 and K is the compressibility of water in N/m^2.

|r|, while the relative wave celerity is calculated from $\phi = \frac{\arctan\left[\frac{\mathrm{Im}(r_{1,2})}{\mathrm{Re}(r_{1,2})}\right]}{kc\Delta t}$. This case is valid when hydraulic resistance is absent, implying that no deformation of the wave due to hydraulic resistance occurs. In Figure 36 and Figure 37 the relative wave celerity and the value for |r| are depicted for several Courant numbers and two values for the weighting factor θ (0.55 and 0.75 respectively).

As can be seen, the accuracy decreases with decreasing value for the Courant number (i.e. at a larger time step). When however, the number of grid points per wavelength increases (i.e. a smaller value for Δx) the value for the relative celerity and the amplification factor is closer to unity. When increasing the value for the weighting factor, these values decrease, implying better stability properties at the price of a lower accuracy. In fact the Preissmann scheme is known to have maximum accuracy when θ=0.50, in which case the scheme is second order accurate. In practice this value cannot be applied since the scheme becomes zero-dissipative (i.e. numerical damping is absent), making the scheme sensitive to the occurrence of non-linear instability. In this sense, a modeller should make choices: better stability at the cost of a lesser accuracy and increased accuracy at the cost of increasing calculation time. It should be well remembered that the graphs shown are only valid for a simplified case with respect to the applied differential equations and the type of boundary conditions (i.e. the Cauchy case). For practical problems the modeller should make some trial runs to find an optimum with respect to Δt, Δx, θ and the resulting calculation time.

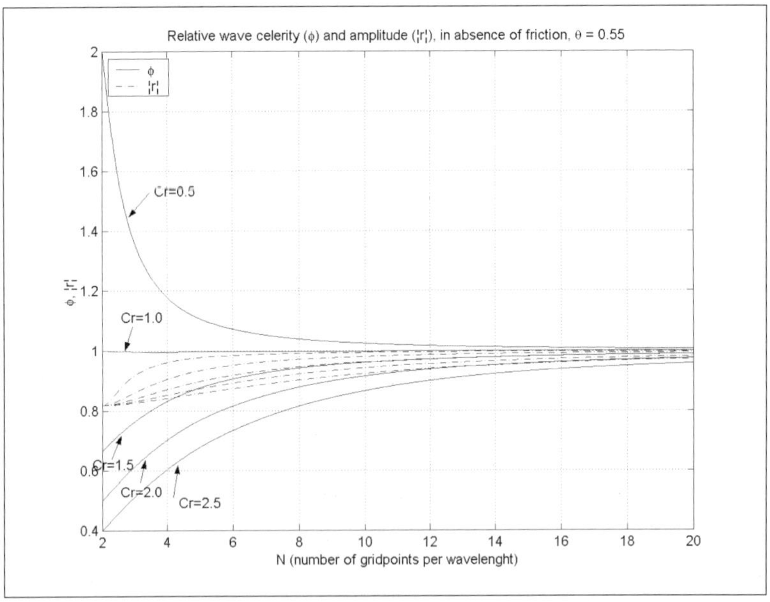

Figure 36: Damping and relative wave celerity for the Preissmann scheme.

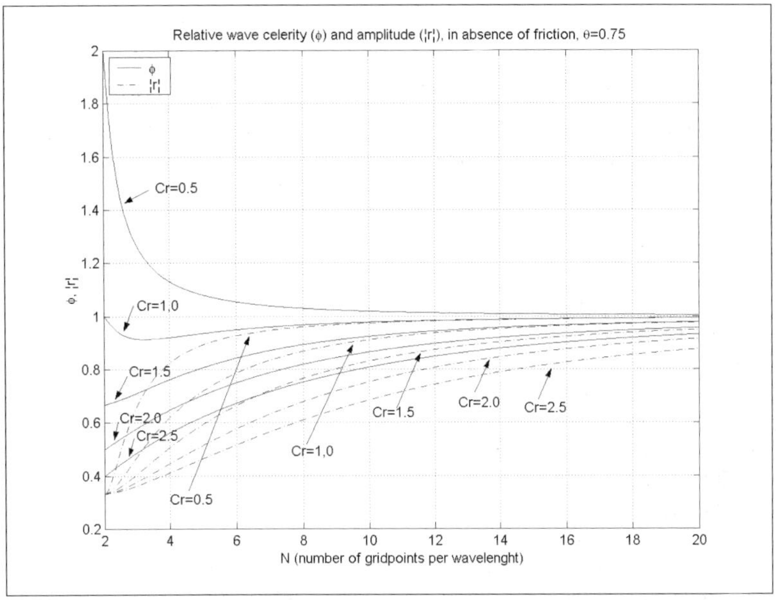

Figure 37: Damping and relative wave celerity for the Preissmann scheme.

2.5 Modelling errors

2.5.1 General

Modelling errors are defined as aberrations in the model due to:

- Incorrect description of processes.
- Numerical errors due to calculation methods applied.
- Simplifying or omitting minor processes.
- Data base errors.
- Effects of programmers solutions, working around difficulties in the implementation of formulas, and numerical difficulties in the software.

In the discussion database errors are not considered for the moment since these will be discussed in Chapter 3. The remaining error sources may be hard to recognise and will differ from software tool to software tool. This is the main reason software is being updated and debugged for several years after its first release. Simplification in process descriptions have been, as far as they are known, described in the previous sections. Errors in process descriptions may only become apparent when comparing model results with experimental data obtained from field measurements.

The simplifications applied in most models as described in this chapter are widely accepted. The modeller however, must be aware of these simplifications in order to be able to make a correct interpretation of the modelling results.

2.5.2. Software induced errors.

Some software-induced errors may only become apparent for special cases escaping the normal control procedures by the manufacturer. A practical manner to get some insight in the potential effect of modelling errors is by using one well-defined database with two calculation models. The constraints put on these software tools are:

- Both software tools are to be developed in two completely separated environments.
- Both tools should be based on identical and commonly accepted processes and methods like the ones described in this chapter.

It is stressed that it is impossible to tell from such a comparison which model is the best. This experiment has been performed using the 'Dutch Didactical Drainage Network' DDDN (see Clemens et al (1998)). The software tools applied are HydroWorks® version 3.3 (Wixcey et al (1992)) and EXTRAN/GM (van Luijtelaar (1993)). Both models showed only small deviations in 10 separate rain events used (the 10 standard rain events defined in the Dutch standard). The layout of the DDDN is shown in Figure 38. Since it is impossible to report all the modelling results for both models, the results were compared with respect to the parameters:

$$\beta = 1 - 2\frac{V_{Hydroworks} - V_{Extran/GM}}{V_{Hydroworks} + V_{Extran/GM}} \qquad \text{(eq. 2.79)}$$

in which V is total volume spilled over each separate external weir during every individual storm, and

$$\tau = 1 - 2\frac{T_{Hydroworks} - T_{Extran/GM}}{T_{Hydroworks} + T_{Extran/GM}} \qquad \text{(eq. 2.80)}$$

in which T is the duration of the spill over each external weir. Furthermore, the absolute volumes spilled and the starting moment of external spills and the location at which flooding occurs are reported. In Table 9 and Table 10, some results are shown. A more elaborate description of the results can be found in the original paper (Clemens et al (1998)). From this study the following conclusions are drawn:

- For storm events showing high rain intensities (events 06-10) the models perform almost equally.
- In cases at which an external weir functions only for a relatively short time or at low discharges, differences occur.
- For external weir 09002, the models give significantly different results. It should be noted however, that the spilled volumes in an absolute sense are insignificant when compared to the total volumes (2 % for storm 2 ranging to 0.03 % in storm 10).

	number standard storm event([19])									
Weir	01	02	03	04	05	06	07	08	09	10
02013	-	-	-1.00	0.97	0.95	0.96	0.96	0.98	1.00	1.00
04003	-	-	-	-	-	0.85	0.93	0.96	1.00	1.00
06004	0.99	1.00	0.99	1.00	1.00	1.00	1.00	1.00	1.00	1.00
06006	0.99	1.00	0.99	1.00	1.00	1.00	1.00	1.00	1.00	1.00
07004	0.96	0.97	0.98	0.98	0.99	0.99	0.99	0.99	1.00	1.00
07005	0.99	0.99	0.99	1.00	1.00	0.99	1.00	0.99	1.00	1.00
08002	-	-	0.92	0.91	0.95	0.94	0.95	0.96	0.98	0.99
09002	-	-	0.75	-1.00	0.85	0.54	0.32	-1.00	0.77	0.60
10003	-	-	-	-	-	-	-	-	0.98	1.00

Table 9: β values.

([19]) The numbers of the standard storm event refer to the events defined by VROM (1995) in the update 1999.

Weir	start Hydroworks (min)	start Extran/Gm (min)	Duration Hydroworks (min)	duration Extran/Gm (min)	τ
02013	20.50	19	27.00	28	0.96
04003	27.00	27	10.75	10	0.93
06004	22.50	22	49.25	50	0.98
06006	22.50	22	49.25	50	0.98
07004	31.50	30	59.50	60	0.99
07005	20.25	21	46.75	46	0.98
08002	32.00	32	42.00	42	1.00
09002	64.75	-	12.25	3	-0.21
10003	-	X	-	-	-

Table 10: Values for τ, storm event no. 07.

As a main conclusion it can be stated that software engineering related sources of errors are present, but stay within acceptable limits. It is suggested to introduce some kind of benchmark (e.g. the DDDN) to quantify the consistency of newly introduced software tools used in the field of hydrodynamic calculations in urban drainage.

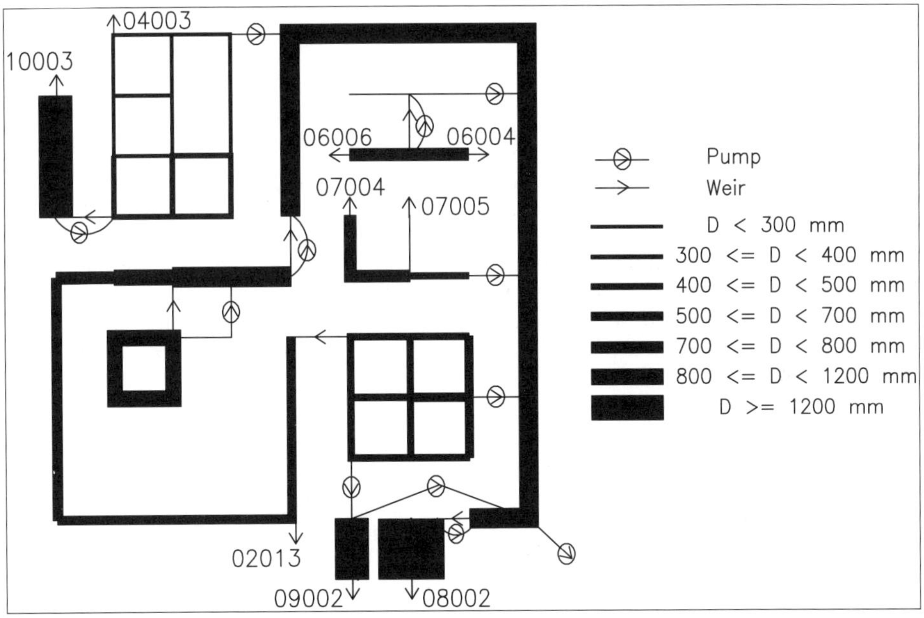

Figure 38: The layout of the Dutch Didactical Drainage Network.

2.5.2 Known errors in process descriptions

Some processes can not, or with great difficulty only, be modelled accurately even in a detailed model. With respect to the hydraulic model and the run-off model these processes have been discussed in the previous sections. In practice however, there are some other effects not accounted for. An example is the fact that house manifolds, gully pots and conduits connecting them to the drainage system are usually not accounted for in the calculations.
In the first place the amount of storage capacity held by these conduits may be significant, estimates up to 10% of the in-system storage are reported (Clemens et al (1999), Foundation

RIONED et al (1998)). Another point, related to this, is that most models assume that all the run-off enters the system at the manholes. In fact, this is certainly not the case; the run-off enters the system at many more inlet points. Theoretically, this can be incorporated in a model at the cost of introducing many more calculation points in the system but the effects are relatively small. An effect though that is mainly forgotten is the retardation of the flow in time due to the hold up in the house manifolds and gully pot conduits. The retardation due to this is in fact incorporated in the run-off model.

2.6 Conclusions

The main conclusions with respect to the topics discussed in this chapter are:

- Applying the most advanced approaches (i.e. a full 3-D dynamic calculation) for a complete network and rain series calculations is not practically feasible in the near future. Therefore, a 1-D simplification is widely applied.
- Due to the 1-D simplification, several extra model parameters evolve related to geometry and corrections for simplification with regard to streaming conditions.
- Numerical techniques must be used to make the equations accessible for computer programs; the numerical procedures however, may introduce restrictions on the applicability.
- For the popular Preissmann scheme it has been shown, using a normal mode stability analysis, that oscillations may evolve. This scheme is widely believed to be unconditionally stable, this however is only true for problems having initial conditions only. When boundary conditions exist (as they will in any practical case) instabilities can occur. A natural reaction of many users of hydrodynamic software is to decrease the time step when confronted with instability; it has been shown that in many cases this will not be a correct remedy. Another remedy to avoid oscillations (or $2\Delta x$ waves), is to increase the hydraulic resistance with decreasing water depth. In some cases this may work. It has been shown however, that in doing so in other cases the problem is even aggravated.
- The choice of parameters related to the numerical solution, like step size (Δx) and time step (Δt), does have an influence on the results obtained.
- A variety of run-off models exists, basically these models show little differences in their description of run-off processes. A common feature of these models however, is the fact that they are all more or less black- or grey box models, having in many cases a large number of model parameters referring to quantities that cannot be measured directly.
- Two software tools for hydrodynamic calculation, developed separately on the principles described in this chapter have been compared with respect to consistency. It is concluded that the differences between the tools are relatively small. There are however differences that are induced by the application of different numerical schemes (in our case an explicit scheme versus the implicit Preissmann scheme) and several programmers solutions.
- When using a numerical model one must be aware of the effects of time step, step size and other model- and software related effects and limitations
- Model parameters (as well for the hydrological models as for the hydraulic model) as reported in literature prove to show a wide range of numerical values, depending on the system studied. Therefore, it is not possible to use parameter values obtained from literature when trying to model a particular urban drainage system with a high degree of accuracy and reliability.

CHAPTER 3: DATA AND DATA ERRORS

3.1 introduction

In this chapter, the focus is on data and data errors and methods to eliminate them. The types of data needed to 'feed' a hydrodynamic model are classified as explained in Section 2.1.2:

- Geometrical and structural data of the sewer system.
- Drained area's (types and magnitudes).
- Model parameters (for the hydraulic and the hydrological model).
- 'Steering' parameters for the numerical processes applied (these are software specific).

In this chapter techniques to detect and remove obvious errors in the databases used in urban drainage are discussed. Furthermore, the overall effects of database errors and data uncertainty on the results of calculations are quantified for two drainage systems.
The latter two types of data (model and steering parameters) are only shortly discussed in this chapter, since Chapter 2 dealt with these, while in Chapters 5 and 6 the focus is on the quantification of uncertainties of model parameters. Therefore, the first two types of data are discussed in some depth here. When referring to 'errors' in a database a clear distinction has to be made between 'blunders' and data inaccuracy. Blunders are completely incorrect information due to typing errors, systematic errors or outliers in measured values. Inaccuracy in data in databases are basically due to measuring inaccuracy or due to a limitation on the number digits that can be stored due to the format chosen. Every class of data types incorporates a certain amount of inaccuracy imposing an effect on the calculation results when using a model. In this respect Price & Osborne (1986) state:

> *'Discrepancies between, say, a WASSP model (NWC, 1982) and reality are more due to faulty data describing the system than to inadequate sub-models'.*

Thereby they stress the inherent importance of obtaining the highest possible level of correctness of the geometrical and structural database. The highest possible level of accuracy can be defined as:

> *Every piece of information in the database complies exactly with the actual physical situation, allowing only deviations due to non-systematic measuring errors.*

Even when doing the painstaking (and costly) job of measuring everything, it is never certain that the database complies with this definition due to typing errors, errors in conversion software etc. Furthermore, one must, due to limited budgets, resort to using information originating from different sources, having several degrees of actuality and correctness.

3.2 general

Many urban drainage networks have been constructed in the pre-computer era. This implies that original information on their construction, geometry and details, is available at best in the form of drawings, maps or written manuscripts. The accuracy and reliability of these sources of information are questionable for several reasons:

- Discrepancies between original plans and actual construction that have not been corrected afterwards in revised plans ('as constructed').
- Loss of original information.
- Poor documentation of construction details.

During the years of large-scale construction (see Chapter 1 of this thesis) the importance of maintaining accurate information was not recognised. The whole concept of what is known as

‘rational maintenance’ was absent. It took until the early 1980’s when it was ‘discovered’ that the technical state of sewer systems had deteriorated alarmingly (see e.g. Oomens (1992)). This was because regular maintenance had been neglected for decades; in fact it was believed maintenance was hardly needed. Another point contributing to the neglect the maintenance is that sewer systems form an unseen, and during a long period unaccounted for, sector in urban infrastructure.
In practice, reliable information on the structure and geometry of a sewer system under study can only be obtained by a detailed field inventory. Due to its high costs however, this is only rarely applied on a full scale. Mostly this is done only for the older parts of a system of which no information at all is available.
Starting in the late 1960’s and early 1970’s application automated calculations in urban drainage started. Databases were built based on available information, in this manner two images of the drainage system exist in parallel:

- The system as represented on drawings and in design reports (master plans).
- The system as it was implemented in computer databases.

As explained, deviations will exist between these images([20]) and between the images and the actual physical system. Decisions to make alterations in the real system are nowadays mainly based on insights gained using the computer image. Therefore, it is of importance to maintain a system of data management to ensure consistency of the data and the real world system. In other engineering disciplines a large amount of literature is available on the topics 'engineering databases' and 'Database integrity' (see e.g. Fernandez et al (1980) or CIAD (1985)). In particular databases used for the computer assisted manufacturing of e.g. integrated circuits and aeroplanes are required to meet high standards. The aspect of database integrity is hardly discussed in scientific literature on urban drainage([21]). In engineering practice though, error elimination is one the most time consuming activities when modelling an urban drainage network. Techniques to recognise and remove database errors have been developed parallel and simultaneously by several consulting engineering firms. The information needed to ‘feed’ a hydraulic model can be summarised as:

- Structure of the system.
- Geometry of the objects in the system.
- Hydraulic parameters.
- Contributing run-off area in the catchment.
- Model parameters for the run-off model.
- Storm event data.
- Dry-weather flow data.

In the remainder of this the emphasis is on structural and geometrical errors in databases

3.3 Structural and geometrical errors

The distinction made between structural and geometrical data is explained as follows:
Data defining the structure of a sewer system are those data that hold information with respect to:

- *The type of objects constituting the system (manholes, conduits, pumps, weirs etc.)*

([20]) In this respect the problem with the geometrical model is similar to that of the process model; several steps are to be taken between reality and database used in calculations, each introducing errors. (See Section 2.1 of this thesis).
([21]) Zeman et al report some basics in Marsalek et al (1998) pages 169-177.

- *The manner in which these objects are related to each other (e.g. a conduit is linked to two manholes, a pumping station acts in a specified manhole.)*

Geometrical data define the physical dimensions (attributes) of each object in terms of time independent values:

- *The location in the x, y, z, space of each object.*
- *The profile of a conduit (e.g. round or egg-shaped).*
- *The shape of a manhole as a function of height.*
- *Switch levels for pumping stations (or Real Time Control rules).*
- *Weir levels and lengths.*

The structure of a sewer system is defined by the connections that exist between the individual elements like nodes (manholes) and discharging elements like conduits, pumping stations, orifices and weirs. Every element has a unique identification and each node has a unique identification, while the conduits (and other discharging objects like weirs and pumping stations) are identified by defining with which nodes they are connected. In this manner the structure of the network is defined and can be used for automated manipulation (see Section 2.4 also).The geometry of the system is defined by the location and the physical shape of the objects in the network. With respect to the nodes the minimum information demand is the street level. In a more sophisticated approach the easting, northing, bottom level and the geometry of the manhole as a function of depth has to be known. Since the minimum information needed is only sufficient when working with models based on stationary process descriptions, in this thesis the sophisticated approach is regarded as the starting point.

In Figure 39 the basic information needed for a node and a conduit is depicted; a more complete treatise on the information need for hydrodynamic calculations is found in either user guides supplied with software([22]), or national standards (e.g. Faber et al (1995,1999)).

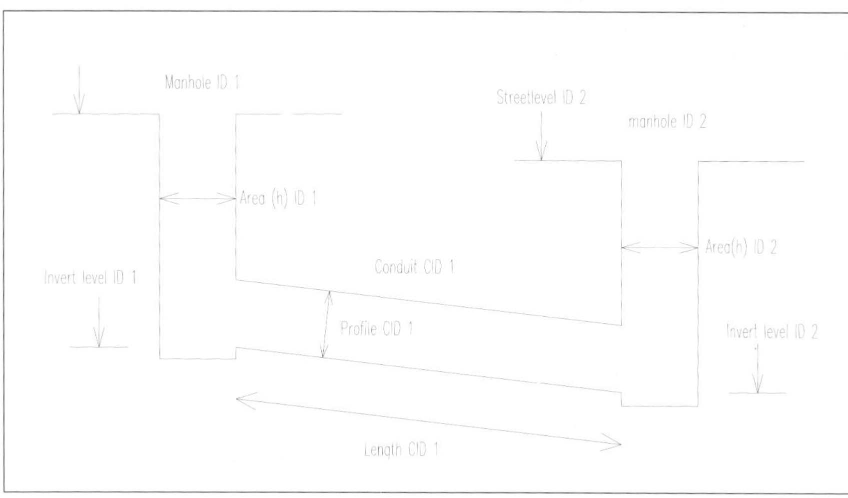

Figure 39: Basic information.

([22]) See e.g. users guide for HydroWorks® or SWMM users manual (1983).

3.4 Methods to recognise and remove geometrical and structural errors in a database

3.4.1 General

A database containing the information needed to use a hydrodynamic model for even a moderate drainage system holds numerous individual data that have to be obtained either from analogue sources (drawings) of from field surveys. Therefore, a database is vulnerable for data errors originating from sources like typing errors, errors in the fieldwork or software bugs in conversion programs. A substantial amount of the time of high-qualified personnel in consulting engineering firms was, and is, used to trace and remove errors from databases used in hydraulic calculations. Formal scientific literature on this subject in urban drainage is almost absent; this is most likely due to the fact that pure scientists hardly ever have to perform the tedious job of debugging large databases. In practice the information needed to build a database is obtained from different sources:

- (Revised) construction drawings.
- Results from field surveys.
- Existing databases.
- Knowledge of (former) construction sites managers.

First, all available data are to be gathered and brought together in a digital database so as to allow for automated manipulation for error recognition. In this section a short overview of methods developed in engineering firms to efficiently trace and remove database errors is presented; in Section 3.5.4. a working sequence is described that can be followed to build databases for use in hydrodynamic calculations. In modern commercial software some of these techniques are imbedded; the origin of most of the techniques, though, is day-to-day engineering practice.

3.4.2 Simple consistency checks

One of the most powerful tools in recognising structural errors is visualising the database; faulty connections are mostly seen immediately. This method became available on a large scale during the 1980's. Many databases built before this period prove to contain serious structural errors when examined. Figure 40 visualises such a database. This particular database was formerly used to check the hydraulic functioning of a drainage system using a computer program for steady state calculations. As can be seen immediately, some conduits are connected between nodes kilometres apart. Furthermore, there are several nodes not connected to the system by conduits at all.
The reasons these aberrations were not recognised in earlier studies may be due to the following facts([23]):

- The length of the conduits was a fixed Figure in the database (no redundancy was introduced).
- No thorough verification against field observations was made. (This must be the case since the single nodes must have shown flooding in that no diversion of water from these nodes is possible).
- Visualisation techniques were not applied.

([23]) It must be mentioned that the problems encountered in practice were ominous; frequent flooding occurred after alterations were effected, based on calculations made using this particular database.

In the early built databases co-ordinates often were left out, see. e.g. Kop (1972), making a correct visualisation virtually impossible.

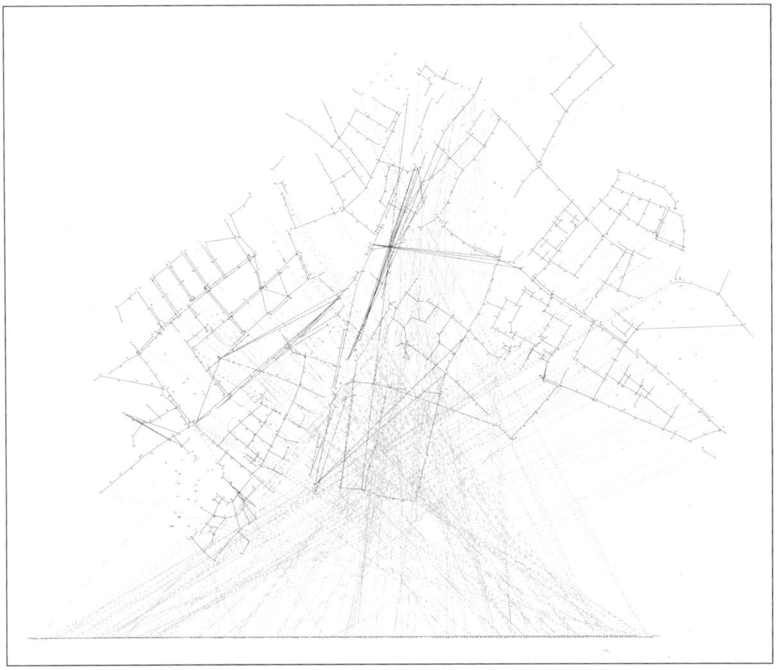

Figure 40: A 'modellers night-mare'.

In modern databases the length of a given conduit is still a parameter to be filled in, though it is simple to calculate the length using node co-ordinates. This calculated length should only be used in control mechanisms. Due to curvature, the actual length of a conduit can deviate substantially from a calculated value. In an automated method for checking the database the following check may be put to effect:

$$\text{If } \left| \frac{L_{database} - L_{coordinates}}{L_{database}} \right| > \varepsilon \qquad \text{(eq. 3.1)}$$

a closer examination (consistency check) of the data involved is to be performed([24]). Here the value of ε is any small positive convenient number, e.g. 0.05. This check is easy to implement in a computer program; combined with a visualisation of the network under study, errors in co-ordinates and conduit lengths are easily tracked and removed.

Another major source of errors is formed by invert levels of conduits, street levels, weir crest levels etc. Generally speaking, all measures of height are assumed to be correct. When studying databases used for steady state modelling, for which the constraints are substantially less than those required for transient models, numerous errors in this respect are present. This

([24]) When calculating the length of a pipe using co-ordinates a correction is to be applied with respect to the dimensions of the manholes on both sides of the pipe, the co-ordinates are valid for the centre of manholes.

is due to the fact that in steady state models the storage term is neglected, meaning that the only levels being crucial are weir levels and street levels, which are relatively easily checked. A first, practical, test on older databases is to look for level data with a value of 99.99 or – 99.99 m, since these values were often used as a default value indicating the actual value was unknown. It is surprising how many errors are traced with this simple test. Another method to trace errors in dimensions of height and conduit profile dimensions is to track soffit levels fulfilling the condition:

$$H_{street} - H_{soffit} < \kappa \qquad \text{(eq. 3.2)}$$

In which κ is a figure set at a value of e.g. 0,4 m. This check is also easily implemented in a computer program. Similar checks can be performed with respect to weir levels and switch levels for pumping stations. In the 1990's PC's became powerful enough to enable 3-D visualisation of entire networks.

Figure 41: An example of 3-D visualisation, after van Capelleveen (1999).

In this manner obvious errors in levels (especially invert levels and street levels) could be recognised immediately (analogous to the 2-D visual checks). At present virtual reality techniques are employed to visualise databases. This can even be done in such a manner that the designing engineer actually can 'walk around' in or 'fly and dive' above and under his database, see e.g. van Capelleveen (1999) and van Capelleveen & Clemens (1999). An example is shown in Figure 41.

Using the simple and more advanced methods outlined in this section many basic errors in databases are effectively recognised. Applying them does not guarantee a 100% correct structural/geometrical database:

- Small deviations in height and length measures may still be present.
- Deviations in profile dimensions (e.g. 300 mm diameter instead of 400 mm) are not detected.

Errors in profile shape (e.g. round instead of egg-shaped) are not recognised.

3.4.3 Lost storage

When having effected the rather trivial checks described in Section 3.4.2, a number of less obvious errors can be present in the data used. An effective method to recognise errors in invert levels, switching level of pumping stations, (internal) weir levels is to look for the amount and the location of what is known as 'lost storage'.

> *Lost storage is the amount of volume held by a given drainage system that cannot be emptied either by gravitationally induced discharge or by pumping stations.*

The principle of lost storage is simple; it is found at locations in the drainage system that can not be emptied entirely. In Figure 42 some examples are given. Due to the fact that the invert level of a pipe is wrong (say 0,5 m beneath all the other invert levels in its direct surroundings), this particular pipe can not be emptied. This may be due to a typing error when building the database or it may be in line with reality; this has to be verified. The principle shown in Figure 42 can also be manifest on a larger scale; it happens that a sub area in the total catchment forms a local minimum in level. In such a case several errors may be the cause: a software bug in a conversion program or the effect of merging together data from different sources using different references with respect to the level. In this case the storage loss also may be real due to e.g. incorrect levelling during construction.
Another form of storage loss is encountered when sub-catchments are linked together. The sub-catchments can be connected by pumping stations and sluice gates. When the switch-off level of the main pumping station is wrong, the system as a whole may show 'storage loss'. In practice this can be a real world problem; in many cases the drainage system is used for avoiding a too high switching frequency of pumps. This must be regarded as bad practice; in doing so pockets of stagnant wastewater for relative long periods of time (i.e. > 30 minutes) are created. Since this reduces the re-aeration process of the water strongly, eventually anaerobic conditions evolve initiating bacteriological processes that result in the formation of H_2S. In a moist atmosphere (as is naturally present in a drainage system), this may lead to the formation of sulphuric acid that corrodes the construction. A detailed study of these processes is reported by Bielecki & Schremmer (1987). The location and amount of storage loss is hard to detect by manual labour, especially in large databases. This is the main reason algorithms have been developed to enable an automated search. In order to locate and quantify lost storage the algorithm described in appendix I.3 may be applied. The algorithm is capable of detecting manholes and conduits that cannot be emptied totally; in other words, it shows where local minima are present in the structure and/or geometry defined in the database. Furthermore, it is capable of quantifying the amount of lost storage. Using simple visualisation techniques one is guided quickly where to look for aberrations. These aberrations can be a real world problem (due to subsidence or poor construction practice) or they may be due to measuring errors or simple typing errors while building the database. Application of this algorithm, which is easy to implement in a computer code, results in information on the locations (i.e. node identification and link identification) where lost storage is present. Furthermore, it calculates the amount of lost storage per location.

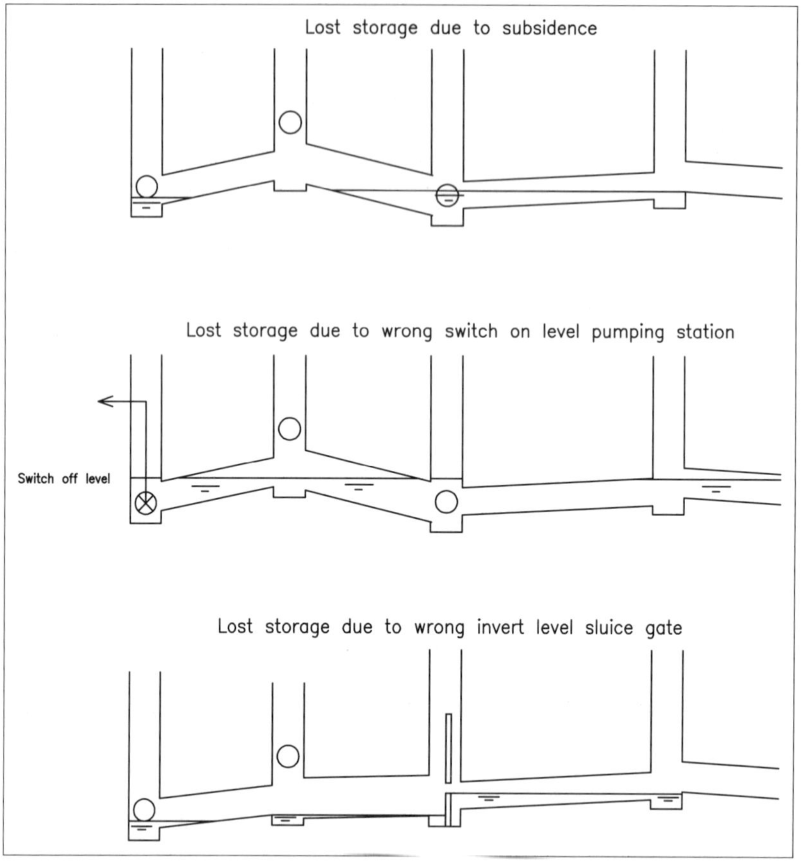

Figure 42: Several forms of storage loss.

An alternative method to locate lost storage, though less elegant and less accurate, is the use of a hydrodynamic model; the following procedure is suggested:

Step 1: Initialise a calculation with a completely filled system.

Step 2: Start the calculation not allowing any water (run-off nor DWF) in the system; due to outlets, weirs or pumping stations the system will be gradually emptied. By inspection, verify whether a stationary situation with non-operating pumps has evolved.

Step 3: As soon as a stationary situation has evolved, inspect the results for water level at the nodes and indicate at which location this level is above the lowest invert level; at these locations lost storage is present. If necessary this is easily depicted using graphical aids.

The latter method has the advantage, from a managerial point of view, that no separate software tool is to be developed and maintained for lost storage calculations. The drawbacks of the illustrated method however, are:

- Due to the shallow water depth problem (see Section 2.4.3.2) software induced errors in the calculation of the volume present in pipes may evolve, due to a layer of 'numerical' water present in the system. The occurrence of this depends on the numerical integration method applied. This results in errors in the volume calculation; the magnitude of these errors depends on choices made by the manufacturer of the program and is therefore seldom accessible for the average user.
- The calculation effort, when compared to the first algorithm, is extraordinary large; this is more so for large catchments (i.e. > 5000 conduits).

Either technique however, is a powerful tool to trace errors with respect to database errors, like incorrect switch levels for pumping stations and errors in invert levels of conduits. The interest in the exact amount of lost storage is mainly focussed on areas in which only small differences in terrain level occur (e.g. The Netherlands, parts of Belgium, Northern Germany) or when regulations are (partly) based on the amount of available effective storage volume in the system. As a tool for tracing database errors, however, detection of locations suffering from lost storage may be employed with success in any sewer system.

3.4.4 Summary

When building a database the following procedure is suggested:

1. Make an inventory of the available data sources.
2. Categorise them with respect to reliability.
3. When necessary convert the data to digital information.
4. Visualise the database in 2-D.
5. Check for discrepancies between actual conduit length and calculated conduit length.
6. Check measures of height.
7. Visualise the database in 3-D.
8. Locate lost storage.
9. Make a hydrodynamic calculation and check for locations with flooding against known locations in practice.

action	Type of errors found
1:Inventory	Missing data
2:Categorise	Lack of overall reliability
3:A-D Conversion	-
4:2-D Visualisation	Missing conduits, wrong coordinates, faulty connections
5:Check lengths	Wrong lengths or co-ordinates.
6:Check measures of height	Gross errors like invert levels above street level, height values set at defaults like 99.99 or –99.99
7:3-D Visualisation	Smaller errors in height measures escaping step 6
8: Locate lost storage	Wrong switch-off levels pumping stations, smaller errors in invert levels or geometrical data on conduits
9:Trial run with hydrodynamic software checking against known flooding locations	Wrong geometrical measures, incorrect pumping capacities, wrong weir width.

Table 11: Successive steps in building a database.

After making an inventory of data sources available, it may be concluded that several data are present in more than one source. In that case a decision has to be made what source is to be used when the data are not identical. An example of this is that a revised drawing says the street level is 1.2 m, whereas the available database says it is 1.3 m. An (indicative) order of decreasing reliability is suggested:

1. Recently measured data.
2. Data from revised construction drawings.
3. Construction drawings.
4. Data in an older database.
5. Information obtained from the (former) construction site manager.

If several sources of information are in contradiction, the only manner to obtain reliable information is to measure the data needed in the field. In any of the steps 4 to 9 errors can show up due to programming bugs in the A-D conversion software used, especially when some kind of systematic deviation is found([25]).

Performing step 9 may reveal errors in the database like missing conduits, incorrect weir- or street levels or errors in conduit geometry. It is stressed that no exact information on the data is obtained in this manner. However, if there is a substantial deviation between the number and the locations at which flooding occurs in the calculation, as compared to field observations, this almost surely indicates the presence of database errors.

3.5 Amount and type of catchment area

3.5.1 General

In order get a complete database, information on the catchment area has to be obtained. As was discussed in Section 2.2, the information that needs to be gathered depends in part on the type of hydrological model applied. For instance, when using a depression storage model in which the terrain slope is a main factor (e.g. the model according to Kidd (1979)) this information has to be obtained in the field. Since terrain slope is relatively hard to measure accurately in an urban area for every individual sub-area, the use of such a model is not considered practically feasible unless default values are applied as can be found in manuals.

The problem of obtaining information on the contributing area is simple from a theoretical point of view;

- A decision has to be made which hydrological model is to be applied; this determines what type of information (apart from location and amount) for each individual area is to be obtained.
- All areas potentially contributing to the runoff are to be quantified.
- For each individual area it has to be decided at which location in the drainage system the run-off will enter.

In practical cases, the first step is the most crucial one; because once a project for building database is under way it is hardly ever feasible to add or change the type of information to be

---

([25])An example encountered in practice is the following: due to a bug in a routine a conversion programme systematically set all street levels between –1.00 and 1.00 to their absolute value, this bug only showed up when a drainage system was modelled that had street levels all between –1.00 m and 1.00 m.

gathered. The choice made on which hydrological model to be applied is mostly determined by the following considerations:

- Which hydrological models are available in the software applied?
- Is it a practical possibility to obtain all data needed (this is largely a matter of available budget)?
- Is the resulting database maintainable in the future?

Since the debate on hydrological models is far from ended, many countries have decided on implementing a 'national standard' on the topic what hydrological model is to be applied, implying a directive for the information to be gathered. This is defendable from a practical point of view. The main drawback however, is that it inhibits a large scale comparison of run-off models, since databases built along the lines of a certain standard are generally unfit for use in conjunction with other models. For example, the Dutch standard uses three categories for depression storage, which implies that the terrain slope is not introduced as a separate item in the databases. In this manner the use of the model defined by Kidd (1979) for depression storage is inhibited. A same problem applies with respect to the routing model; in the Dutch standard the run-off length is not a parameter while in several runoff models this is a crucial parameter.

The effects of choices made in the early stages of database construction are felt throughout any project involving hydrodynamic calculation for urban drainage systems. With respect to the main issue of this thesis, calibration of hydrodynamic models, the effect is obvious. It is senseless to try to calibrate a software tool in which a hydrological model is applied that needs data that are not available in the database.

On the other hand, the work of e.g. Van de Ven (1989) led to the opinion that relatively simple hydrological models perform well enough for practical uses compared to more complicated models (see also Section 2.2. or Foundation RIONED (1992)). Topics on which more consensus among specialists is obtained are quantification of areas and determining the location at which the run-off enters the drainage system.

3.5.2 Quantification of areas

Many techniques are available to quantify the area of an object generating run-off. In most cases, detailed maps are the basic source for this information. Full digital, mixed analogue-digital (or hybrid techniques) and full analogue techniques are distinguished.

Many older databases are built using fully analogue techniques with detailed maps as a basis and the well-known planimeter as a tool. In most cases these databases suffer from obvious errors like typing or calculation errors since building these databases involves a substantial amount of manual labour. These techniques are hardly applied anymore and therefore will not be discussed any further here.

At present, situations occur in which only analogue maps are available while a digital database has to be obtained. Under these conditions hybrid techniques are applied. These techniques aim at minimising the use of highly qualified personnel. An example of such a technique is manually adding a colour to individual surfaces of equal signature (depending on differences defined by the model applied). The resulting map is digitised in such a manner that each colour gets a unique code attached to a 'pixel' representing a certain amount of area and located by a set of x,y co-ordinates. Using GIS techniques (see e.g. Prodanović et al (1998) or Măksimović (1998)) it is easy to calculate the total amount of every surface type. Furthermore, techniques like Thiessen polygons can be employed to assign a given surface to a manhole or conduit at which it discharges run-off. It should be noted that such an assigning technique will not result in a distribution as found in actual practise. It is however, a practical impossibility to find out at which location every individual sub area discharges into the drainage system.

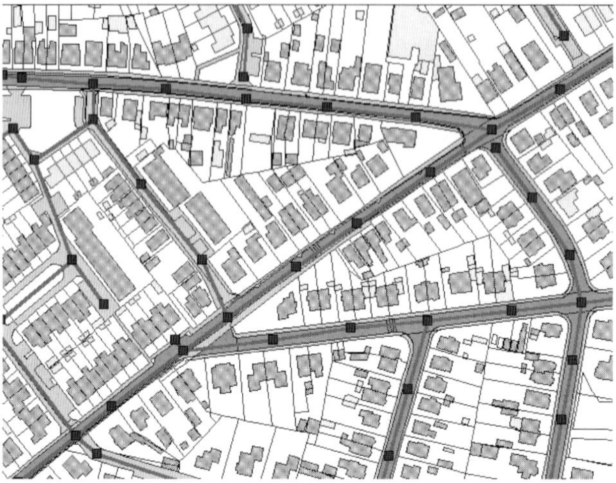

Figure 43: An example of the result of applying GIS techniques for obtaining information with respect to amount and type of surfaces contributing to run-off generation. Different shades of grey indicate different types of surface. A technique like e.g. Thiessen polygons (also known as Dirichet tiles or Sobolov spaces) is used to assign individual surfaces to the manholes at which they are supposed to discharge under storm conditions.

Garb & Send (1994) describe a method to incorporate local terrain slope in determining the location at which a particular sub-area discharges. The demand put on the accuracy of the data however, is prohibitive for application in practice on a large scale. For large sub-areas it is of importance to find out the discharging location by means of a field survey.
Another, more sophisticated, method is based on the use of fully digitised maps. In many cases the areas represented on these maps have an attribute added determining the type of surface involved. In Figure 43 an example of the resulting map is shown.

A serious drawback of these methods is that, without a detailed field survey, the obtained results suffer from inaccuracies regardless of the 'exact' impression they make. In most cases impervious area on private terrain is not included in the maps.
Especially terraces, paths and small roads in private gardens may contribute substantially. In a few cases such an inventory was made (See e.g. Witteveen+Bos (1997 and 1999)), indicating that the possible accuracy of GIS based techniques is between 3 and 10%. This Figure depends largely on the quality and actuality of the original information rather than on the techniques applied.

3.6 Storm event data

The question what storm event to use largely depends on the purpose for which a model is applied. When designing a newly built system mostly standardised storm events are used to test the design with respect to safety against flooding. A large body of literature exist on the issue on what storm events to use (see e.g. Berlamont (1997)); in any case some statistical property (return period) is assigned to a design storm. The choice of which return period is used depends on the acceptable failure frequency; in different countries different approaches are adopted. In the European norm the following differentiation with respect to return period is suggested:

Pressurised flow should occur with a return period larger than or equal to

- T=1 year in rural areas
- T=2 year in residential areas
- T=2 to 5 years in commercial or industrial zones
- T=10 years for tunnels

Flooding should occur with a return period larger than or equal to:

- T=10 year in rural areas
- T=20 years in residential areas
- T=30 years in commercial or industrial areas
- T=50 years for tunnels

These figures are, especially with respect to pressurised flow by no means achievable in flat areas like the Netherlands and parts of Belgium. Therefore other, nationally defined acceptable return periods are adopted. In the Netherlands for instance, a return period of T=1 year or T=2 years with respect to flooding is used in designs.

Normally so-called IDF relations are used to decide which rain intensity over a certain period is to be used. IDF relations are Intensity-Duration-Frequency relations; these relations are derived from a time series of observations by means of partial series analysis (see e.g. Vaes et al (1996)). An example of these relations is shown in Figure 44.

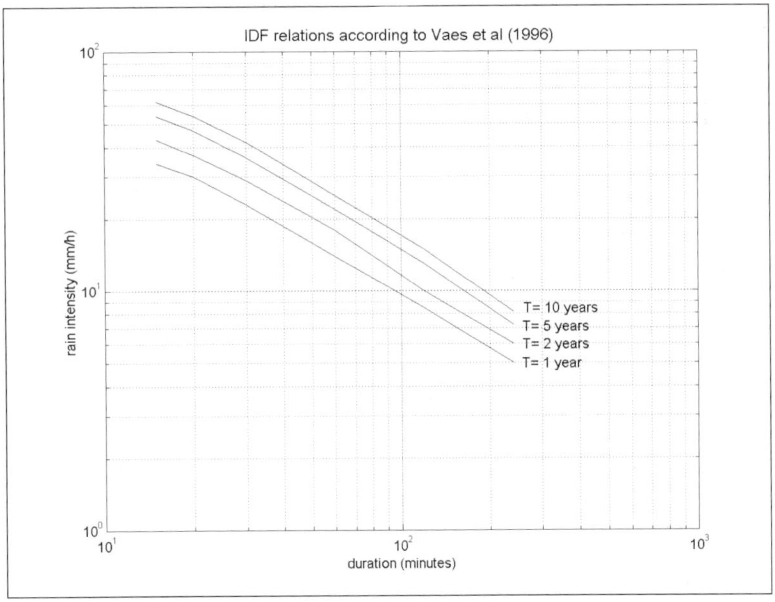

Figure 44: An example of IDF-relations.

Making a choice with respect to return period and duration results in the rain intensity to be applied. However, this allows for a first rough design only; when using hydrodynamic models variation with time has to be supplied also. With respect to the shape of a design storm no definite method has been presented so far. The main problem is that the shape of the hyetograph combined with the hydraulic behaviour of a given drainage network defines the flooding frequency; this implies that basically for every individual network design storms should be derived. A far better way, from a theoretical point of view, is to apply a rain series calculation(e.g. 10 years or storm data) and analyse the results. This approach is known as 'statistics a posteriori' and is not a practically feasible option for larger systems at present. Therefore standardised hyetographs are derived like in the Netherlands (van Luijtelaar & Rebergen (1997) or in Belgium (Vaes et al (1996)), or by the British Meteorological Service. An example of the Dutch standardised storms is shown in Figure 45.

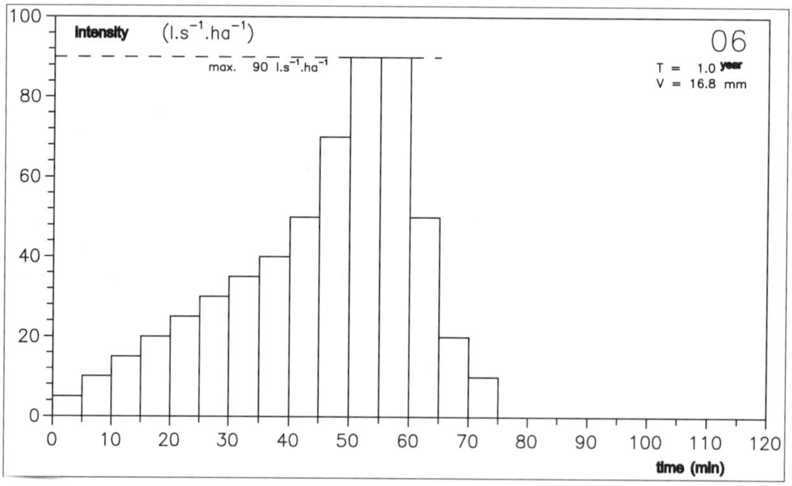

Figure 45: Standard storm event no. 6 (Faber et al, 1995,1999).

It must be remembered that these design storms do not take into account the spatial variability: this can be a factor to be counted with when interpreting measuring data; see e.g. Willems (1998).

3.7 Dry-weather flow data

Dry weather flow is due to several sources:
-Leakage.
-Industrial wastewater discharges.
-Domestic wastewater discharges.

In this case also, it is of importance to take notice of the purpose of the modelling activity; for design purposes normally standardised figures for municipal wastewater and assumptions for industrial discharges and leakage are adopted.
For design purposes, a leakage term of 1 l per km pipe per second is mainly applied. In older systems however, especially when high ground water tables are present, leakage forms a noticeable factor in many cases. This may add up to 50% of the DWF as encountered in

several municipalities in the Netherlands. Similar figures are reported for drainage systems in other parts of the world (see e.g. Murray (1987)).
Data on industrial wastewater discharges can be obtained from the industries themselves (if exactly known) or may be estimated by studying the specifications in the discharge permits. It should be realised that depending on the type of industry it is also of importance to get information on the discharge pattern in time. If for instance a batch process is applied, this may result in short peak loads; a continuous process, on the other hand, will result in a stationary load.
The DWF from households is often directly related to the drinking water consumption. The figures obtained in this fashion, however, cannot be exact. This is because a portion of the water will not reach the drainage system at all; e.g., water used for garden sprinkling and car washing is infiltrated in the ground or evaporates. In any case, the drinking water consumption pattern over the day will certainly differ from the DWF discharge patter due to retardation of the discharge and temporary in-house storage (e.g. in bathtubs).
Therefore obtaining unbiased information on the DWF with respect to amount and daily variation is no routine task. In Chapter 6 of this thesis a method is described to obtain DWF patterns based on information of the pumping actions of a main sewer pump. This more precise information is of importance when redesigning a system in which DWF is an important factor (e.g. purely DWF systems) or for calibration purposes.
In any case, exact information for every individual discharge into a system is practically impossible to obtain; therefore, often averaged figures are applied (i.e. mean hourly discharge per inhabitant of e.g. 12 l/h). Depending on the size and the characteristic (large urban catchment or a small rural catchment) differences are observed in the discharge pattern.

3.8 The effects of database errors on the modelling results([26])

3.8.1 General

After applying the techniques to track and remove errors from the model input as outlined in the previous sections in this chapter, several uncertainties remain present in the model input.
In practice, it is known that the data set applied is never 100% perfect. Imperfections in geometrical data can be distinguished into two groups:

- Structural errors: layout failures, missing links and nodes, data of non-existing structures, wrong profile codes, completely wrong profile dimensions;
- Errors in descriptive geometrical parameters, like diameters (measures vary within tolerance limits of the production process), measures of level, pipe lengths;

Serious structural errors are mostly well recognised (see Section 3.4); smaller errors, e.g. missing a few smaller links, escape the control procedure as presented in Section 3.4.4. Errors in geometrical data, e.g. a diameter of 400 mm instead of 300 mm, egg shaped profile instead of a round profile are less detectable, and will therefore be present in almost any database. Based on the practical experience over a period of more than 10 years it is estimated that at least 1 percent of the geometrical data is suffering from these kinds of errors even after applying the error recognition techniques described. In addition to the geometrical data, it is necessary to estimate local hydraulic loss coefficients, hydraulic roughness etc. In practice it is hardly ever feasible to quantify these parameters exactly by means of measuring data.

([26]) This section is based on, and a wider extension of, two congress papers: Ummels & Clemens (1998) and Clemens & von der Heide (1999).

3.8.2 Research technique

The goal is to obtain a first insight into the relative effect database errors have on the overall reliability of hydrodynamic calculations. To this end the effect of database errors is compared to the effect of other sources of uncertainty. The effects of variation in rain intensity in space is left out of the analysis. Willems (1998) studied these effects, using a multi-reservoir model in which much of the hydrodynamic effects are lost. He concluded that spatial variability in large systems accounts for circa 29% of the overall variability in results. In this research the focus is on the effect of the use of the hydrodynamic models only and the effects of geometrical and structural errors in the database only.
In order to quantify the effects database errors have on calculation results it is suggested to regard the applied parameters as random picked variables, defined by an average value and a standard deviation. The average values are based upon generally accepted figures (Faber (1995,1999).

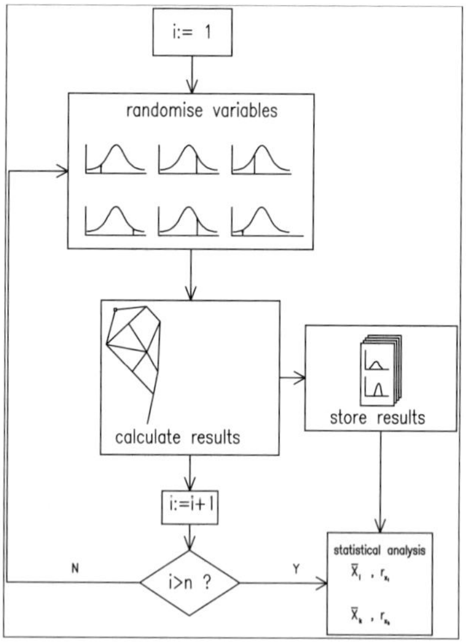

Figure 46 : Monte Carlo technique.

In Figure 46 the basic research method is outlined. An urban drainage network is schematised into a set of data. Applying a randomiser on several parameters, 'n' slightly different data sets are generated. Using these data sets 'n' hydrodynamic calculations are performed (using the software package HydroWorks®). In this manner 'n' data sets containing the results, which are slightly different in each individual calculation, are obtained. The calculation results are stored and analysed after completing the calculations. The parameters analysed are:
- Total flow from spills and weirs.
- Maximum flows from spills and weirs.
- Maximum water levels in manholes.
- Number of locations where flooding occurs.

In this research five groups of data, which are in practice linked in the manner of acquisition, are distinguished:

A/ Runoff parameters.
B/ Catchment area.
C/ Hydraulic parameters.
D/ Geometry (variance within measuring errors of level of profile dimensions) of the system.
E/ Structure of the network.

In order to obtain estimates for the influence of uncertainties in every group of parameters a set of Monte Carlo simulations is carried out as outlined in table 5.
In total 16 sets of calculations are made for two urban drainage systems, the sets II through VIII give information on the influence of uncertainties of every group of data (A through E) separately, while the results of set I will give information of the influence of uncertainties in the whole data set. In the sets VI, VII and VIII the percentage of structural errors was respectively 1, 10 and 20 %. The rain event used in every calculation is standard event no. 06 as defined in the Dutch guidelines for hydrodynamic calculations (Faber (1995, 1999)).
The number of simulations within one Monte Carlo simulation has been varied, in order to investigate the effect of the number of simulation results using $n \in [1,..1000]$ are shown. Generally, the number of simulation increases depending on the object of the study. In this study the main focus is on obtaining estimates for the first and second moment (i.e. mean and variance) allowing a limited number of Monte Carlo simulations (order of magnitude of 10^2). When, however, estimates for third and fourth moments (Skewness and Kurtosis, respectively) are to be obtained the number of simulations rapidly increases (order of magnitude 10^3-10^4).
The calculations are carried out using a Pentium II 300 Mhz based computer. The software is tested extensively with respect to the correctness of the random generator (Kleywegt (1992) and Kleywegt & Clemens (1993)) and the effect of truncation errors in HydroWorks®. The latter software package was run 1000 times using system I, applying standard rain event no. 06. The calculation results were binary compared and no aberrations were found. An alternative to the Monte Carlo simulation is a first order approximation in which the non-linear functions describing the hydraulic behaviour are linearised (i.e. by Taylor series expansion). As stated by Stevens & Haan (1993), this approach is only feasible under certain restraints on the input function. In the case of a full hydrodynamic model the non-linearity and the occurrence of discontinuities is prohibitive for the use of a first order approximation, therefore the Monte Carlo simulation is applied in this research.
In order to get an idea of the type of probability distribution for the results obtained the following possibilities were investigated:

- The normal (Gaussian) distribution
- The Weibull distribution
- The Gumbel distribution

The parameters in these distributions were estimated by the methods of the central moments, details can be found in annexe IV.

3.8.4 The drainage systems used

The method is applied on two urban drainage catchments; these are shown in Figure 47 and Figure 48.

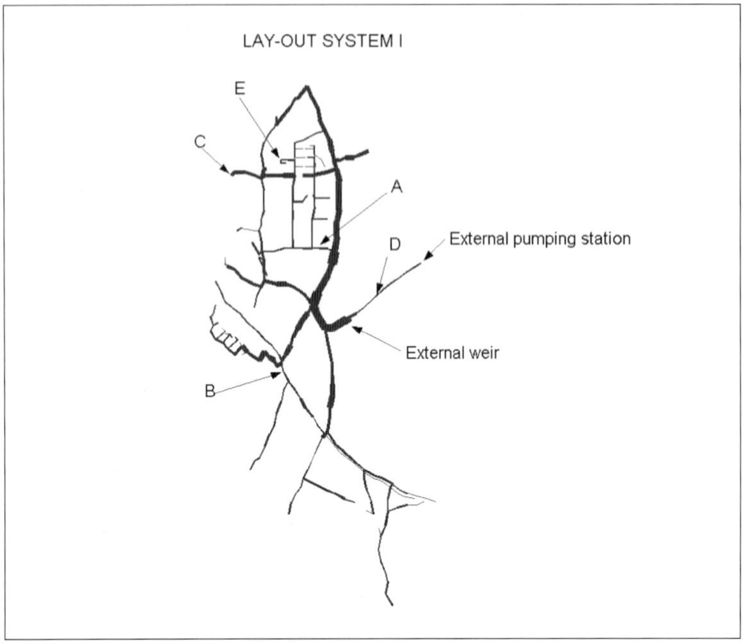

Figure 47: Layout system I.

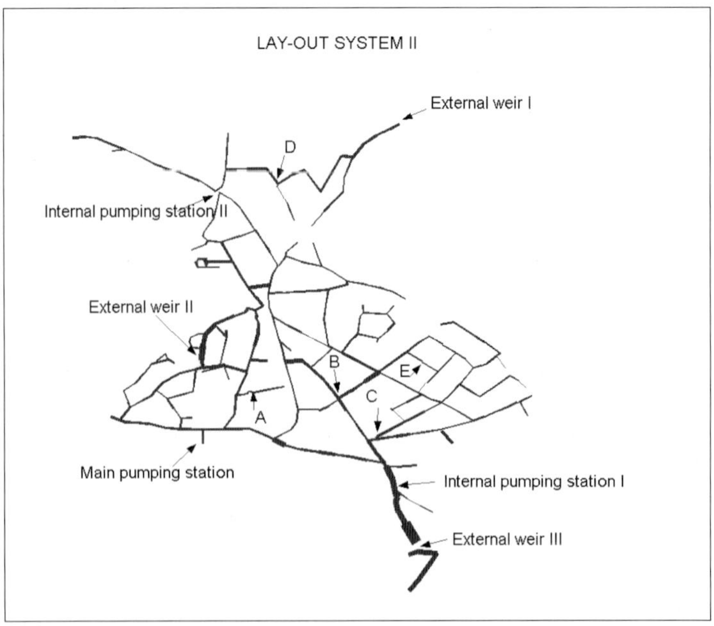

Figure 48: Layout System II.

The main characteristics of the systems used are summarised in Table 12; system I is a rather small, simple system incorporating only one external weir and one pumping station in a mildly sloping area.
System II is somewhat larger, it contains 3 external weirs, two internal weirs and one external pumping station and is located in a flat area.

system	paved area (ha.)	Total length (m)	Diameters (m)	Inhabitants	ground level (m)	storage volume (m^3)
1	15.8	9157	0.2-1.25	2050	17.8-28.6	834
2	25.7	13178	0.2-1.00	4600	8.4-10.6	1726

Table 12: Main characteristics of the research systems.

In both systems, apart from the weirs, five locations were selected to track the effect of parameter variation and geometrical errors on discharge and water level variation.

	A	B	C	D	E
System I	Looped	Branched	Periferical	Main stream	End pipe
System II	Periferical	Main stream	Main stream	Branched	Periferical

Table 13 : Main characteristics of the location studied within the systems.

In the study the errors introduced have a Gaussian probability distribution with mean values and standard deviation as summarised in Table 14.

Parameter	Description	Data group	Average	Standard deviation
H, h_2 ([27])	Invert levels at up- and downstream end	D	-	0.05 m
H_c	Weir crest level	D	-	0.01 m
H_g	Street level	D	-	0.01 m
K_n	Hydraulic roughness	C	3 mm	1.0 mm
c	Weir coefficient	C	1.4	0.35
D ([28])	Profile diameter	D	-	0.5 %
P_{in}, P_{out}	Switch on Switch off level pumping station	D	-	0.01 m
Q_{pump}	Pumping capacity	C	-	5%
ξ	Local loss coefficient	C	1.2	0.4
i	Infiltration rate	C	2 mm/h	30%
c_r	Linear reservoir constant	A	0.2	30%
b	Initial and depression losses	A	0.5 mm	30%
A	(sub) catchment area	B	-	5%

Table 14: Mean values and standard deviations used for database parameters.

The introduction of structural errors was done using a computer program, randomly introducing these types of errors as a user defined percentage in the geometrical and structural data in the database applied. The errors introduced are:
- Profile errors (e.g. a round pipe instead of an egg-shaped pipe).
- Nominal dimension errors, i.e. 1 or 2 nominal profile steps.
- Conduits left out.

These errors were introduced in such a manner that:
- The database remained consistent.
- The main DWF route was not broken up (this would not escape standard checks).

([27]) Bloemen (1997).
([28]) Nijmeegse betonindustrie De Hamer (1998).

- *Main routings towards pumping stations and external weirs were not significantly reduced in their hydraulic capacity (this should not escape the attention of an experienced modeller).*

The calculation results are analysed with respect to the statistical distribution; nominal value and standard deviation are calculated. In particular the following results are analysed:

- Number of locations on which flooding occurs in the calculation, N and σ_N
- The volumes spilled over external weirs V and σ_V.
- The maximum discharge over external weirs Q_{max} and σ_{Qmax}.
- The standard deviation of the maximum water level (σ_{Hmax}) in a number of manholes at different points in the system.

The motivation for these choices is:

- The maximum water depth is an important criterion by which a designing engineer judges his design.
- Water depth is easily measured in practice.
- Spilled volume is an important parameter with respect to the design in relation to environmental effects.
- Maximum spilled discharge is an important parameter with respect to the design of receiving surface waters.
- The number of flooded manholes is an important parameter used when judging a design with respect to flooding.

Set	Run-off	Area	Hydraulics	Geometry	Structure
I	Varied	Varied	Varied	Varied	Varied (1%)
II	Varied	Fixed	Fixed	Fixed	Fixed
III	Fixed	Varied	fixed	fixed	Fixed
IV	Fixed	Fixed	Varied	fixed	Fixed
V	Fixed	Fixed	Fixed	Varied	Fixed
VI	Fixed	Fixed	Fixed	Fixed	Varied (1%)
VII	Fixed	Fixed	Fixed	Fixed	Varied (10%)
VIII	Fixed	Fixed	Fixed	Fixed	Varied (20%)

Table 15: Characteristics of the eight calculation sets.

3.8.5 Results system I

The results are presented in some detail annexe IV; here some characteristic results are shown. In Figure 49 and Figure 50 the results for mean value and standard deviation for the water levels are shown as a function of the number of Monte Carlo simulations made. As can be seen the mean value reaches a nearly constant level after about 100 simulations whereas for the standard deviation it takes about 250 simulations to reach a more or less constant level. From these results it was concluded that using 250 runs gives enough information for further analysis; therefore the results presented in the tables are based upon 250 simulations.

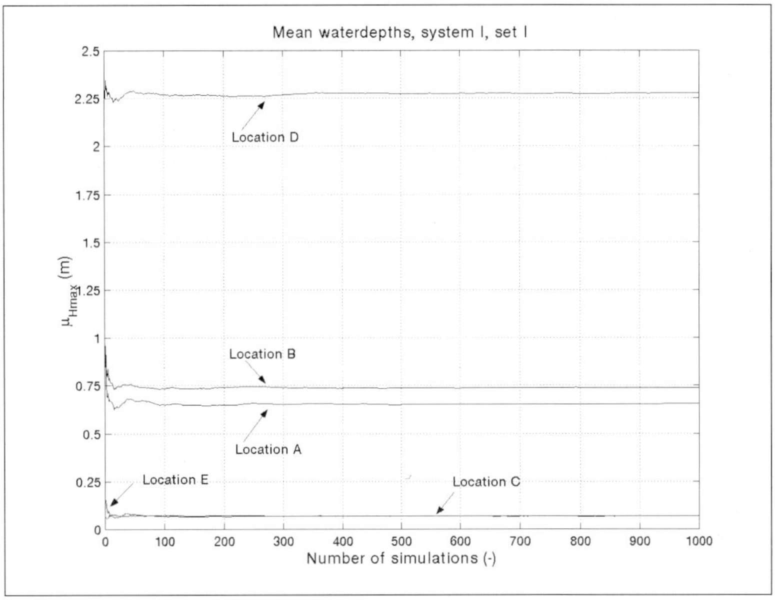

Figure 49: Mean value for the maximum water depths as function of the number of Monte Carlo simulations.

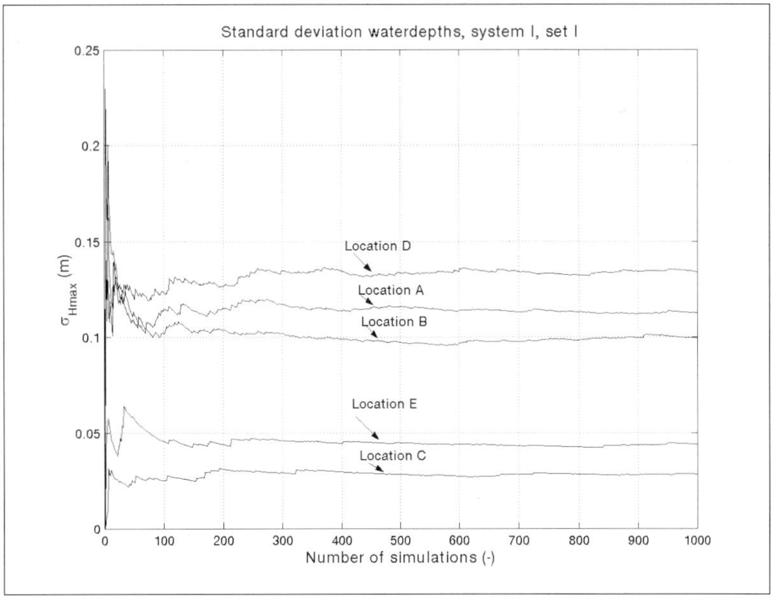

Figure 50: Standard deviation in maximum water depths as function of the number of Monte Carlo simulations. After circa 250 simulations the results are almost stable.

In Table 16 and Figure 51 the correlation between the results are shown. As can be seen the correlation is insignificant for most combination of results, except for the maximum discharge and the spilled volume over the external weir.

	$H_{max,A}$	$H_{max,B}$	$H_{max,C}$	$H_{max,D}$	$H_{max,E}$	$Q_{max\ I}$	V
$H_{max,A}$	1	0.0447	0.0481	0.2694	0.3806	0.1467	0.5331
$H_{max,B}$		1	0.0821	-0.2814	0.0796	0.0629	0.4000
$H_{max,C}$			1	-0.0481	0.0461	0.0486	0.0558
$H_{max,D}$				1	0.2542	0.1786	0.1023
$H_{max,E}$					1	-0.0631	0.1129
Q_{max}						1	0.7503
V							1

Table 16 : Correlation matrix for system I, set I. As can be seen only the correlation between the volume spilled over the weir and the maximum overflow discharge show a mutual correlation of any significance. The other results prove to be almost independent of each other.

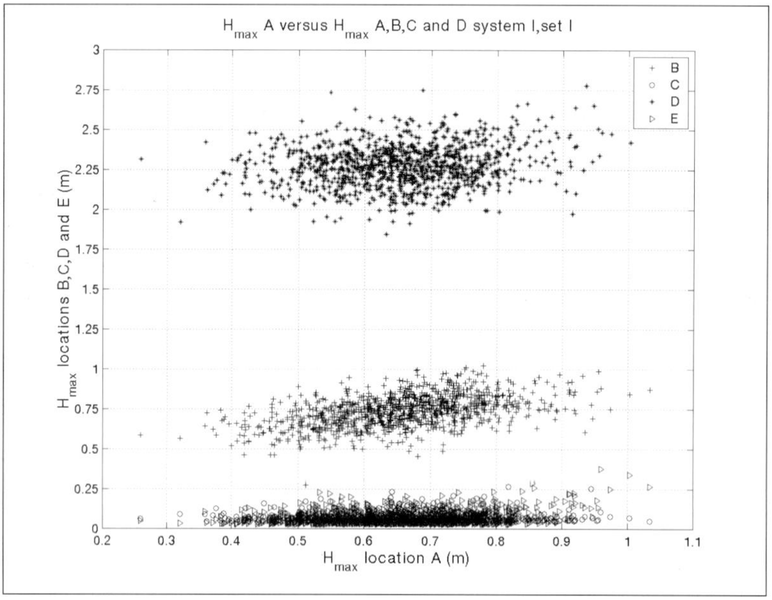

Figure 51: H_{max} location A versus H_{max} at locations B,C,D and E. This Figure illustrates some of the correlations presented in Table 16.

The probability density functions for the maximum water levels, discharges and spilled volumes can be characterised as being either Gaussian, Weibul or Gumbel for most cases. In some cases however, the distributions shows several sharp distinct peaks.

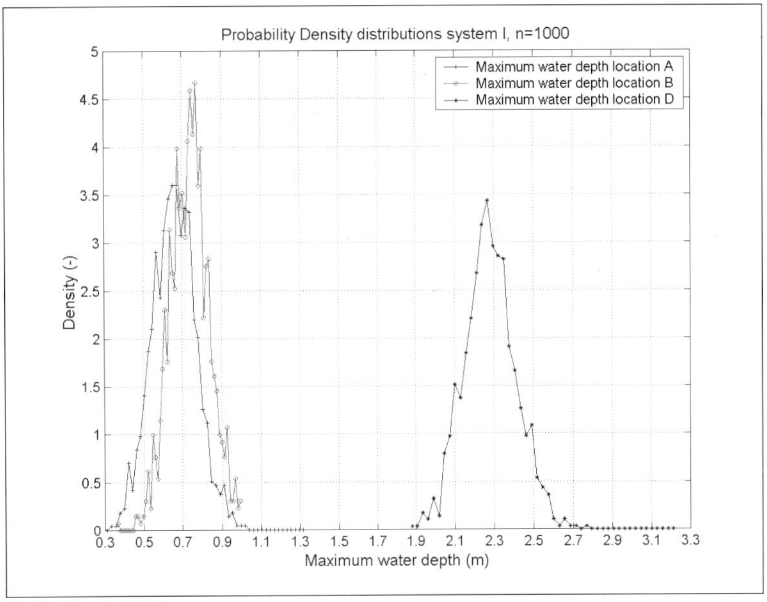

Figure 52: Density distributions maximum water depth location A, B and D, set I, system I.

For the locations A, B and D the probability distribution for the maximum water depths is by good approximation Gaussian (see Figure 52), whereas locations C and E are best described by an extreme value distribution (either Weibull or Gumbel), see Figure 53.

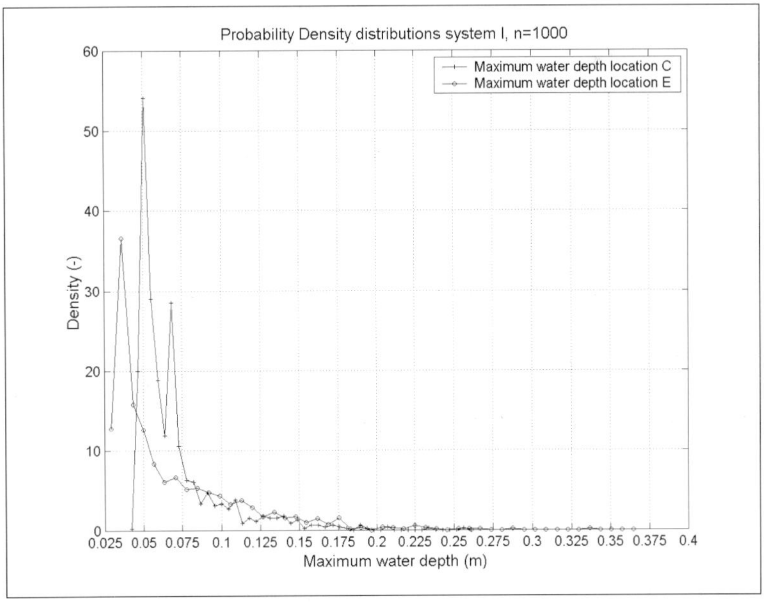

Figure 53: Density distribution maximum water depth location C and E, system I, set I

Globally the results obtained for the calculation sets in which the parameters groups run-off, area, hydraulic parameters or geometrical dimensions were varied, one of the continuous distributions best describe the density functions obtained. When the parameter group 'Structure' is varied solely, the distribution functions show sharp (distinct) peaks (see e.g. Figure 54. It must be noted, however that for the case in which only 1% of structural errors was introduced no distinct peaks evolved.

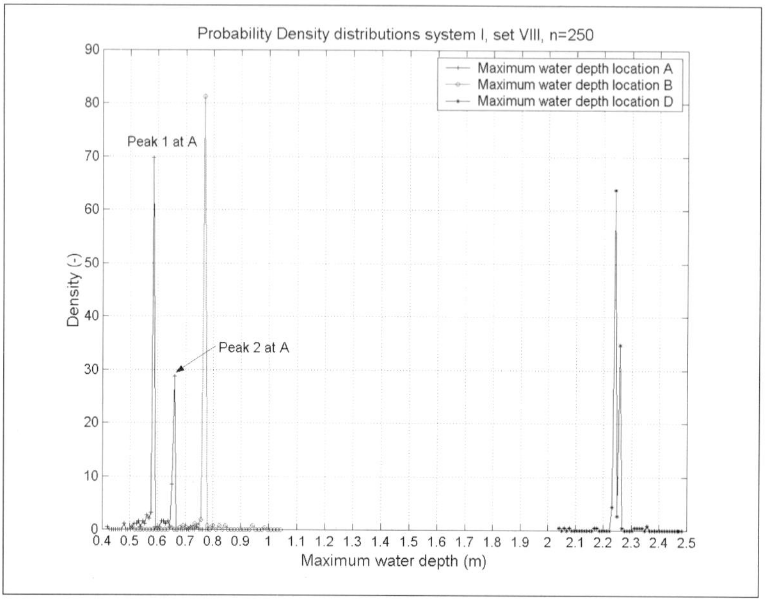

Figure 54: Density distribution maximum water depth, locations A, B and D, set VIII (20% structural errors), system I. As can be seen at location A two distinct peaks evolve, though at a relative small mutual distance.

3.8.6 Results system II

For system II also the results are presented in some detail in annexe IV,

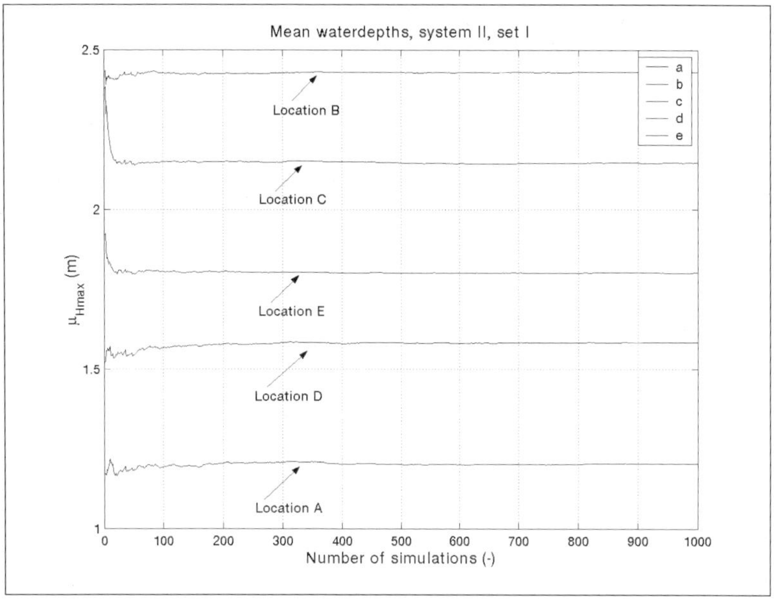

Figure 55: Mean water depths as a function of the number of simulations.

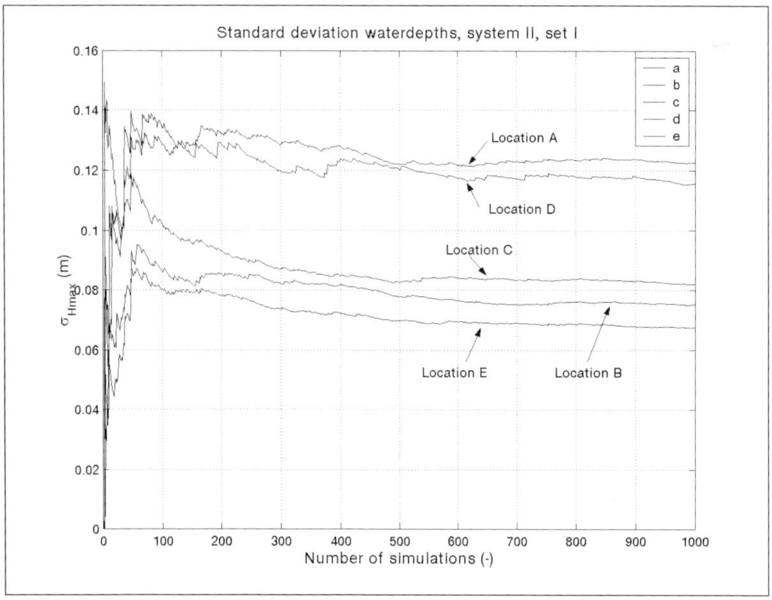

Figure 56: Standard deviation as a function of the number of simulations.

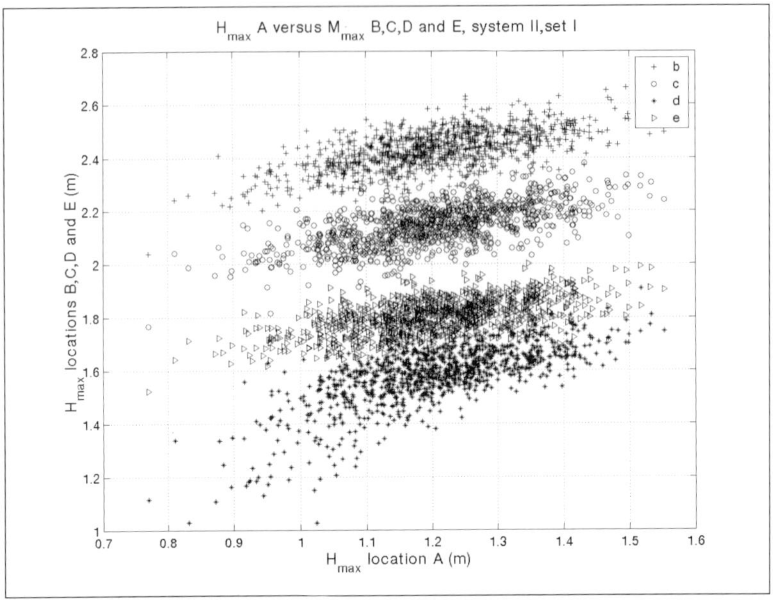

Figure 57: H_{max} at location A versus H_{max} at locations B, C, D and E. This Figure illustrates some of the mutual correlation presented in Table 17, as can be seen there is a larger correlation between water depths in system II than found for system I (see Figure 51).

	$H_{max,A}$	$H_{max,B}$	$H_{max,C}$	$H_{max,D}$	$H_{max,E}$	$Q_{max\ I}$	$Q_{max,II}$	$Q_{max,III}$	V_I	V_{II}	V_{III}
$H_{max,A}$	1	0.6367	0.6071	0.6928	0.5622	0.1111	-0.0800	-0.1078	0.6086	0.6512	0.6511
$H_{max,B}$		1	0.5304	0.5531	0.5002	0.1073	-0.0903	-0.1126	0.4918	0.4770	0.4748
$H_{max,C}$			1	0.4802	0.4742	0.0063	-0.2699	-0.2862	0.4408	0.3441	0.3422
$H_{max,D}$				1	0.5336	0.4062	0.1556	0.1771	0.7438	0.7397	0.7402
$H_{max,E}$					1	0.1614	-0.0390	-0.0609	0.5014	0.4685	0.4691
$Q_{max\ I}$						1	0.6886	0.8495	0.9708	0.6236	0.6205
$Q_{max\ II}$							1	0.8116	0.3185	0.7082	0.4333
$Q_{max\ III}$								1	0.4006	0.5079	0.4321
V_I									1	0.8889	0.8738
V_{II}										1	0.6236
V_{III}											1

Table 17: Correlation matrix for system II, set I. As can be seen the correlation between water levels, maximum discharges and spilled volumes are one order of magnitude larger than for system I (see Table 16) although the values found indicate that (except for the shaded relations) no prediction for e.g. an overflow volume can be based on a maximum water level. The larger correlation values in system II compared to system I are caused by the characteristics of this system. System II is relatively flat, implying that the system functions more or less uniformly over the separate elements of which it is composed.

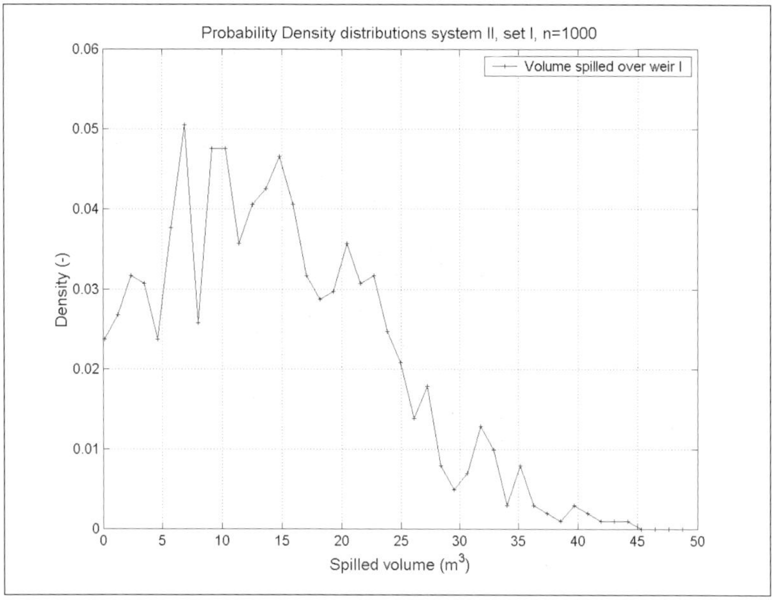

Figure 58: Density distribution spilled volume, weir I, set I, system II. As can be seen there is a substantial density at Q=0, this implies that there is realistic chance that no volume is spilled at all.

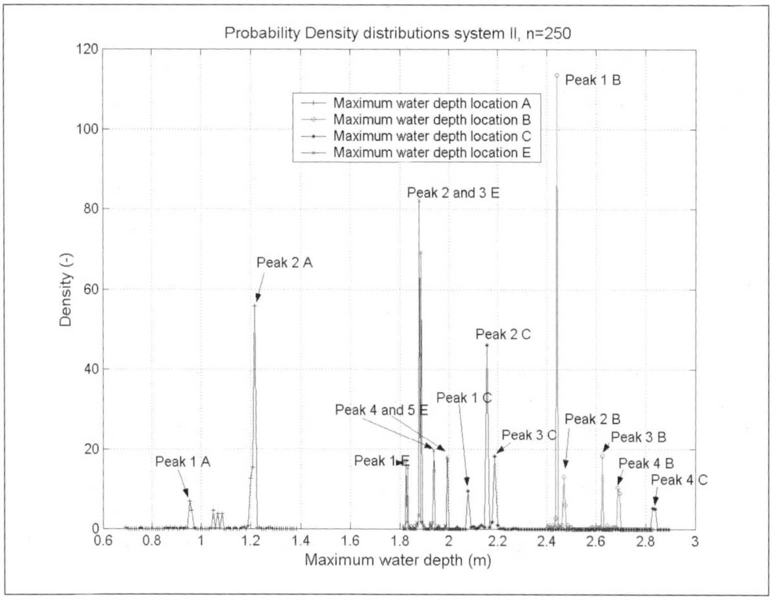

Figure 59: Density distributions for maximum water depths locations A, B ,C and E set VIII (20% structural errors), system II. When compared with the density distributions for 1%, several distinct peaks evolve.

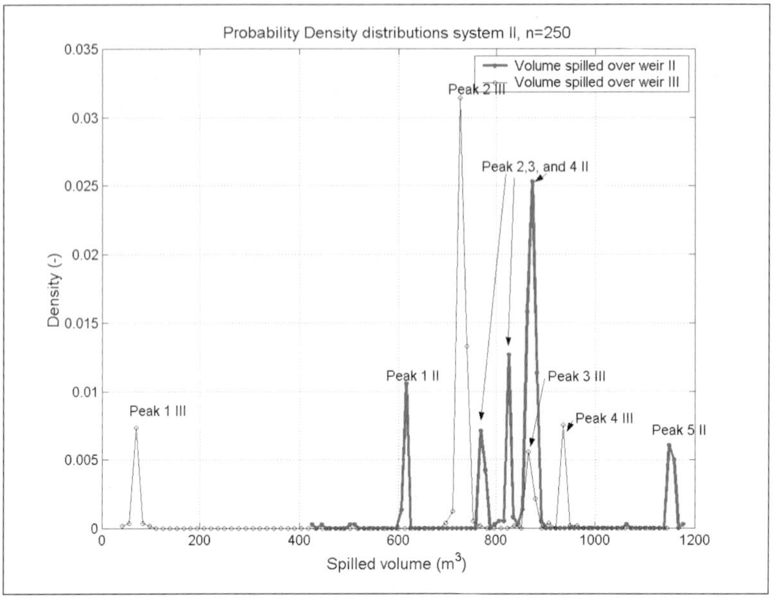

Figure 60: Density distributions for spilled volumes over weirs II and III, set VIII (20% structural errors), system II. As can be seen several distinct peaks evolve so-called 'modes', associated with some specific structural errors in the database.

3.8.7 Analysis results system I

In Figure 49 and Figure 50 respectively the mean and standard deviation for the maximum water depths as function of the number of Monte Carlo simulations are shown. As can be seen, estimates for the mean values are obtained after a limited number of simulations (N>50), while an estimate for the variance is obtained only after circa 250 simulations. In the further analysis for system I 250 simulations are used.

Based on the central limit theorem it would be expected that the probability distribution of the results would be Gaussian to a good approximation. The distributions for H_{max} at the locations C and E are best reproduced by a Weibull distribution. All other distributions (water depths, discharges and spilled volumes) are best described by a Gaussian distribution in compliance with the expected result, at least for those calculation sets that do not involve variation in the structural database properties.
In the probability distributions for the sets VI, VII and VIII (1, 10 and 20 % geometrical errors), several peaks are present. The occurrence of these peaks is evident when the percentage of the geometrical errors is 10 or 20 %. At 1 %, the effect of geometrical errors is relatively small when compared to the effect of the other parameter groups.

The reason the water levels at location C and E follow an extreme value distribution may be explained from their position in the system. Both locations are situated at the border of the system as a top-end and are at a relatively high level in the system. In Figure 61 the typical situation is sketched, at location B the maximum water depth is distributed Gaussian. At the given load (rain and DWF) the flow from manhole A to B is at such a low rate that hardly any noticeable effect is present in the water level at manhole A due to this. The only effect on the water level in manhole A is due to a backwater effect caused by the water level at manhole B. This implies that the probability distribution of the water levels at manhole A is mainly influenced by the upper part of the (e.g. Gaussian) distribution of the water level at manhole B.

Figure 61: Backwater effect on probability distribution.

As can be seen in Table 16, there is no significant correlation between the maximum water depths at the 5 locations studied. The maximum discharge and the spilled volume are positively correlated; this is also the case for the water depths at A and B and the spilled volume. This is easily explained from the structure of the network; both A and B are in main transporting routes towards the external weir, whereas C and E are top-ends.

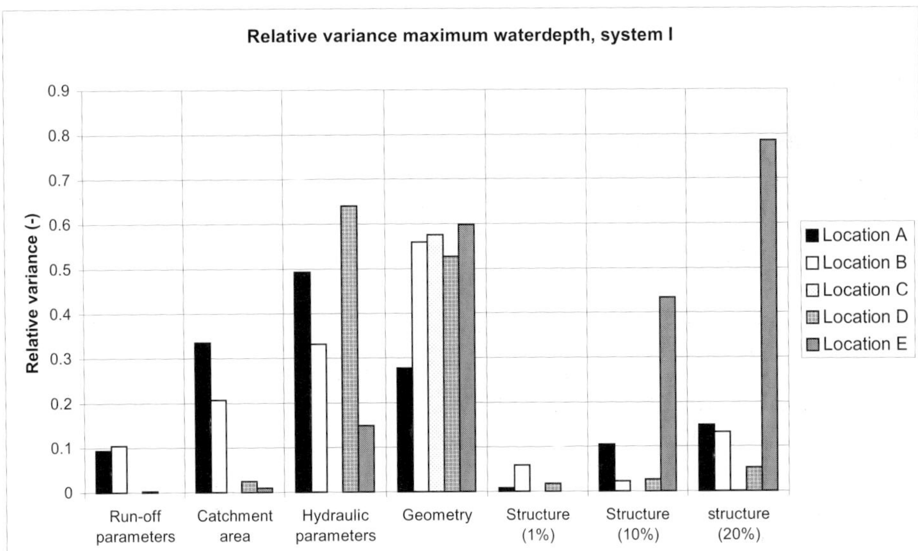

Figure 62: Relative variance in maximum water depths in system I.

var	σ_A	$\left(\frac{\sigma_i}{\sigma_1}\right)^2_A$	σ_B	$\left(\frac{\sigma_i}{\sigma_1}\right)^2_B$	σ_C	$\left(\frac{\sigma_i}{\sigma_1}\right)^2_C$	σ_D	$\left(\frac{\sigma_i}{\sigma_1}\right)^2_D$	σ_E	$\left(\frac{\sigma_i}{\sigma_1}\right)^2_E$
1	0.114	1.00000	0.099	1.00000	0.029	1.00000	0.135	1.00000	0.044	1.00000
2	0.035	0.09425	0.032	0.10448	0	0	0.007	0.00269	0	0
3	0.066	0.33518	0.045	0.20665	0	0	0.021	0.02420	0.0044	0.01
4	0.080	0.49246	0.057	0.33149	0	0	0.108	0.64000	0.017	0.14927
5	0.060	0.27700	0.074	0.55872	0.022	0.57551	0.098	0.52697	0.034	0.59711
6	0.010	0.00769	0.024	0.05877	0	0	0.018	0.01778	0	0
7	0.037	0.10534	0.015	0.02296	0	0	0.022	0.02656	0.029	0.43440
8	0.044	0.14897	0.036	0.13223	0.001	0.00119	0.031	0.05273	0.039	0.78565

Table 18: Relative variance in maximum water depth, system I.

In Figure 62 the relative([29]) variance in the water levels for locations A,..,E for the calculation sets I,...,VIII are shown. In set I all parameters are varied, and this set is therefore used as reference by which to measure the effect of the individual parameter groups.
As can be seen the sets 4 and 5 (hydraulic parameters and geometrical measures respectively) show a relatively large relative variance. Variation of the parameter values in the hydrological model has a smaller influence. For the locations A and B the variation in catchment area and run-off parameters do have a significant influence also. For location C geometrical properties seem to be most important.

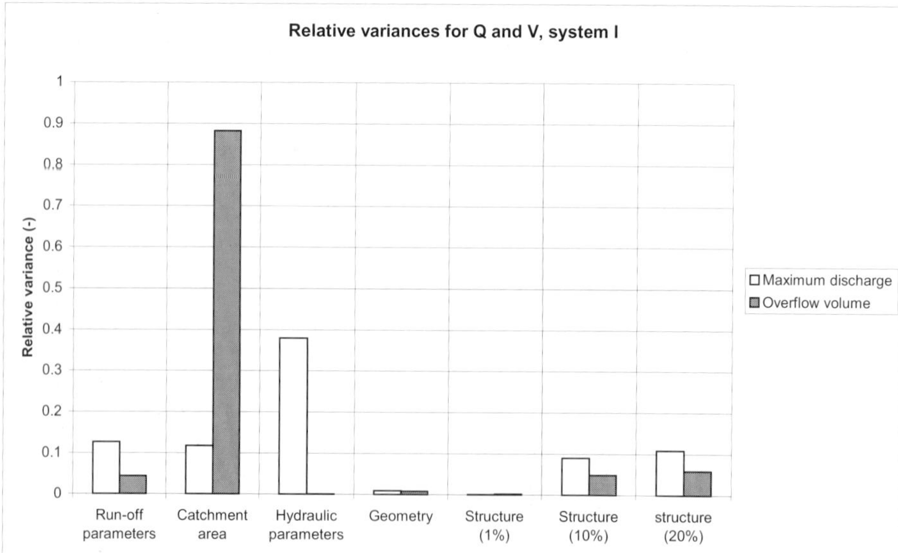

Figure 63: Relative variances for overflow volume and maximum discharge, system I.

([29]) The relative variance is defined as $\left(\frac{\sigma_i}{\sigma_1}\right)^2$, in which σ_1 is the standard deviation in calculation set I (variation in all parameter groups) and σ_i is the standard deviation due to variation in one parameter group as defined in calculation set i.

variation in:	σ_Q	$\left(\frac{\sigma_i}{\sigma_1}\right)^2_Q$	σ_V	$\left(\frac{\sigma_i}{\sigma_1}\right)^2_V$
All parameter groups	0.073	1.0000	132.12	1
Run-off parameters	0.026	0.1269	27.97	0.0448
Catchment area	0.025	0.1173	124.11	0.8824
Hydraulic parameters	0.045	0.3800	4.09	0.0010
Geometry	0.007	0.0092	11.75	0.0079
Structure (1%)	0.002	0.0008	5.20	0.0015
Structure (10%)	0.022	0.0908	29.20	0.0488
Structure (20%)	0.024	0.1081	32.25	0.0596

Table 19 : Relative variance for maximum discharge and spilled volume, system I.

Furthermore it is seen that the introduction of 1% structural errors results in only a minor relative variance in the water levels. When introducing 10% or 20% errors the relative deviation is noticeable, but still within acceptable limits when compared to set 1 (i.e. <1.0). Location E is most vulnerable for structural errors at error levels at or above 10%.

In Figure 63 and Table 19 the relative variance for the overflow volume and the maximum overflow discharge is shown for the calculation sets. With respect to the maximum overflow discharge, variation in hydraulic parameters is most important, while geometrical variation is of minor influence. The effect of structural errors is only significant for 10% and 20% errors. The overflow volume is influenced mainly by variation in catchment area. Variation in structure however, has a minor influence even when 10% or 20% errors are introduced.

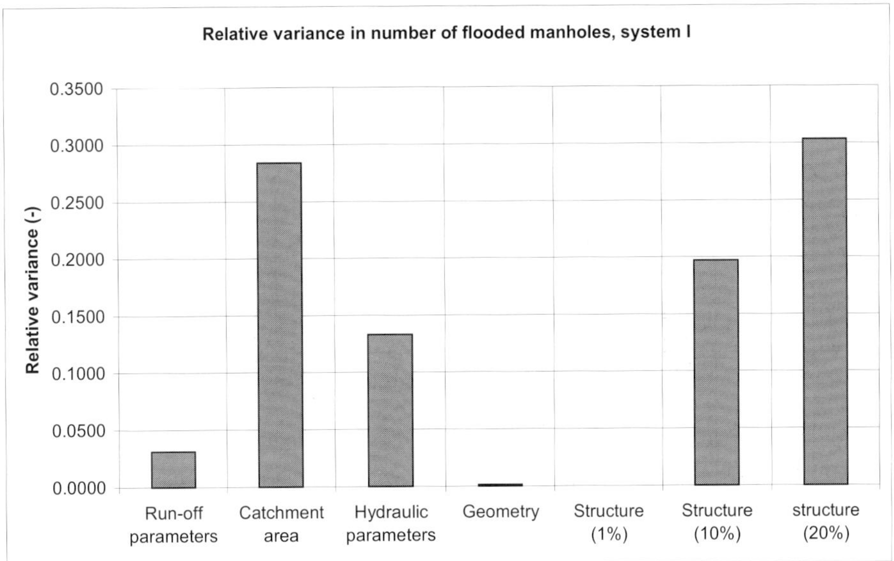

Figure 64: Relative variance in the number of flooded manholes, system, I.

variance in	σ_N	$\left(\frac{\sigma_i}{\sigma_1}\right)_N^2$
All	1.066	1.0000
Run-off parameters	0.188	0.0311
Catchment area	0.568	0.2839
Hydraulic parameters	0.389	0.1332
Geometry	0.039	0.0013
Structure (1%)	0.000	0.0000
Structure (10%)	0.473	0.1969
structure (20%)	0.587	0.3032

Table 20: Relative variance in the number of flooded manholes, system I.

In Table 20 and Figure 64 the relative variance in the number of flooded manholes is shown. In particular the catchment area has an important influence of the results; structural errors only become of significant importance when >10%.

3.8.8 Analysis system II

The effect of an increasing number of Monte Carlo simulations on the results is shown in Figure 55 (estimates for the mean value) and in Figure 56 (estimate for the standard deviation). As can be seen the estimate for the mean water level is almost constant for N>200), while the estimate becomes stable only for N>700; the differences however with the estimates obtained with N=400 are relatively small. For reasons of calculation time the presented results are based upon N=400.

For all parameters, except for the water level at location D, the results match with a Gaussian distribution. In this case, the system is rather flat, which implies that situations like those occurring in system I at location C and E, leading to an extreme value distribution, do not occur. The distribution for the water level at location D in system II can only be understood as the effect of a branched system versus a looped system, since location D was the only location situated in a branched part. The goodness of ´fit´, however, for the Normal distribution or the Gumbel distribution is only small.
In Figure 65 the relative variance in maximum water levels is shown for different calculation sets. Clearly, variation in catchment area and hydraulic parameters has the most significant influence on the calculation results. Furthermore, location A and D respond in a different manner to variation in the level of structural errors when compared to the other locations. Location D seems to be relatively insensitive for structural errors; 1%, 10% and 20% errors in structure show about the same relative variance. In contrast, location A shows a relative large deviation even at 1% errors, but the effect of increasing the percentage of errors does not result in a sharp increase of the deviation as seen in locations B, C and E. Locations A and D are a top-end and a location in a branched (sub) system respectively, while the other locations are in the looped (sub) system.

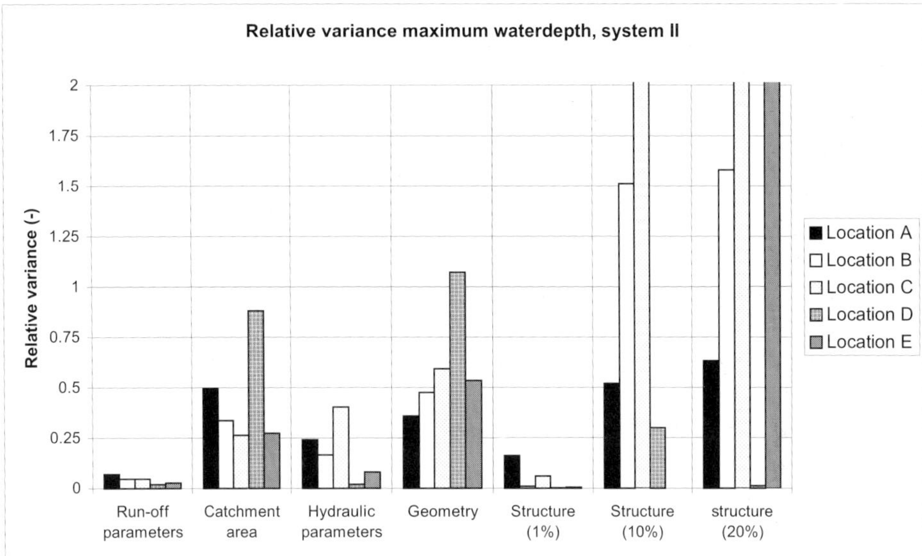

Figure 65: Relative variances in maximum water depths in system II.

var	σ_A	$\left(\frac{\sigma_i}{\sigma_1}\right)^2_A$	σ_B	$\left(\frac{\sigma_i}{\sigma_1}\right)^2_B$	σ_C	$\left(\frac{\sigma_i}{\sigma_1}\right)^2_C$	σ_D	$\left(\frac{\sigma_i}{\sigma_1}\right)^2_D$	σ_E	$\left(\frac{\sigma_i}{\sigma_1}\right)^2_E$
all	0.1220	1.0000	0.0740	1.0000	0.0740	1.0000	0.1150	1.0000	0.0670	1.0000
run-off parameters	0.0320	0.0688	0.0160	0.0467	0.0160	0.0467	0.0160	0.0194	0.0110	0.0270
area	0.0860	0.4969	0.0430	0.3377	0.0380	0.2637	0.1080	0.8820	0.0350	0.2729
Hydraulic parameters.	0.0600	0.2419	0.0300	0.1644	0.0470	0.4034	0.0160	0.0194	0.0190	0.0804
Geometry	0.0730	0.3580	0.0510	0.4750	0.0570	0.5933	0.1190	1.0708	0.0490	0.5349
Structure 1 %	0.0240	0.0387	0.0070	0.0089	0.0180	0.0592	0.0060	0.0027	0.0040	0.0036
Structure 10%	0.0880	0.5203	0.0910	1.5122	0.2210	8.9191	0.0630	0.3001	0.0590	0.7754
Structure 20 %	0.0970	0.6322	0.0930	1.5794	0.2230	9.0813	0.0120	0.0109	0.0460	0.4714

Table 21: Relative variances in maximum water depth, system II

It must be noted however that in the analysis of the results for increasing variance in structural errors, the relative variance does not tell the whole story. For practical purposes it is of more importance to note that with an increasing error percentage the density distributions show 'peaks' that are more distinct with increasing error percentage. This implies that in a database with a relative high error percentage (>10%) the probability for obtaining incorrect results increases.

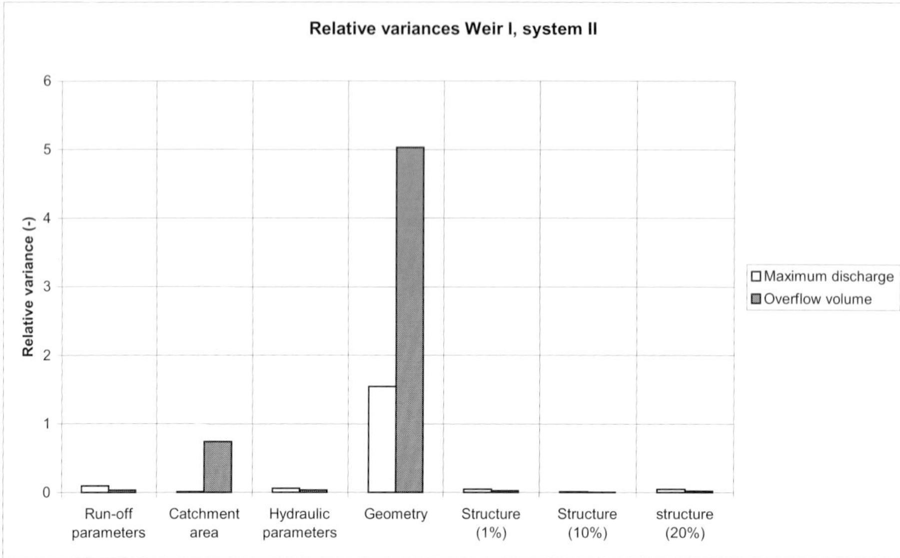

Figure 66: Relative variance in overflow volume and -discharge, weir I system II.

	σ_Q	$\left(\frac{\sigma_i}{\sigma_1}\right)^2_Q$	σ_V	$\left(\frac{\sigma_i}{\sigma_1}\right)^2_V$
All	0.0177	1.0000	9.00	1.0000
Run-off parameters	0.0054	0.0931	1.66	0.0340
Catchment area	0,0018	0.0103	7.73	0.7377
Hydraulic parameters	0.0044	0.0618	1.63	0.0328
Geometry	0.0220	1.5449	20.19	5.0325
Structure (1%)	0.0040	0.0511	1.47	0.0267
Structure (10%)	0.0020	0.0128	0.60	0.0044
Structure (20%	0.0040	0.0511	1.31	0.0212

Table 22: Relative variance for maximum discharge and volume, weir I, system II.

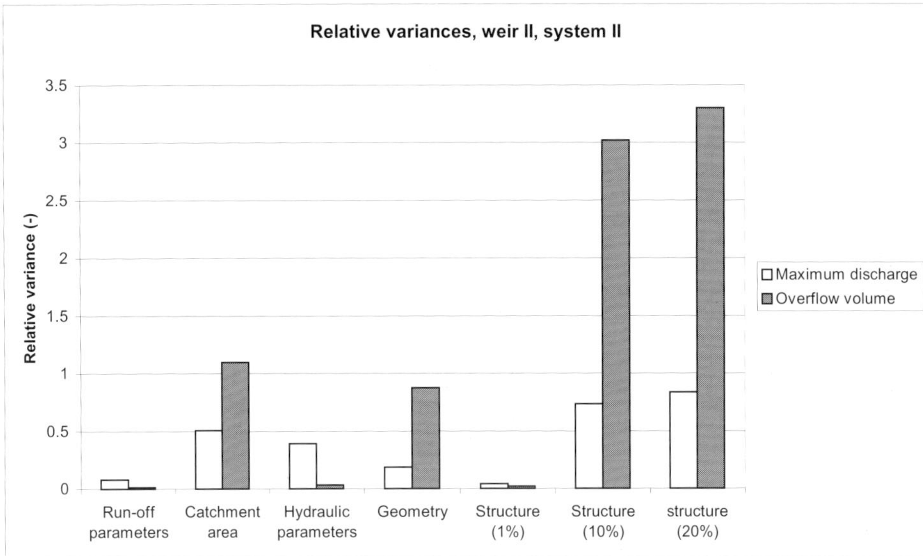

Figure 67: Relative deviation in overflow and discharge, weir II, system II.

	σ_Q	$\left(\frac{\sigma_i}{\sigma_1}\right)_Q^2$	σ_V	$\left(\frac{\sigma_i}{\sigma_1}\right)_V^2$
All	0.0350	1.0000	75.83	1.0000
Run-off parameters	0.0100	0.0816	8.81	0.0135
catchment area	0.0250	0.5102	79.48	1.0986
Hydraulic parameters	0.0220	0.3951	14.05	0.0343
Geometry	0.0150	0.1837	70.90	0.8742
Structure (1%)	0.0070	0.0400	10.90	0.0207
Structure (10%)	0.0300	0.7347	131.82	3.0219
Structure (20%)	0.0320	0.8359	137.75	3.2999

Table 23: Relative variance for maximum discharge and volume, weir II , system II.

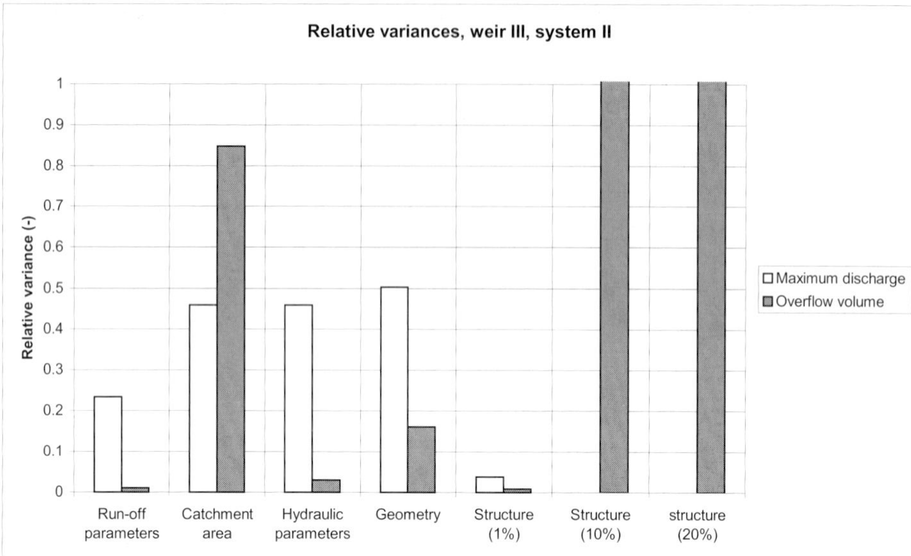

Figure 68: Relative variance in overflow volume and -discharge, weir III system II.

	σ_Q	$\left(\frac{\sigma_i}{\sigma_1}\right)_Q^2$	σ_V	$\left(\frac{\sigma_i}{\sigma_1}\right)_V^2$
All	0.0310	1.0000	77.98	1.0000
Run-off parameters	0.0150	0.2341	8.04	0.0106
Catchment area	0.0210	0.4589	71.84	0.8487
Hydraulic parameters	0.0210	0.4589	13.51	0.0300
Geometry	0.0220	0.5036	31.32	0.1613
Structure (1%)	0.0060	0.0375	7.21	0.0085
Structure (10%)	0.1400	20.3954	233.23	8.9455
Structure (20%)	0.1420	20.9823	239.11	9.4022

Table 24: Relative variance for maximum discharge and volume, weir III , system II.

In system II there is a significant difference between the behaviour of the weirs II en III on the one hand and weir I on the other. The standard deviation shown by weir I is, on the average, 6-8 times more pronounced then by the other weirs. The general trend is that the influence of the total run-off area is by far the most important. Though when looking at the maximum discharge, the effect of the hydraulic parameters (i.e. in this case the overflow coefficient) is also a major factor influencing the overall error propagation. From Table 22, Table 23 and Table 24 it becomes clear that especially weir III is sensitive to structural errors. In Table 17 the correlation matrix for calculation set I is shown, as can be seen the maximum water depths at all locations show a relative strong positive correlation (i.e. circa 0.50) when compared to the correlation-values found for system I.

In Table 25 and Figure 69 the relative variance for the number of flooded manholes is shown for system II. In system II all types of parameters have a noticeable influence. The introduction of 1% structural errors does not have a large effect. With an increasing error percentage however, the effect increases.

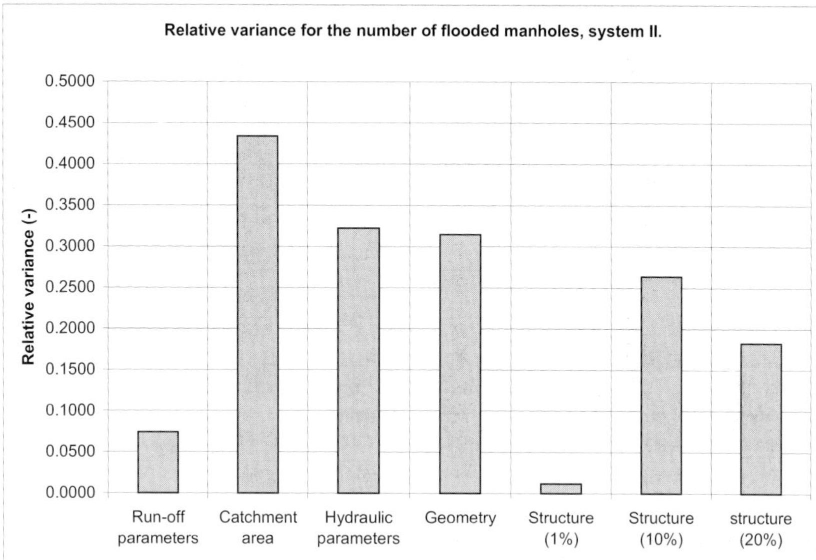

Figure 69: Relative variance for the number of flooded manholes, system II.

	σ_N	$\left(\frac{\sigma_i}{\sigma_1}\right)_N^2$
All	7.79	1.0000
Run-off parameters	2.12	0.0741
Catchment area	5.13	0.4337
Hydraulic parameters	4.42	0.3219
Geometry	4.37	0.3147
Structure (1%)	0.84	0.0116
Structure (10%)	4	0.2637
structure (20%)	3.33	0.1827

Table 25: Relative variance in the number of flooded manholes, system II.

3.8.9 Discussion

The number of Monte Carlo simulations required to obtain estimates for the first and second moment (mean and variance) depends on the system under study. This is illustrated by the fact that for system I 250 simulations were sufficient and in system at least 400 simulations had to be done. Characterising system I as a 'steep' system and system II as a flat (or typically Dutch system), the analysis in the previous sections shows that variance (or uncertainty) in model parameters in the hydraulic model has a more pronounced influence on the results in a steep system than in a flat system. This is especially so with respect to the maximum water depths occurring. This easily understood since in steep systems the flow velocities can become > 1 m/s, whereas in the flat system the occurrence of these high velocities is rare. This implies that local head losses are a noticeable factor in steep systems when compared to flat systems. Another observation with respect to maximum water depths is that the effect of parameter uncertainty depends on the exact location in the system that is studied. In both systems variance in the geometry of the system is of major importance, while errors in the structure of the system only become significant when the error percentage is >10%.

When comparing the correlation matrices for both systems (Table 16 and Table 17) it is seen that the correlation between maximum water depths at various locations in systems I is significantly less than those in system II. This may be explained from the different characteristics of both systems. It is intuitively understood that in a flat, looped network (system II) water depths are more heavily correlated than in a relatively steep network.
The characteristics of the probability distributions when introducing only structural errors, deviate from the distributions found when varying other parameter groups.
The main difference is that the density distributions show peaks. For both systems there is only one peak when the error percentage is 1%. This implies that the results are located in a small band of possible water values. When the errors percentage increases the density distributions tend to show several peaks. This implies that the results are located in several small bands of values. These bands are referred to as 'modes'. Recalling the fact that the type of structural errors are such that they escape the normal error recognition protocols, the occurrence of these 'modes' is a danger to be kept in mind when redesigning a drainage system. When such an error is present in a database and it does not lead to a direct warning like flooding, it is easily missed by the designing engineer. The effect of not recognising the error has an unpredictable effect on the quality of the resulting design. An example is seen in Figure 60, the spilled volumes over external weir III in system II can be in the 'mode' of peak I (circa 70 m^3) but with an almost equal probability in peak 4 circa 930 m^3. This kind of differences can have an important effect on the suggested measures to be taken to correct the functioning of a system under study.
In fact, this is an example of a 'false computer image' of a sewer system.
The overall conclusion of the analysis is that there is no single group of parameters that dominates the variance in results of a hydrodynamic model. Therefore, all parameter groups should be known with the highest accuracy practically achievable. Furthermore, it is concluded that the errors in the database with respect to structure should preferably not exceed 1 %. The only manner to achieve this degree of accuracy in the database is to maintain a good system of data management. One manner to maintain a good database is to perform cross-validations whenever possible. For instance, when an in-sewer inspection on the technical state is carried out, one is well advised to verify the available data of the construction under inspection also. The over-head costs of this verification are negligible, while it helps to avoid unnoticed errors in modelling.

3.9 Conclusions chapter 3

Obtaining a 100 % perfect database for an urban drainage system is hard to achieve in practice. Techniques like visualisation in 2 dimension or projections of 3-D images are effective in tracking and removing obvious errors (blunders) in geometrical structure and measures of height.
Smaller errors can be recognised when studying the so-called ‘lost storage’ by means of an algorithm as discussed and presented in Section 3.5.3. The only safe way to be certain that the database used is in line with the real structure in practice, is to obtain field data for every

object in the database. Due to the high costs involved with field survey's however, this is seldom done. Especially in large catchments the practical problems in this respect are obvious. When applying the data-error recognition techniques as described in this chapter no guarantee for an error-free database is obtained.
Based on some assumptions with respect to possible errors in measures of height and the introduction of geometrical and structural errors in the database escaping the attention of an experienced modeller, a sensitivity analysis for two different catchments is made.
When studying the effects of data errors and parameter errors on calculation results the following conclusions are drawn:

- When putting the consistency checks to effect, the influence on the results is in such a magnitude that the calculation may safely be used for design purposes.
- When studying an existing system, the minimum to be done is a cross validation between observation (like flooding areas) and model predictions. This does not make a detailed measuring campaign necessary.
- The number of Monte Carlo simulations needed to obtain estimates for mean and variance depend on the system under study and are in the order of magnitude of 10^2. If the focus is on a more detailed insight in the extremes of the probability distributions (as will be for design purposes) the number of simulations needed will increase.
- In any case the type of distribution obtained for the results in a Monte Carlo simulation has to be studied, since it cannot always be expected to be Gaussian. This is especially so in systems having large differences in construction levels.
- When the number of geometric and structural errors is limited (circa 1%) their influence on the results is also limited. With an increasing percentage, the effects overwhelm other uncertainties. This is a further proof for the correctness of the postulated effect of these errors on model results stated by Price & Osborne (1986) (see also Section 3.1).
- The possible effects on water levels calculated may be up to decimetres; this is too large for obtaining a correct calibration as described in chapter 5 of this thesis. When setting out to calibrate an urban drainage system, the constraints put on the applied database are more severe than normally applied in practice.
- Branched networks are more sensitive for structural and geometrical errors in the database when compared to looped systems.
- The amount of area in a catchment contributing to the run-off is of major influence and should therefore be as correct as possible.
- For (re) design purposes a small percentage of structural and geometrical errors in the database can be acceptable since the effect on predicted maximum water levels does not result in grossly over- or underestimating the occurrence of flooding.
- In Section 3.1, the highest possible level of correctness for a database was defined as:

  '*Every piece of information in the database complies exactly with the actual physical situation, allowing only deviations due to non-systematic measuring errors.*'

  As has become clear from the analysis in Section 3.8, it is not necessary to obtain this utter level of correctness in practical cases. A certain maximum error level can be allowed as long as the effects of this error level are relatively low when compared to the other sources of uncertainty in the calculations made. Although no general conclusions are allowed to be drawn on the limited set of systems analysed, the allowable error level with respect to structural errors in the database is estimated to be circa 1%. Prerequisites are however, that the database has been checked by every possible means as described in the Section 3.5 in this chapter. A further increase in the accuracy and reliability of the model results can only be obtained when measuring data are used to calibrate the model applied.

CHAPTER 4: MEASURING PROGRAMS

4.1 Introduction

4.1.1 General

Using a model (see Chapters 2 and 3) is one way to study the hydraulic behaviour of an urban drainage system. This method has the advantage that at relatively low cost and in a relatively short time a detailed insight is gained in this hydraulic behaviour. The main drawbacks however, are the uncertainties in the results obtained due to data errors (see Chapter 3) and the fact that several process parameters must either be estimated or are to be obtained from generally accepted default values.
Complementary to using models are measuring programs. By measuring information on the actual hydraulic behaviour is obtained circumventing the drawbacks of a model. A measuring program is hardly ever able to give information as detailed as a model can; this is because at a limited number of locations information can be obtained over a limited span of time only.
So, a measuring program results, like a model, in just an other 'image' of the hydraulic behaviour of the real system under study. The characteristics however, are different:

- The information obtained is directly related to the real processes.
- In practice, the resolution in time and space is limited.
- The data acquisition is real time, resulting in long waiting periods.
- No 'what-if' design questions can, practically, be answered.

The required information that is to be obtained by a measuring programme varies with its purpose. For instance, for research purposes the reliability of the information must be at a high level while the duration of the programme is mostly limited. On the other hand, for monitoring purposes the duration may be unlimited and a somewhat lower level of reliability must accepted if the programme is to be economically feasible. A more rigorous discussion of the aspects related to measuring projects can be found in STOWA (1996[a,b]).
In this chapter the focus is on techniques to judge designs of measuring networks by their effectiveness. As will become clear, the effectiveness of such a network depends on the following variables:

- The measuring frequency.
- The measuring accuracy.
- The duration of the measuring programme.
- The inherent variability in time and space of the process under study.

Since this thesis focuses on water quantity aspects only, measuring programs aiming at the quantification of pollution loads from CSO's are not discussed. The techniques presented however can be applied on this measuring objective also.
The discussion is limited to:

- Water level measurement.
- Discharge measurement.
- Rain intensity measurement.

4.1.2 Measuring programs

A measuring programme comprises different aspects and activities that are interrelated. This is schematically shown in Figure 70. After translating the measuring objective into a criterion by which a network design can be judged, either a model or historical data (or a combination) is applied to decide on the variables to be measured, measuring locations and sampling frequencies. An important step is to decide where in the system data are to be obtained. Due to practical reasons (like inaccessibility because of traffic), only a limited number of locations are suitable. Furthermore, an increase of measuring locations leads to a sharp increase in the investments and the operational costs involved. Therefore, this selection has to be done with care. Intuitively it is clear that the locations must be spread over the system ([30]). Without going into details here, a (preliminary) model of the system is applied to judge the information content of several practically achievable measuring setups of which the best one is chosen.

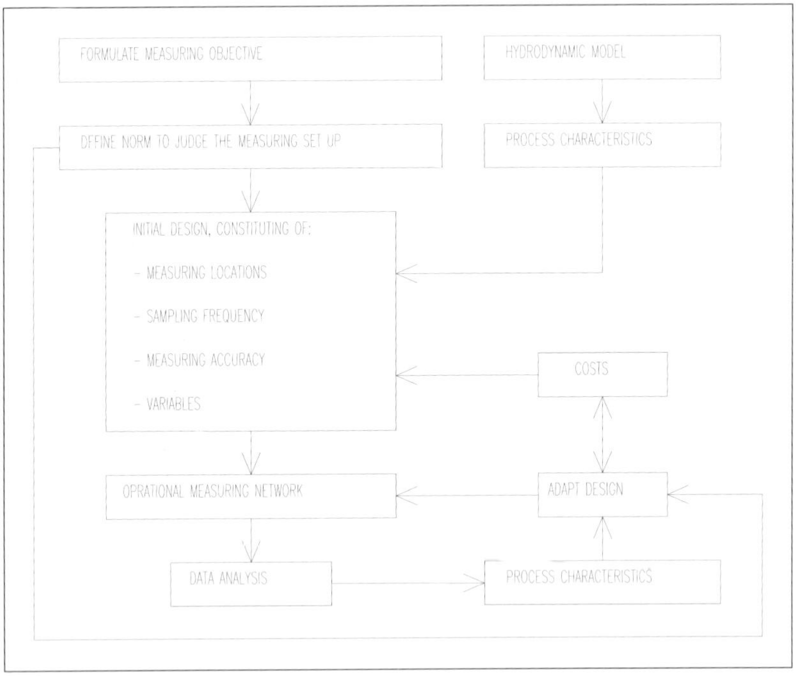

Figure 70: The iterative process in measuring network design and adjustments.

The total amount of information obtained from a measuring setup is a function of the measuring accuracy, measuring frequency and the spatial density of the measuring locations. Too large a density does not contribute to the information content while too sparse a system results in too little information. Once the measuring locations have been chosen a measuring program can be set up, resulting in a set of requirements with respect to measuring accuracies, -frequencies and aspects of data handling. Based on this program, equipment is obtained meeting the requirements set. After installation the measuring campaign is started. It is of

([30]) This is easily seen when considering the fact that 'n' data loggers on one location monitoring the same parameter do not give additional information compared to just one logger (assuming this logger functions correctly).

importance to accurately keep record in a logbook of the events taking place ([31]). During the measuring campaign, it may occur that information is found which leads to necessary adaptations of the model database. In some cases this can lead to the necessity to change the selection of the measuring location. The results from a quick data analysis should become available on a regular, frequent basis (e.g. every 2 weeks), since it can reveal problems with the functioning of the equipment like zero drift or synchronicity problems.

4.2 Measuring network design

4.2.1 General

When designing a measuring network it is of importance to obtain a good estimate for the sampling density in time and space. Such an estimate must be based on the reliability of the information obtained from the measurements related to the goal of the measuring campaign. This implies that a relation has to be defined between sampling frequency and some criterion by which the information from the measuring data can be judged. For almost any goal this relation is defined by four variables; the characteristic scales of the process, the variability of the process, the measuring period and the obtainable measuring accuracy. To obtain estimates for the characteristic scale of the process studied, historical data can be used in cases of existing networks. For newly built systems an initial estimate is usually derived from models. Eventually, the sampling frequency can be optimised by iteration as soon as data become available. For scientific ends, usually rather short measuring periods are applied, normally limiting the possible application of iterative refinement of the measuring network, though a high degree of reliability is desired. When implementing a measuring network for monitoring purposes (like e.g. for routine control of discharge permits) measuring continues during long periods. In such cases the degree of reliability is decided upon the results obtained from multi-criteria analyses between costs and benefits (see e.g. Terstriep (1986)). The main goal of interest in this thesis with respect to measuring networks is that of obtaining information that can be applied for parameter estimation of hydrodynamic models. It is possible to optimise measuring locations, sampling frequencies and measuring accuracy simultaneously. For instance Kalman filtering has been applied with success to this end on models in other modelling sectors (see e.g. van Geer (1987)). The use of Kalman filtering however, is beyond the scope of this thesis because it asks for massive calculation power, even when recently developed efficient algorithms are applied (see e.g. Verlaan (1998)). An important criterion in this thesis is that all techniques described can be performed on a personal computer even for relatively large networks (i.e. > 5000 nodes). The effect of choosing measuring locations is discussed briefly in Chapter 5 (Section 5.11), in which some effects of choosing different locations on the quality of the parameter optimisation process are illustrated. In this section the focus is on defining the sampling frequency in time for measuring programmes in urban drainage using relatively simple techniques. To this end the following issues are discussed:

- Estimating relevant time scales.
- Trend detection.
- State reconstruction.

The availability of prior estimates for relevant time scales are of importance for a designing engineer to check the results of more complicated methods. In order to show the influence the

([31]) In a logbook information like problems encountered with the measuring installation, operational information, weather conditions are recorded. This information is of importance when performing data analysis and in the calibration process.

measuring goal has on the measuring frequencies to be applied, trend detection and state reconstruction are discussed.

4.2.2. Relevant time scales in urban drainage

4.2.2.1 General

In Chapter 2 the processes taken into account when modelling an urban drainage system are described. In the next subsections, simple techniques to estimate characteristic time scales for some of these processes are presented and illustrated with examples. In this section the decimation time is used as a measure. This is a somewhat different approach compared to the measure used in Section 4.2.3.3 in which a time scale measure is based upon auto-correlation functions. The main difference is that the decimation time is based upon a (simplified) description of the physical process, whereas the auto-correlation function has a statistical basis. The latter approach is practical when either a complicated combination of processes is significant or when only measuring data are available. The order of magnitude of the characteristic time scale however, is equal in both approaches.

4.2.2.2 DWF conditions

It is well known that DWF shows periodic fluctuations, in which daily and weekly patterns are present. In small rural areas patterns with a typical time scale of 4 to 6 hours are found, whereas in large urban conglomerates such pattern hardly exist (see e.g. Koot (1979)).
In some cases seasonal patterns can be recognised due to changes in ground water table.

4.2.2.3 Overflow mode

In order to obtain an estimate for the time scale for a system during an external spill (overflow mode) a simplified eigen-process (i.e. no excitation is working on the system) is studied.
To this end the mass balance of a tank spilling over a weir is modelled:

$$\frac{dV(h)}{dt} - \alpha h^{\beta} = \frac{\partial V(h)}{\partial h}\frac{\partial h}{\partial t} - \alpha h^{\beta} = 0 \qquad \text{(eq. 4. 1)}$$

Assuming the derivative of the volume with respect to the water depth to be constant, this may be simplified into:

$$A\frac{dh}{dt} - \alpha h^{\beta} = 0 \qquad \text{(eq. 4. 2)}$$

In which:

h	water depth above weir crest	(m)
A	area of the free surface	(m^2)
V	volume	(m^3)
α	weir constant	($m^{1.5}s^{-1}$)
β	power constant normally 1.5	(-)

The solution for (eq. 4.1) is easily found by separation of variables;

$$t = \frac{A}{\alpha}(1-\beta)\left[h^{(1-\beta)} - h_0^{(1-\beta)}\right] \qquad \text{(eq. 4. 3)}$$

As a measure for the characteristic time scale of the system the decimation time is used (i.e. the time the system needs to reach h = 0.1 * h(t=0) = 0.1*h_0). This implies that the characteristic time of the system depends on system variables (A, α and β) as well as on the initial condition h_0. (Note that the characteristic time is process as well as system dependent). Apart from the simplifications and the necessity to use estimated values for the systems variables. It is clear that it is impossible to obtain an accurate estimate of the characteristic time-scale for any practical case, since the initial state varies strongly. However, a lower limit for $t_{0.1}$ can be obtained by setting h_0 at its expected maximum value (i.e. the difference between weir level and street level). Neglecting buoyancy and resistance effects, the value for A can be estimated as the free surface area in the manholes and pipes in the drainage system, while textbook values (unless better estimates are available, see Section 4.4.3) for the values of α and β are used.

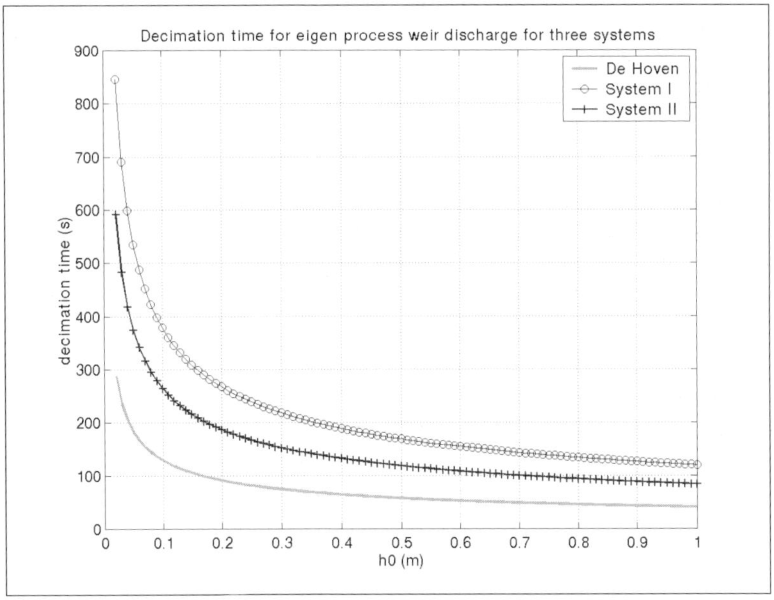

Figure 71: The decimation time for three drainage systems as a function of the initial water depth above weir crest. In this case β=1.5 and $\alpha = L\sqrt{gm}$, L= weir length, m=1.5 [-].

System	Free surface area (m^2)	α ($m^{1.5}$/s)	A/α ($m^{0.5}$/s)
De Hoven	161	4.27	37.70
System I	249	2.25	110.67
System II	696	9.00	77.33

Table 26: Relevant characteristics of the decimation time shown in Figure 71.

4.2.2.4 Emptying of a system

After a storm event a system is emptied gradually. Neglecting variation in DWF altogether, this process is described (heavily simplified) by:

$$\frac{dV(h)}{dt} = \frac{\partial V(h)}{\partial h}\frac{\partial h}{\partial t} = A(h)\frac{dh}{dt} = -Q_p \qquad \text{(eq. 4. 4)}$$

The general solution of equation 4.4 is:

$$\int_{h_0}^{h} A(h)dh = Q_p * (t - t_0) \qquad \text{(eq. 4. 5)}$$

This implies that when A(h) is known (either as an analytical function or in the form of tabulated values) equation 4.5 can be solved. For practical cases A(h) is derived using the relation between storage volume and water level for a given drainage system. This latter relation is calculated using the principles as outlined in annexe I. In Figure 72 an example of such a relation is shown.

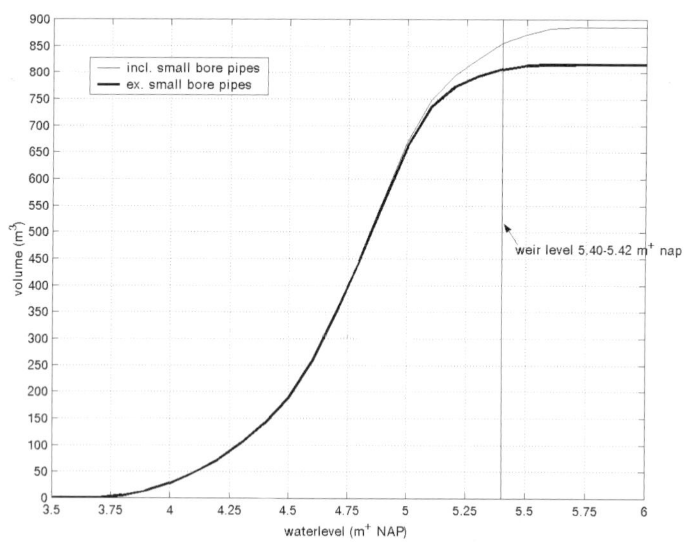

Figure 72: Volume-water level relation for 'De Hoven'.

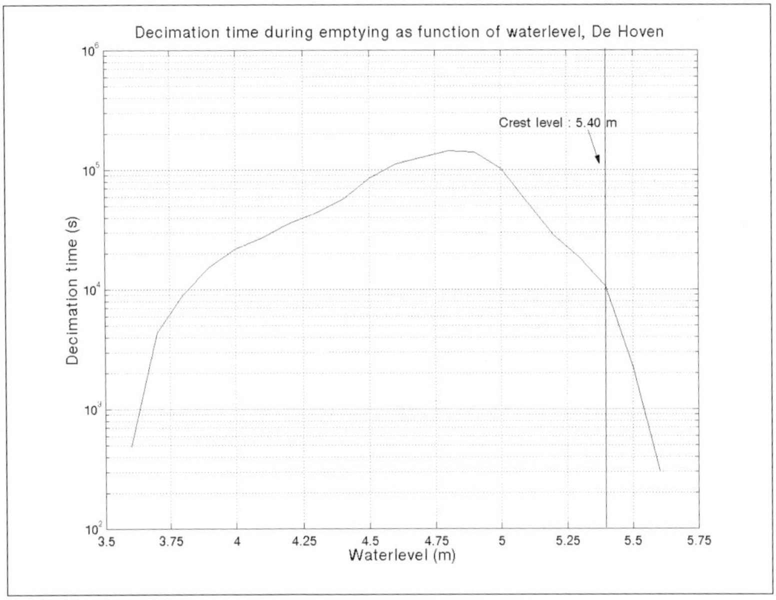

Figure 73: Decimation time, derived using the volume-water level relation shown in Figure 72. The maximum decimation time is 143350 seconds (at a water level of 4.80 m) while the minimum decimation time (found when the system is almost completely emptied) is 500 seconds.

4.2.2.5 Filling of a system

When a systems gets filled (due to run-off) exactly the same reasoning is followed as was done in Section 4.2.2.4, with the exception that the discharge entering the system is much larger. To this end the value of Q_p is replaced by;

$$Q_p^* = Q_p - A_c i_{max} \qquad \text{(eq. 4. 6)}$$

In which

A_c	Catchment area	(ha.)
i_{max}	Maximum rain intensity	($l.s^{-1}ha^{-1}$)

In this approach effects of run-off and infiltration are neglected, unless moderate intensities are applied and it is know that run-off effects are a major factor influencing the results. As an example the maximum and minimum decimation time shown in Table 27 for the system of 'De Hoven' (see Chapter 6) are found for the 10 standard storms as defined in the Dutch standard. When studying the results in Table 27, it is seen that the standard storms result in an underestimate for the characteristic time when compared to the historical storms. This is because the standard storms are originally meant to be used for design purposes with respect to flooding. These design storm are relatively short and show relatively high rain intensities forcing any system into short characteristic time scales.
Applying the results in measuring network design, however results in a 'safe' (i.e. too high) sampling frequency. As can be seen, the characteristic time scale during overflow is significantly less than during the filling process. This implies that if an overflow occurs the

sampling interval must be decreased in order to obtain enough information to describe the overflow process accurately

Storm	maximum rain intensity ($l.s^{-1}.ha^{-1}$)	range decimation time during filling (s)	minimum decimation time during overflow (s)
01&02	50	580-7080	111-101
03&04	70	400-5130	88-81
05&06	90	315-4022	82-77
07&8	110	256-3310	77-74
09	160	175-2290	70
10	210	133-1750	68
Storm August 25 1998	55.7	520-7014	90
Storm September 2 1998	13.35	23800-34644	-
Storm October 7 1998	13.35	18850-34664	-
Storm October 9-13 1998	40.04	5360-9950	-
Storm October 24 1998	30.77	980-13222	127

Table 27: Range of the decimation time during the filling process of 'De Hoven' for the standard storms according to the Dutch guidelines. As can be seen the value for the characteristic time varies with about a factor 4 between storm events (almost linearly with the maximum rain intensity) and varies with a order of magnitude of 10^2 due to variation in the storage-water level relation of the system under study.

4.2.2.6 Concluding remarks on characteristic time scales

In the previous sub-sections it was shown that by applying basic process descriptions estimates for characteristic time scales of several processes taking place during DWF-mode as well as during storm conditions can be obtained. It was also made plausible that these estimates depend largely on the initial condition assumed. Therefore, estimates obtained using the results in the previous section are to be handled with care. Nevertheless, at least they offer some basis beyond 'expert judgement' when a preliminary design of some monitoring network in urban drainage is to be made.

As was made clear, a wide variation is to be expected in characteristic time scales, depending on the drainage system, the initial conditions and the relevant process occurring. This implies that no general estimate can be given for sampling frequencies, since they depend heavily on the actual situation under study. However, the following orders of magnitude for characteristic time scales can be used as a first indication:

- DWF mode : 1000-5000 s
- Overflow mode (without excitation) : 50-100 s
- Emptying of a system : 1000-50000 s
- Filling of a system : 100- 10000 s

Depending on what processes are considered relevant in a given case, the sampling frequency must be in the same order of magnitude as the corresponding characteristic time scale.

4.2.3 Trend detection

4.2.3.1 General

Examples of measuring goals related to trend detection are:

- Detection of an increase in mean DWF over time (e.g. due to infiltration of groundwater).
- Detection of sudden changes in water level (e.g. caused by a sudden extra DWF load like the discharge of a groundwater extraction into the drainage system).

In cases like this, the locations at which to measure are decided upon based on some suspicion with respect to illegal discharges or based on evidence obtained from visual inspection of the system. Therefore, the design of the measuring network concentrates around the following issues:

- The sampling frequency.
- The accuracy of the measurements.
- The duration of the measurements.

These three parameters are linked together as will be explained in Section 4.2.3.2.
It is noted that the time scale taken into consideration defines what is to be regarded as a trend. For instance when analysing a short time series (e.g. 3-4 hours) of DWF it is likely to observe an (almost) linear trend. According to the theorem of Shannon, a continuous signal can be reconstructed without loss of information using a series of discrete samples if the sampling interval (in space- or time dimension) is less than or equal to the smallest time or distance scales present in the process studied. This implies that in practice a decision has to be made what degree of accuracy is required in relation to the smallest relevant time scale. Therefore, the problem to be solved is not so much to avoid information loss, but merely what amount of information loss is acceptable. The information loss induced by a choice for a certain sampling interval depends on the correlation structures of the process studied. The correlation function is the relation between two variables x and y as a function of the mutual time- or space distance. Furthermore, the intrinsic variability of the process at a fixed point in time or space is of importance. The variability is quantified by the variance σ^2.
The correlation structure and the variance together define the covariance structure.
The information content that can be extracted from measuring data generally:

- Decreases with increasing variance.
- Decreases with an increasing ratio between the dominant process scale and the sampling interval.
- Increases with increasing ratio between measuring duration and dominant process scale.

4.2.3.2 A criterion for design in trend detection

A trend in a process is in fact a variation of the processes mean value in time and/or place. Lettenmaier (1976) approaches trend delectability in terms of statistical tests. A null hypothesis H_0 stating 'no trend has occurred' is tested against the Hypothesis H_1; 'a trend did occur'. Based on the information obtained from measuring data either H_0 or H_1 is accepted as the 'truth'. When applying a statistical test the power and the confidence for the choice made are quantified and are used as a norm by which to pick a sampling frequency.
In Table 28 the four natural possibilities are shown, two types of errors can occur;

- A trend is not present, but is indicated by the analysis of the measuring data; a type 1 error.
- A trend is present, but is not indicated; a type 2 error.

The confidence of a test is the probability for a correct indication of 'no trend occurrence', while the power is defined as the probability that an indicated trend is in line with reality.

	test indication H_0	test indication H_1
'real' state: H_0	No error $P=1-\alpha$ (Confidence)	Type 1 error, $P=\alpha$
'real' state: H_1	Type 2 error, $P=\beta$	No error, $P=1-\beta$ (Power)

Table 28: The four possible outcomes for testing a hypothesis.

Both confidence and power are to be at a high level, or reversibly, the probability for type 1 and type 2 errors (α and β respectively) is to be at a low level. In Figure 74 the step trend and linear trend are schematically shown. In the case of a linear trend, the time scale of interest is of importance. E.g., over a short period (i.e. order of magnitudes of 10 minutes- 1 hour), the occurrence of a storm may, without knowledge of rain intensities, be interpreted as an increasing trend in the DWF mode. Therefore when analysing the measuring results a data analysis has to be made in order to eliminate the effect of storms in the time series when trying to detect a trend in DWF. Furthermore, the DWF as function of time is assumed to be a stationary stochastic process, the criterion for design with respect to trend detectability does not result in a sampling frequency needed for reproducing the DWF patterns.

There are several statistical tests that can be applied to the problem outlined; for trend detection the following two cases are of importance:

- Detection of a step trend (a more or less sudden change in mean value).
- Detection of an increase or decrease (modelled as a linear trend in time).

The statistical tests most suited for these trends are the Mann-Whitney test and Spearmans ρ test respectively (Conover (1971)). Lettenmaier (1976) has shown that, based on these two statistical tests, at a given measuring duration T and a fixed sampling interval Δt the power for a trend Tr to be detected in a process having a variation σ_p^2 depends on:

$$\psi(\Delta t, T, Tr) = \frac{1}{2}\frac{Tr}{\sigma_p}\sqrt{N_{equi}(\Delta t, T)} \qquad \text{(eq. 4. 7)}$$

for the step trend and

$$\psi(\Delta t, T, Tr) = \frac{Tr}{\sigma_p\sqrt{12}}\frac{\sqrt{N_{equi}(\Delta t, T)(N_{equi}(\Delta t, T)+1)(N_{equi}(\Delta t, T)-1)}}{N_{equi}(\Delta t, T)} \qquad \text{(eq. 4. 8)}$$

for the 'linear' trend.

The power and confidence of the tests applied are related by:

$$1-\beta = \Phi\left[\psi(\Delta t, T, Tr) - \xi\left(\frac{\alpha}{2}\right)\right] \qquad \text{(eq. 4. 9)}$$

In which Φ denotes the standard normal distribution and $\xi(\alpha/2)$ denotes the quantile at $\alpha/2$. $N_{equi}(\Delta t, T)$ is the equivalent number of samples, the subscript 'equi' refers to the fact that a set of N samples showing mutual correlation (implying they hold redundant information) contains information equivalent to N_{equi} non-correlated samples

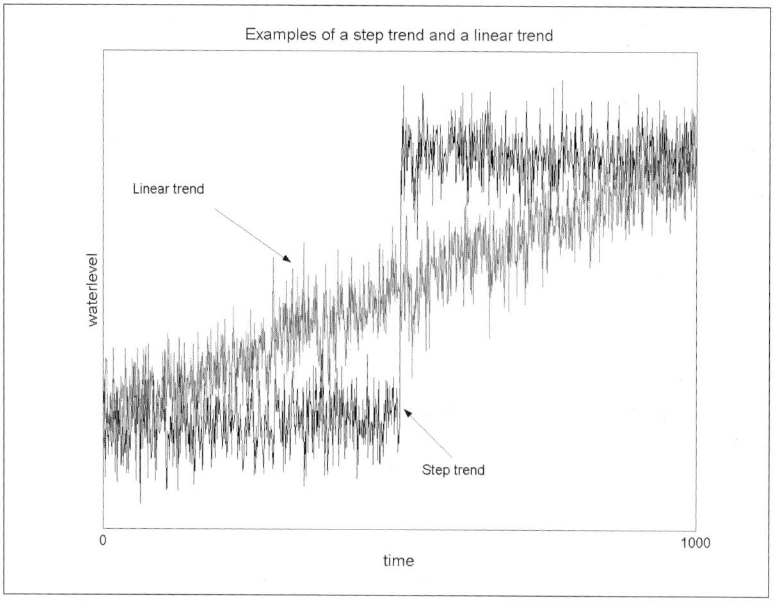

Figure 74: Step and linear trend (schematically).

The relation between N and N_{equi} is (Bayes & Hammersley (1946)):

$$N_{equi}(\Delta t, T) = \frac{N(\Delta t, T)}{1 + 2\sum_{i=1}^{N(\delta t, T)}\left[\left(1 - \frac{i}{N(\Delta t, T)}\right)\rho_x(i\Delta t)\right]} \quad \text{(eq. 4. 10)}$$

$\rho_x(i\Delta t)$ is the correlation function. In the case of independent samples, i.e. $\rho_x(i\Delta t)=0$ for all $i>0$, N_{equi} $(\Delta t,T)=N(\Delta t,N)$. This only occurs when the sampling interval is much larger than the characteristic scales of the processes under study. When all samples are 100% correlated then $N_{equi}=1$, regardless of the number of samples taken, indicating an inefficient sampling frequency. In practical cases, the correlation lies between these two extremes. So, when α, $\rho_x(i\Delta t)$ and Tr are known, Δt can be calculated.

4.2.3.3 Correlation functions for urban drainage systems

For measuring systems in more or less continuous processes in homogeneous media (like e.g. groundwater flow, or flows in large water-bodies) it is possible to use covariance structures for optimisation of the information content of a measuring system (see e.g. Schilperoort et al (1982). The main problem when trying to implement this method on measuring networks for urban drainage is that the covariance structures are hard to detect and change in character between DWF and storm conditions. The latter observation was illustrated by Stigter (2000). He studied the covariance patterns of a measuring network in a large catchment in Deventer with 13 water level gauges. He concluded, after analysis of the measuring data, that the correlation between distinct measuring locations strongly depends on the water levels occurring. Therefore, it is not simple to set some prior assumption of the shape of the correlation functions valid for every condition that can occur. When a numerical model is available, this can be used to obtain an estimate for the correlation function between several locations in a system. An important issue when working with space distances in drainage networks is that the physical distance cannot be applied as a measure. This is because two distinct points in a system may be connected in different manners depending on the water level. This is illustrated in Figure 75. As long as the water levels are <H_{nvert} the water levels in manhole A and B have no mutual correlation; but on the other hand if the water levels are >H_{invert} a strong relation between them (and thus a large correlation) exists. The main conclusion is that with respect to the time scale the sampling frequency can be determined. With respect to the spatial density of the network, a design can only be based upon either existing measuring data or on a (numerical) model.

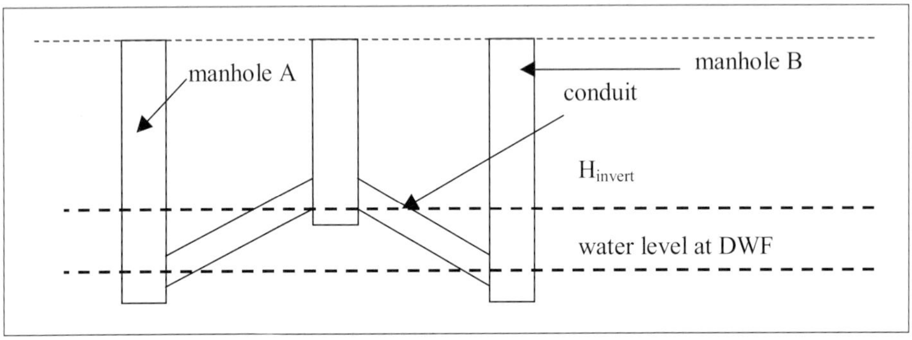

Figure 75 : During DWF the cross-correlation between water levels in manholes A and B is small (if present it is caused by a more or less similar DWF load). If the water level rises above H_{invert}, the cross-correlation will become significant. (Figure after Stigter (2000)).

The characteristic scales of the process are obtained from its covariance structure. The correlation function for the observations is defined as:

$$\rho_x(i\Delta t) = \frac{\gamma_x(i\Delta t)}{\sigma_p^2 + \sigma_m^2} \qquad \text{(eq. 4. 11)}$$

in which σ_p^2 is the process variance (or inherent variance) and σ_m^2 is the variance due to measuring errors([32]).
The covariance-function is defined by:

$$\gamma_x(i\Delta t) = \begin{cases} \sigma_p^2 + \sigma_m^2 & i = 0 \\ \sigma_p^2 \rho_p(i\Delta t) & i \neq 0 \end{cases} \qquad \text{(eq. 4. 12)}$$

The characteristic time-scale of a process studied is (intuitively) defined as the time scale at which the covariance function is zero([33]).
The time scales for an urban drainage networks are influenced by several variables:

- The 'character' of the system (in terms of e.g. slope and structure)
- The flow mode (DWF-pattern).

([32]) The summation of the variances in the nominator is only valid when the measuring error is independent from the process.
([33]) It is noted that the time scale as defined here has no direct relation with the time scales as defined in Section 4.2.2. In the later case the time scales were based on physical processes, whereas using the covariance function the time scale is based on either observation or modelled results.

In fact, no single characteristic time scale for an urban drainage network can be defined. For a given network, these scales must be quantified either from analysis of historical data or by using a model. To illustrate this the auto-correlation function in DWF mode for two systems are shown:

- 'De Hoven' DWF-mode , municipality of Deventer (see also Chapter 6 of this thesis).
- 'De Maten-zuid' DWF mode, municipality of Apeldoorn (see Witteveen+Bos, (2000)).

For 'De Hoven' and 'De Maten-zuid', measuring data were available, offering the possibility to show the discrepancies between the results of a hydrodynamic model for initial design and the 'real world'.

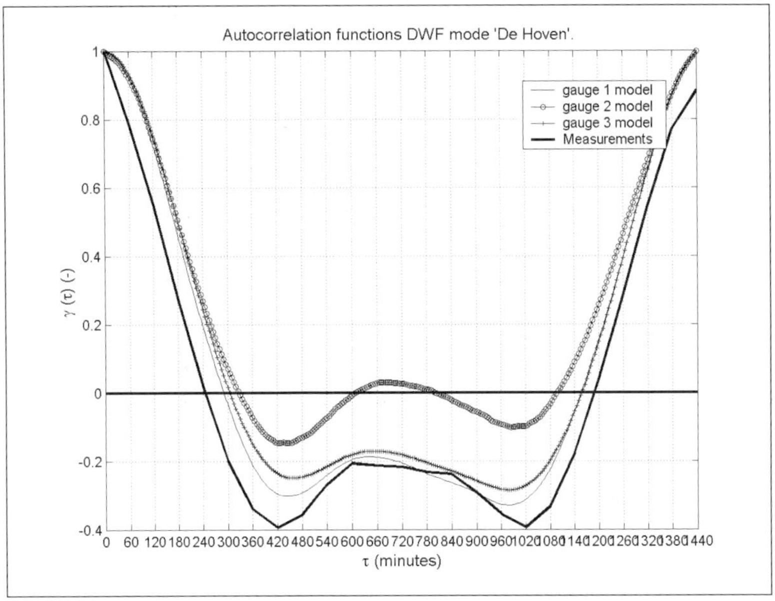

Figure 76: Auto-correlation functions, measured and modelled for 'De Hoven'.

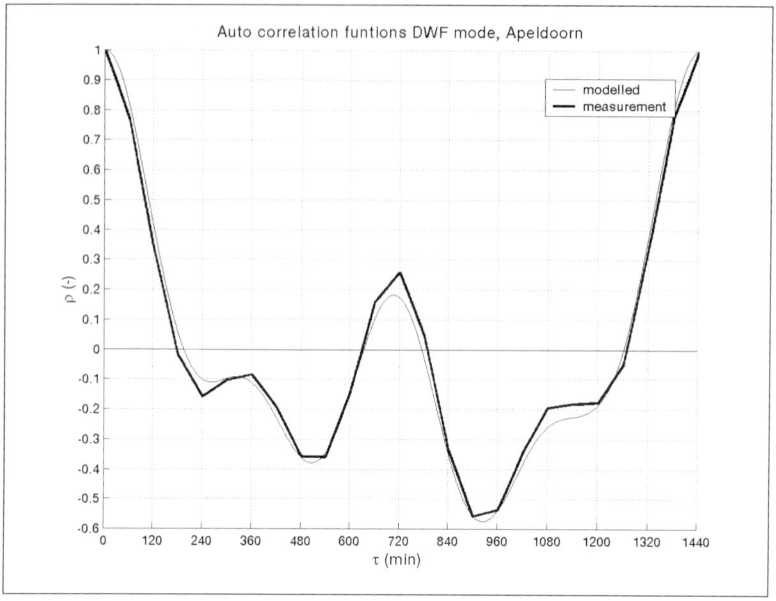

Figure 77: Auto-correlation function for 'De Maten-Zuid', Apeldoorn, DWF mode, measured and modelled.

4.2.3.4 Application of a hydrodynamic model for initial design

In order to demonstrate the applicability of a hydrodynamic model for measurement an example is discussed for which measurements were available. The system used is the same as presented in Chapter 6 of this thesis in the case study for calibration.
Since the co-variance function is used for the determination of the sampling frequency the norm for the correctness of the estimate obtained using test cases is defined as:

$$R = \left. \frac{(1-\beta)_{model}}{(1-\beta)_{historical}} \right|_{\alpha=0.05,\delta t=60\,minutes} \qquad \text{(eq. 4. 13)}$$

R is the ratio between the power of the test for trend detection obtained at a given confidence level (1-α) for the model at the one hand and the historical data at the other. For safety, when using a model this ratio should not deviate too much from unity, and preferably be < 1, implying a conservative estimate by the model.
In Figure 78 and Figure 79 the value of N_{equi} as function of the sampling interval Δt is shown (the period T=48 h). As can be seen there is a clear difference between the gauging point applied in the model on one hand and the results obtained when using measured values. Furthermore, the value for N_{equi} as derived from measurements is substantially larger than the results obtained using the model. This is due to the deviation in the modelled covariance functions and the covariance function as derived from the measurements (see Figure 76).

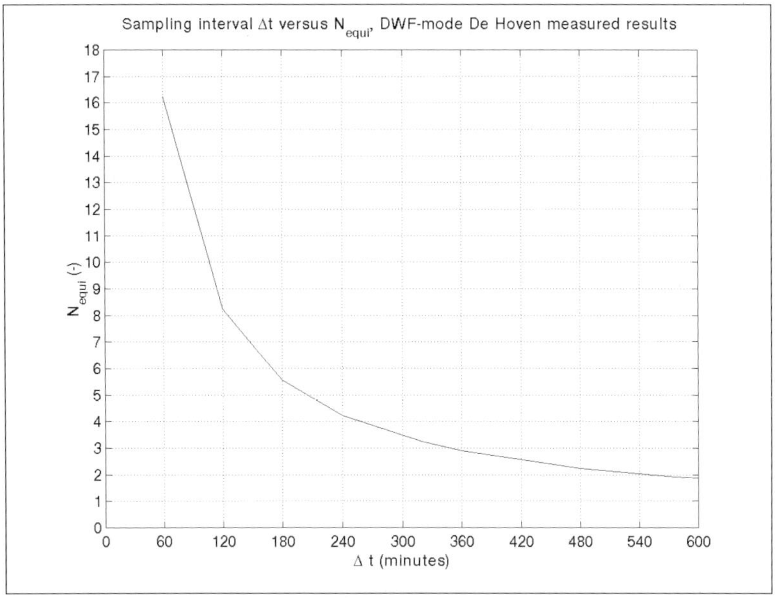

Figure 78: Equivalent number of sampling points as function of the sampling interval based on measurements, drainage system 'De Hoven'.

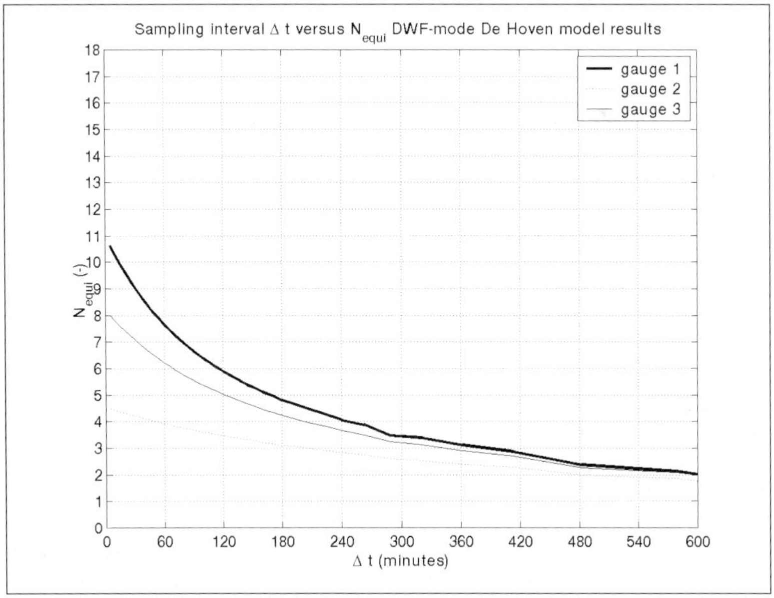

Figure 79: Equivalent measuring points as function of the sampling interval, model results, drainage system 'De Hoven', the measuring period referred to is 48 hours.

When either model results or measuring data are used for the design (or redesign) of the measuring setup, differences occur.

In Table 29 the value for R for gauges 1,2 and 3 are shown for δt=60 minutes, together with the chosen value for δt in the case of application of measured data, the values applied for α and β being 0.05. This latter choice implies the confidence and the power of the test applied to detect a step or linear trend are both 0.95.

	T_r/σ_p=1.0	T_r/σ_p=2.0	T_r/σ_p=3.0	δt (1.0)	δt (2.0)	δt (3.0)
R Gauge 1	0.52(0.57)	0.81(0.56)	0.98(0.72)	- (-)	- (-)	25 (-)
R Gauge 2	0.32(0.38)	0.52(0.31)	0.85(0.41)	- (-)	- (-)	- (-)
R Gauge 3	0.57(0.50)	0.72(0.46)	0.96(0.61)	- (-)	- (-)	80 (-)
Measured	-	-	-	- (-)	72 (-)	171 (-)

Table 29: Model versus historical data in choice of sampling interval for 'De Hoven'. (Values between brackets are for the linear trend), '-' indicates that the design criterion could not be met. As can be seen application of the model results in this case in a conservative estimate (R<1).

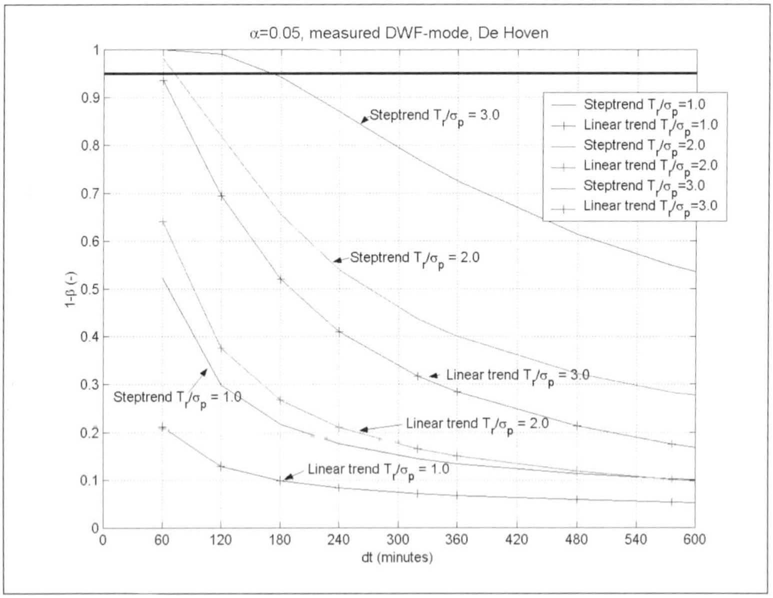

Figure 80: Power for tests on step and linear trends in DWF mode, based on measurements.

In Figure 80 and Figure 81 the power of a test for trend detection are shown, α=0.05, for modelled results and measured results respectively. As can be seen a step trend is more easily detected than a linear trend. If $\beta=\alpha$ is required, it is clear that only a relatively large trend (relative to the inherent variance of the process) can be detected (T_r/σ_p=3.0 meeting such a demand).

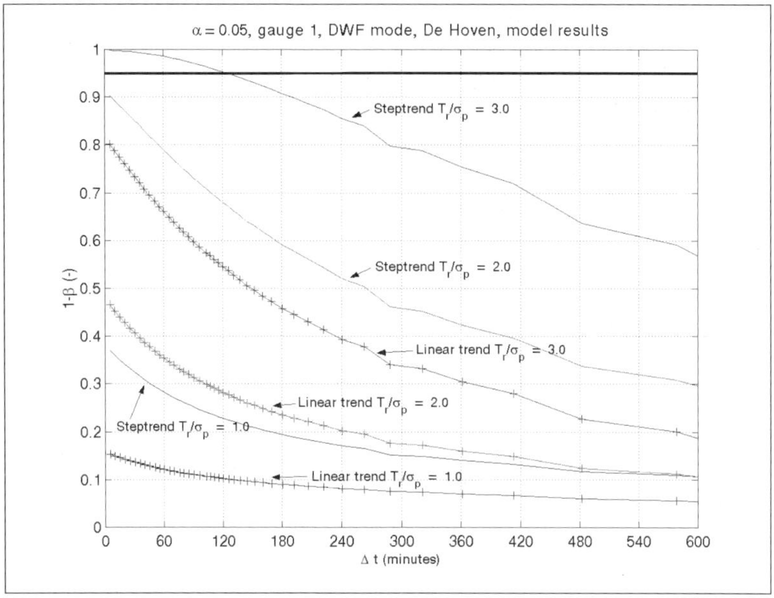

Figure 81: In this case it is possible to detect a step trend as well as a linear trend, even for Tr/σ_p=1.0.

	T_r/σ_p=1.0	T_r/σ_p=2.0	T_r/σ_p=3.0	δt (1.0)	δt (2.0)	δt (3.0)
R Model	0.99(0.99)	1.01(1.00)	1.00(1.00)	26(8)	100(35)	231(71)
Measured	-	-	-	24(6)	99(34)	234(77)

Table 30: Model versus historical data in choice of sampling interval for Apeldoorn. (Values between brackets are for the linear trend). As can be seen application of the model results in this case in an estimate that resembles the results obtained based on historical data almost perfectly.

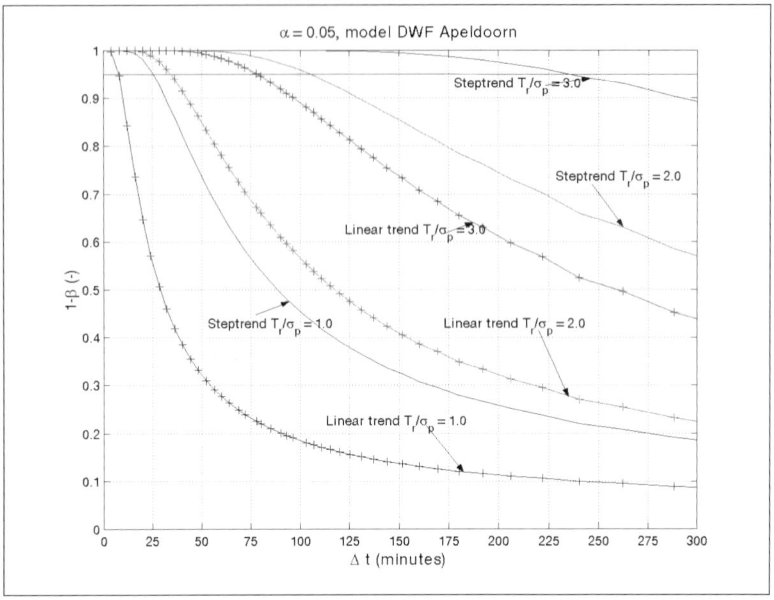

Figure 82: Power for tests on step and linear trends in DWF mode, based on a model, Apeldoorn.

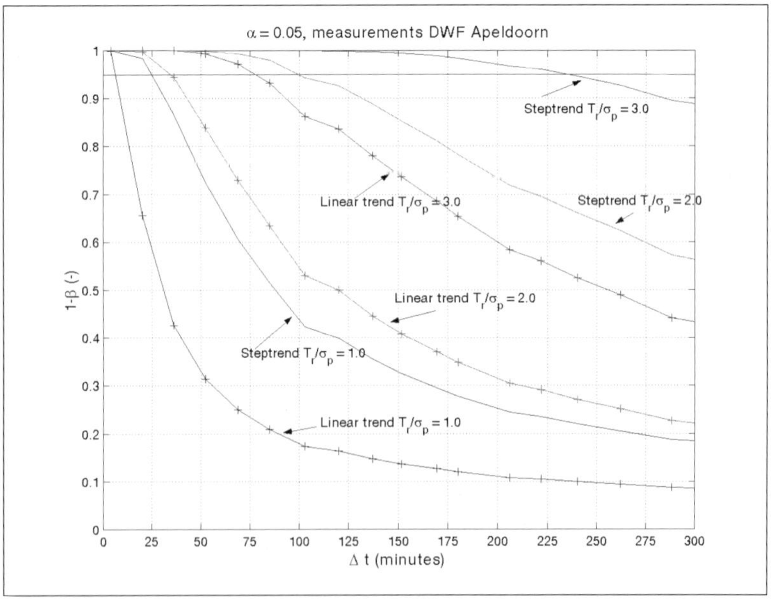

Figure 83: Power for tests on step and linear trends in DWF mode, based on measured values, Apeldoorn.

4.2.4 State reconstruction

4.2.4.1 General

In state reconstruction, the objective is to define a sampling frequency in such a manner that the actuality of a process can be reconstructed at a certain moment in time at which no samples have been taken by using adjacent samples. This implies that the moment in time for which the reconstruction is done must be inside the time-interval between which samples are available. As will be become clear, the sampling frequency to be applied for state reconstruction depends on the characteristics of the signal to be sampled and on the measuring accuracy of the device used for sampling. The expected characteristic time scales can easily be quantified using the results described in Section 4.2.2. Basically, there are two ways to analyse the problem of sampling frequency; in the time domain or in the frequency domain (see e.g. Hamilton (1994)). In the former case the analysis is based upon the statistical effects the sampling frequency has on the information obtained, in the latter case an upper limit for the sampling frequency yielding information is obtained. In the next two sections both methods are discussed and illustrated with some examples.

4.2.4.2 Frequency domain analysis

According to Shannon's theorem, any signal can be expanded into a series of relatively simple functions. This is done using a finite Fourier transform. The finite Fourier transform is analogous to the more general Fourier transform applied to continuous functions, see e.g. Farlow (1982):

$$F(k)=\frac{1}{N}\sum_{j=1}^{j=N} f(j)\omega_N^{(j-1)(k-1)} \qquad \text{(eq. 4. 14)}$$

and its reverse transform is

$$f(j)=\frac{1}{N}\sum_{k=1}^{k=N} F(k)\omega_N^{-(j-1)(k-1)} \qquad \text{(eq. 4. 15)}$$

In which N is the number of discrete equidistant elements in a time series and ω_N is defined as:

$$\omega_N = e^{\frac{-2\pi i}{N}} \qquad \text{(eq. 4. 16)}$$

In which i is the imaginary unit. F(k) with (k=1,…N) is easily computed using a Fast Fourier Transform algorithm (see e.g. Price et al (1992)), the power density spectrum is calculated by multiplying F(k) with its complex conjugate. Now two cases are considered:

- No measuring data are available (the monitoring network has still to be designed).
- Preliminary measurements are available (preferably at a high sampling frequency) and are to be used for making a decision to change the sampling frequency.

In the first case the designer is free to decide upon the measuring device he/she will apply, but no 'real world' information on the signal to be sampled is available. Instead, a model is applied to simulate such a signal. In the second case, the measuring device has been chosen and a decision has to be made to either decrease the sampling frequency or obtain an other sampling device depending on the objective of the monitoring project.
For both ends the following reasoning is suggested:

Suppose the signal to be sampled g(t) is composed out of two signals, a 'real' signal f(t) and a noise term ε(t) originating from the measuring inaccuracy (see Section 4.3) of the sampling device:

$$g(t) = f(t) + \varepsilon(t) \quad \text{(eq. 4. 17)}$$

In the first case g(t) is simulated and ε(t) can be decided upon by the designer, whereas in the second case g(t) is known and either the distance between two successive samples or ε(t) can be influenced. Now ε(t) is stated to be gaussian white noise. This implies the signal is completely random with a gaussian distribution with a mean value of zero and a standard deviation of σ_m. The power density function E(ω) depends strongly on the specific apparatus applied. For simplicity a constant value is assumed (this implies that the measuring uncertainty is assumed to be white noise). If the transform of a time series x(t) is denoted by X(ω) then the power density function is defined by([34]);

$$X_p(\omega) = \sqrt{X(\omega)\overline{X}(\omega)} \quad \text{(eq. 4. 18)}$$

The maximum sampling frequency that yields information of interest (i.e. information on the actual process f(t)) is defined by:

$$G_p(\omega) \approx E_p(\omega) \quad \text{(eq. 4. 19)}$$

Imposing any higher frequency than the frequency for which (eq. 4.19) holds, results in recording noise, or measuring errors.

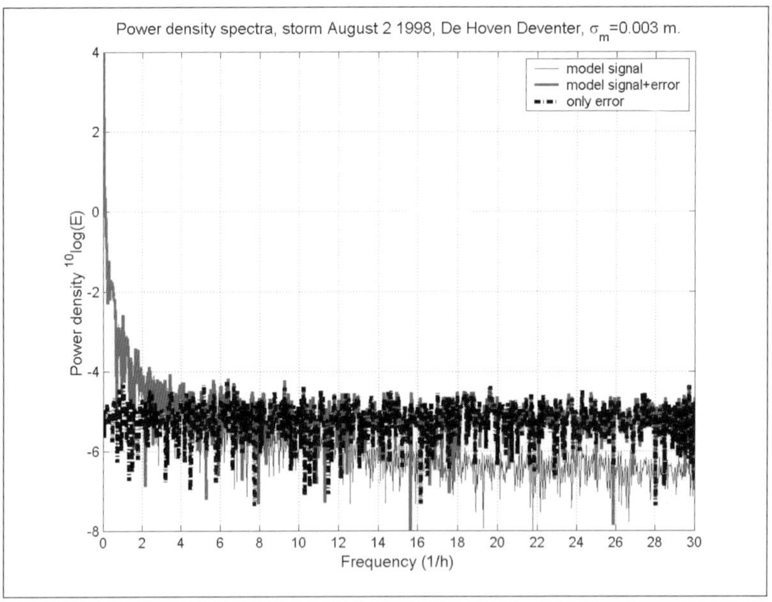

Figure 84: Power density spectrum, storm August 2, 1998 De Hoven Deventer.

Therefore, roughly, when studying the power density spectrum of a high frequency sampled signal, its is stated that when this spectrum is almost constant, higher frequencies do not add

([34])Over-lined symbols define the complex conjugate.

information on the real process. In practice though, it is hard to obtain an objective norm based on the observation outlined, since the power density function shows a highly irregular shape on a smaller scale (see e.g. Figure 84). For example, based on the spectra shown in Figure 84 a maximum recognisable process frequency of 4-8 times per hour would be decided upon resulting in a sampling frequency of 8 to 16 times per hour. A more exact quantification is impossible without further analysis. In order to obtain a criterion by which the maximum process frequency (ω_0) can be quantified, the following equation based on the power density spectrum is proposed:

$$G_p(\omega) < 2E_p(\omega) \quad \text{for all } \omega > \omega_0 \qquad \text{(eq. 4. 20)}$$

In which:

$G_p(\omega)$ Power density function of the original signal
$E_p(\omega)$ Power density function of the measuring inaccuracy signal
ω_0 cut-off frequency

In order to avoid the problem of the small scale variation of the power density functions as shown in Figure 84, the following smoothing technique was applied:

$$\hat{G}(\omega_j) = \sum_{m=-h}^{h} \kappa(\omega_{j+m}, \omega_j) * G(\omega_{j+m}) \qquad \text{(eq. 4. 21)}$$

In which κ is a kernel that indicates how much weight is given to each frequency in the band width h, the kernel sum being unity. The kernel applied here was proposed by Hamilton (1994):

$$\kappa(\omega_{j+m}, \omega_j) = \frac{h+1-|m|}{(h+1)^2} \qquad \text{(eq. 4. 22)}$$

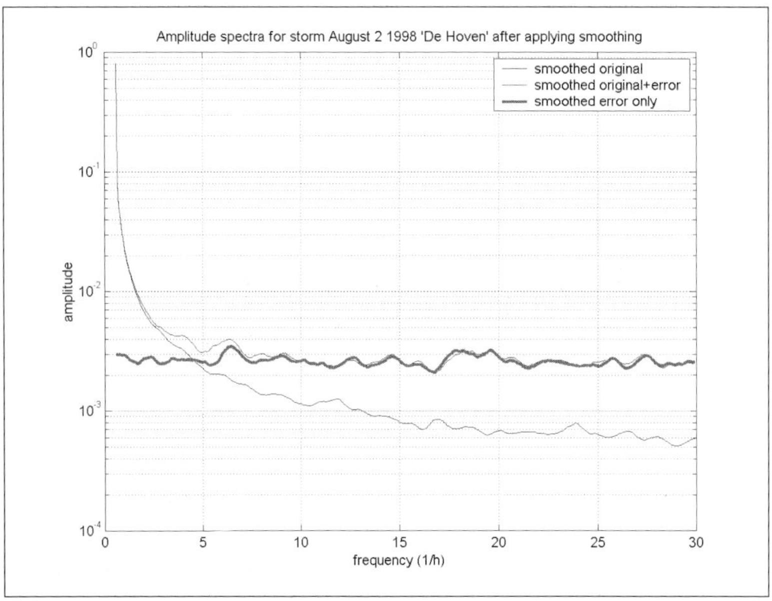

Figure 85: Smoothed spectra, using a bandwidth of 15, for storm on August 2 1998 'De Hoven'.

With respect to the bandwidth h Hamilton (1994) (page 167) states that one must:

> *'rely on subjective judgement to choose the bandwidth that produces the most plausible estimate.'*

For the examples discussed hereafter a bandwidth of 15 frequencies was applied, based on the subjective judgement that applying this value produces a fairly smooth spectrum (see e.g. Figure 85).

Example

This example illustrates the case of choosing a measuring device when no 'real world' data are available. The signal to be studied (f(t)) is obtained using a model; to this end the water level at a certain location in a sewer system is calculated. In fact, the system used is 'De Hoven', also used for the case study in Chapter 6, using standard storm no. 01 through no. 10 and five historical storm events. The measuring device is supposed to have a measuring accuracy([35]) of 0.6% of the measuring range (2 m), implying σ_m=0.003 m. So the signal f(t) was derived from a model and $\varepsilon(t)$ was added to it to obtain the signal g(t) simulating a 'real world' signal.

storm	remarks	δt (s) point 1	δt (s) point 2	δt (s) point 3
storm 01	Overflow	158	135	114
storm 02	Overflow	171	63	60
storm 03	Overflow	167	72	75
storm 04	Overflow	130	60	60
storm 05	Overflow	95	64	65
storm 06	Overflow	134	60	60
storm 07	Overflow	61	89	79
storm 08	Overflow	105	74	72
storm 09	Overflow	60	60	60
storm 10	Overflow	60	60	60
August 25 1998	Overflow	199	280	375
September 2 1998	No overflow	654	675	686
October 7 1998	No overflow	1196	658	1082
October 9-13 1998	No overflow	469	527	505
October 24 1998	Overflow	329	300	146

Table 31: Sampling intervals obtained using frequency domain analysis. The sampling frequency is twice the value of ω_0 since it takes at least two samples per period to reconstruct the underlying process correctly. Therefore the sampling interval is defined by $\delta t=(2\omega_0)^{-1}$.

As can be seen in Table 31, the sampling interval varies over the different locations in the system. Furthermore, the sampling interval in storms leading to an overflow is significantly smaller than those in storms that do not lead to an overflow. This is caused by differences in the characteristic time scales occurring in both types of storm events. (See also Section 4.2.2).

4.2.4.3 Time domain analysis

Schilperoort (1986) has derived a theoretical framework for determining the number of sampling locations for a one-dimensional network of gauging stations (e.g. gauging stations

[35] The measuring accuracy (m_e) of a device is defined as the 95% probability interval, the standard deviation follows from $\sigma_m=0.25*m_e$.

along a river branch). A similar approach is adopted in order to obtain a measure for sampling frequencies with respect to state reconstruction in time at a gauging station in an urban drainage network. Suppose the process state (e.g. water level, discharge or flow velocity) at $t=t_\tau$ ($t_{i-1} < t_\tau < t_i$) is obtained from a simple linear interpolation:

$$\hat{x}_\tau = \alpha x_{i-1} + \beta x_i \qquad \text{(eq. 4. 23)}$$

α and β are weighing factors in the interpolation. Furthermore the process has a variance σ^2_p, a mean value of μ_p and the correlation function is known ($\rho(\tau)$, see Section 4.2.3.3). As measure for the effectiveness of the sampling frequency the mean squared error is used:

$$\left.\begin{aligned} mse = \frac{1}{n}\sum_{k=1}^{k=n}(\hat{x}_\tau - x_\tau)^2 = \\ \sigma_p^2\left[1+\alpha^2+\beta^2-2\alpha\rho(t_{i-1},t_\tau)-2\beta\rho(t_i,t_\tau)+2\alpha\beta\rho(t_{i-1},t_i)\right]+\mu_p^2\left[\alpha+\beta-1\right]^2 \end{aligned}\right\} \qquad \text{(eq. 4. 24)}$$

The last term in equation 4.24 is the bias term, which vanishes if $\alpha+\beta=1$. Minimising *mse* for α and β results in the optimal values for α and β. Assuming $\alpha+\beta=1$, and imposing the following condition:

$$\frac{\partial mse}{\partial \alpha} = \frac{\partial mse}{\partial \beta} = 0 \qquad \text{(eq. 4. 25)}$$

results in:

$$\alpha = \frac{1}{2} + \frac{\rho(t_{i-1},t_\tau)-\rho(t_i,t_\tau)}{2(1-\rho(t_{i-1},t_i))}, \text{ and } \beta = \frac{1}{2} - \frac{\rho(t_{i-1},t_\tau)-\rho(t_i,t_\tau)}{2(1-\rho(t_{i-1},t_i))} \qquad \text{(eq. 4. 26)}$$

storm	remarks	Δt (s) point 1	Δt (s) point 2	Δt (s) point 3
storm 01	Overflow	767	394	544
storm 02	Overflow	660	303	391
storm 03	Overflow	561	472	339
storm 04	Overflow	547	331	332
storm 05	Overflow	419	444	394
storm 06	Overflow	548	336	341
storm 07	Overflow	365	398	392
storm 08	Overflow	466	409	345
storm 09	Overflow	327	368	382
storm 10	Overflow	324	345	370
August 25 1998	Overflow	739	838	849
September 2 1998	No overflow	1619	1802	1414
October 7 1998	No overflow	2524	2095	2298
October 9-13 1998	No overflow	1490	1422	1628
October 24 1998	Overflow	1153	1046	1564

Table 32: Sampling intervals obtained using time-domain analysis for 'De Hoven'. As can be seen the standard storms result in significantly higher sampling frequencies when compared to the historical storms studied. Detailed information on the storms applied is found in annexe VI.

This leads to the following formula for *mse*:

$$mse = \frac{1}{2}\sigma_p^2\left[3+\rho(t_{i-1},t_i)-2\rho(t_{i-1},t_\tau)-2\rho(t_i,t_\tau)-\frac{[\rho(t_{i-1},t_\tau)-\rho(t_i,t_\tau)]^2}{1-\rho(t_{i-1},t_i)}\right] \quad \text{(eq. 4. 27)}$$

A simple case is obtained when interpolation is only done exactly halfway between the two adjacent samples, in that case $\rho(t_{i-1},t_\tau) = \rho(t_i,t_\tau)$, resulting in:

$$mse = \frac{1}{2}\sigma_p^2(3+\rho(t_{i-1},t_i)-4\rho(t_{i-1},t_\tau)) \quad \text{(eq. 4. 28)}$$

As can directly be seen from equation 4.28, if the correlation function is constant and equal to unity the *mse* becomes zero, implying that interpolation does not introduce information loss. On the other hand, if the correlation is (almost) equal to zero (a situation that occurs if the sampling distance is equal to or larger than the characteristic time scale of the process), the *mse* is equal to 1.5 times the inherent variance of the process. In that case it contributes significantly to efficiency loss. In order to estimate an upper limit for the sampling time the following criterion is applied:

$$mse \leq \sigma_m^2 \quad \text{(eq. 4. 29)}$$

If this criterion is applied for a set of storms on the system of De Hoven with σ_m=0.003 m for water level measurement, maximum sampling intervals for gauging point 1, as shown in Table 32 are obtained.

4.2.4.4 Choosing a sampling frequency

In Sections 4.2.4.2 and 4.2.4.3 two methods were discussed to obtain limits on sampling frequencies. Both methods were illustrated by some examples. Frequency domain analysis gives a clue as to what sampling frequency is to be regarded as the limit above which no additional information is obtained on the process under study. On the other hand the sampling frequency obtained from the time domain analysis gives a minimum sampling frequency; if a lower frequency is used the information is regarded to be too sparse. When preparing a measuring campaign both limits can be quantified using either standard design storms or, when available, historical storms combined with a hydrodynamic model and the measuring accuracy of the equipment to be applied. When the limits of the sampling frequency obtained conflict (i.e. the sampling frequency obtained by time domain analysis is larger than the sampling frequency obtained from frequency domain analysis), either the criteria should be adapted or an other type of measuring device is to be applied. In both cases, however, a well-defined basis for changes in design is applied.
For the example 'De Hoven' presented in the previous section, the following analysis is made. As can be seen (see Table 33), there is a clear distinction between storms that led to an external overflow (August 25 and October 24) and storms that did not with respect to the maximum and minimum sampling frequency. This is easily explained from the fact that the characteristic times of the process differ in these two cases (see Section 4.2.2). From historical data a sampling interval of 3 to 10 minutes would be adopted (based on the storm at August 25) while based on the standard storm events a sampling interval between 1 and 4 minutes would be picked. This implies that using the standard storms a safe choice can be made with respect to the sampling interval.

storm	Δt (s) minimal maximum	Δt (s) maximal minimum	range char time
01	394	158	7730-111
02	**303**	171	7730-101
03	339	167	5526-88
04	331	130	5526-81
05	394	95	4260-82
06	336	134	4260-77
07	365	89	3468-72
08	345	105	3468-74
09	327	**60**	2366-70
10	324	60	1795-68
August 25 1998	739	375	7014-90
September 2 1998	1414	686	34664-23800
October 7 1998	2096	1196	34664-18850
October 9-13 1998	1422	527	9950-5360
October 24 1998	1046	329	13222-127

Table 33: Maximum and minimum sampling intervals for 'De Hoven'. The shaded values of 303 s and 60 s indicate the domain in which the sampling interval has to be picked. A value of 300 s would be practical, this would imply however, that in storms 01,...,10 not all information available is obtained. On the other hand, for the historical storms 'too much' information is obtained. This indicates that, at least for this case, application of the standard storms results in a safe preliminary choice of the sampling interval.

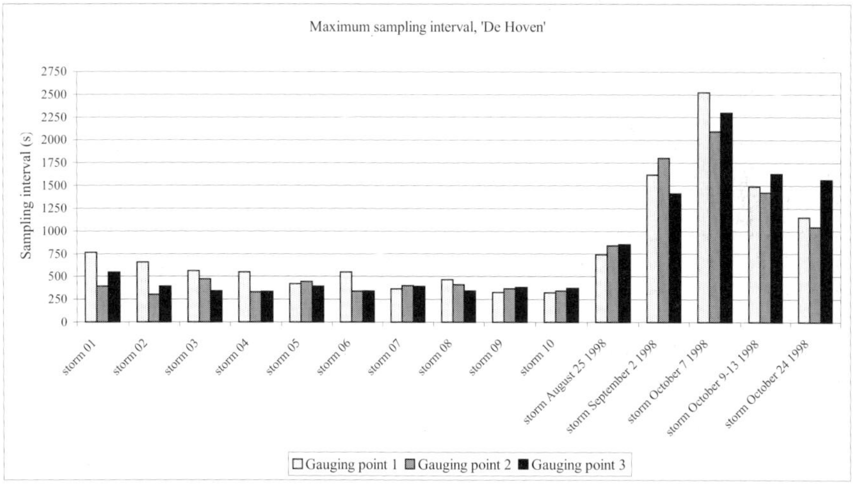

Figure 86: Maximum sampling times for 'De Hoven' for three gauging stations. As can be seen application of the 10 standard storms results in a conservative estimate for the maximum sampling interval when compared to 5 historical storms. Furthermore, it is seen that the maximum sampling interval varies with the storm characteristics as well as with the location in the drainage system. This implies that, prior to implementing a measuring programme, every potential gauging location has to be analysed with respect to maximum and minimum sampling interval. For practical purposes, a uniform sampling interval should be chosen. In this case, a first estimate for the maximum sampling interval (based on storms 1 through 10) would be based on gauging station no. 2 storm 02 at 303 seconds (5 minutes).

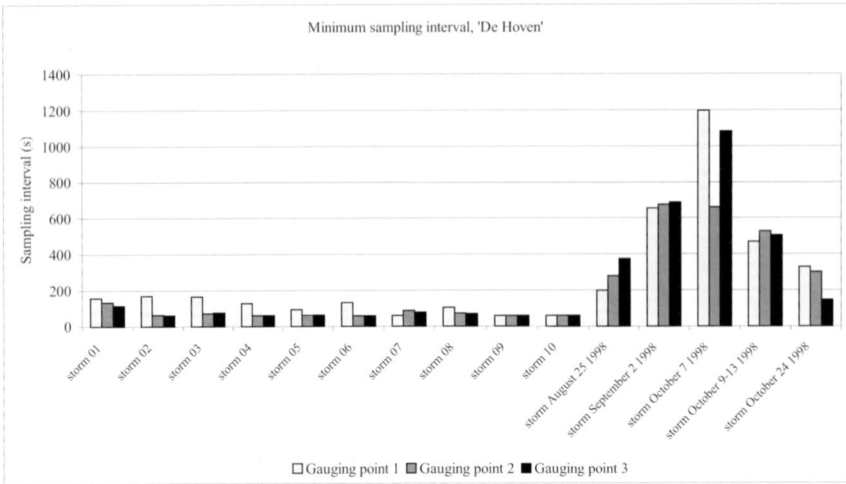

Figure 87: Minimum sampling interval for 'De Hoven'

To illustrate the influence of the characteristics of the drainage system, the sampling interval for Systems I and II as presented in Chapter 3 (Section 3.8.4) obtained using frequency domain analysis and time domain analysis are discussed. The results are shown in Table 34 and Table 35.

System I

In system I, the points C and E deviate significantly from the other points. This is because these points show hardly any variation under the load of most storms while the maximum water depths are small (about 0.1 m). The points C and E are located at the top-ends in the system, as was explained in Section 3.8.7.

storm	minimal maximum Δt (s)	maximal minimum Δt (s)
storm 01	948	487
storm 02	1025	670
storm 03	603	346
storm 04	425	316
storm 05	**405**	268
storm 06	506	405
storm 07	493	252
storm 08	567	256
storm 09	716	**174**
storm 10	707	181

Table 34: Minimum and maximum sampling intervals for System I (based on locations A,B and D). The shaded values of 405 s and 174 s define the range in which a sampling interval can be picked. A practical value would be 300 s(5 minutes). It is to be accepted that too many samples are taken for some storms. On the other hand, for storms 06,..,10 more information than could be extracted is obtained.

Therefore, the processes that can be monitored at locations C and E hold information on these locations only (i.e. the water-level variation is only influenced by a small subsystem). This is the reason such locations are normally not taken into consideration when designing a measuring network. Therefore, locations A and E are omitted from further analysis.

System II

Storm	minimal maximum Δt (s)	maximal minimum Δt (s)
Storm 01	730	476
Storm 02	647	420
Storm 03	608	258
Storm 04	535	390
Storm 05	472	153
Storm 06	558	166
Storm 07	482	186
Storm 08	475	134
Storm 09	335	**91**
Storm 10	**250**	141

Table 35: Minimum and maximum sampling interval for system II. The shaded values (250 s and 91 s) give the domain in which the sampling frequency can be picked. A practical value would be 240 s (or 4 minutes). This would imply that for storms 05,...,10 less information is obtained than is possible.

To avoid data storage problems at high sampling frequencies, several possibilities exist:

- The sampling frequency is made dependent on the occurrence of a storm (this implies that communication between the rain gauge and the water level or discharge gauging station has to be established).
- The sampling frequency is made dependent on the actual water level.
- The sampling frequency is not defined as a fixed value but is made dependent on the occurrence of a minimum difference with respect to the previous sample.

In the latter case a measurement is recorded when the measured water level (or discharge) meets the following condition:

$$\left| h_{actual} - h_{previous\_recorded} \right| > \kappa \qquad \text{(eq. 4. 30)}$$

In which κ is set at any convenient small value larger than the anticipated measuring error. A drawback of measuring values obtained in this manner is that they call for substantial post-processing in order to make them accessible for further analysis. Estimating the value for κ is done in the same manner as was followed in the previous section with respect to a fixed sampling interval in time and allowing for a differentiated sampling interval with respect to the water level. The analysis does not change, only the axes (x and t) are interchanged.

4.3 Measuring accuracy and reliability

In the previous sections the term measuring error or measuring accuracy was used several times without a strict definition. Any physical measurement is characterised by its accuracy and its reliability, several possible definitions for measuring (in)accuracy are applied parallel (see e.g. Taylor (1982)):

- The root mean squared error (rmse).
- The standard deviation.

In this chapter the measuring accuracy is defined as the 95% reliability interval assuming a gaussian probability distribution. This implies that the measuring error is equal to $4*\sigma_p$, in which σ_p is he standard deviation that is directly related to the measuring method and the measuring equipment applied. The reliability of a measurement is influenced by systematic differences between the actual state and the state as measured in a process. A low reliability can be masked by a low accuracy (i.e. a high value for σ_p). Systematic errors are caused either by human errors during the measuring process (like incorrect readings or an incorrect definition of a reference level) or can be caused by systematic errors in the measuring principle or the apparatus applied (e.g. drift of the reference level in water level measurements).

4.4 Measuring methods used in urban drainage

4.4.1 General

When limiting the discussion to water quantity parameters, the following basic parameters are to be measured:
- Water levels.
- Discharges.
- Rain intensities.

Since these parameters will vary in time, it is of particular importance to measure them as a function of time. In the next sections some measuring methods used in urban drainage systems are discussed. For any measuring device used in an urban drainage system the following demands are to be met:
- Robustness in a mechanical sense.
- Good chemical resistance since the conditions in especially DWF and combined systems can be corrosive because of the presence of H_2S gas.
- A high degree of water tightness, this especially holds for devices containing electrical circuits.
- A low maintenance rate (i.e. less than about 4 times per year).

In the past devices originally designed for groundwater level measurements were applied; these proved to be unsuitable for application in sewers. Over the last decade however, several commercial products have been introduced meeting the mentioned demands enabling monitoring on a wider scale.

4.4.2 Water level measurement

4.4.2.1 Pressure head measurement

An important assumption when using pressure head measurement is that the pressure is hydrostatic, allowing direct quantification of the water depth h (see Vaterlaus, (1989)):

$$h = \frac{\Delta p}{g(\rho_w - \rho_a)} - \frac{(\Delta p)^2}{2gK(\rho_w - \rho_a)} \qquad \text{(eq. 4. 31)}$$

In which:

h	Water depth	(m)
g	Gravitational acceleration	(m/s^2)
K	Compression modulus	(N/m^2)

ρ_w	Density of water	(kg/m^3)
ρ_a	Density of air	(kg/m^3)
Δp	Difference between water- and atmospheric pressure	(N/m^2)

The second term in equation 4.31 compensates for the compressibility of water, this only becomes of interest when accurate measurements are needed at large depths. The measuring ranges encountered in urban drainage (1-10 meters) allow for neglecting this compensation([36]). If the value for the density of the medium and the value for the gravitational acceleration constant are known h is easily quantified. By good approximation the value for g is 9.813 m/s^2, while the density of sewerage is somewhat higher that that of pure water (a value of about 1002 kg/m^3 is usually assumed). The density of sewage is influenced by:

- Temperature.
- Total Suspended Solids (TSS) content.
- Gas concentration.

The temperature of wastewater in a sewer system normally does not vary extremely and is almost constant in a range between 15 °C and 20 °C, except for special cases like industrial discharges. Because of TSS and dissolved gasses the density varies (see e.g. Michelbach et al, (1992)) between 999.6 kg/m^3 and 1005.8 kg/m^3. The gas concentration is usually low and no noticeable influence is to be expected. However, when selecting a measuring location the absence of air entrainment (due to sluices, internal weirs etc.) in the direct vicinity of the measuring location is to be verified. Uhl (1993) reports the results of an elaborate study into the accuracy and stability of several commercially obtainable pressure sensors. The main conclusions he drew were:

- An accuracy of 0.1% to 0.6% of the measuring range is achievable under practical conditions.
- The long-term([37]) stability (zero drift) is about 0.5% of the measuring range.

4.4.2.2. Ultra sound methods

An alternative for the pressure sensor is a measuring device using sound reflection as a principle for distance measurement. The principle is based on the measurement of the travelling time of a short (ultra sonic) sound pulse. Such a pulse is send down and reflects on the water surface. The time lag between the moment the pulse was sent and the time the reflection reached the sensor is measured. The distance between sensor and water surface is then easily calculated from $\Delta h = \frac{c\Delta t}{2}$, in which c is the speed of sound and Δt is the measured travelling time of the pulse. The accuracy of this type of measurement is limited by the accuracy of time measurement (which is in practice hardly ever prohibitive) and the value for c. The speed of sound depends on temperature, atmospheric pressure, convection and moisture of the air. In the absence of wind, at a temperature of 0 °C, this speed is 331 m/s, while under the same condition but at 20 °C this is 343 m/s. The influence of the temperature

([36]) The pressure at a depth of 10 m is circa 10^5 N/m^2, neglecting the density difference between water and air, estimating g=10 m/s^2 and K=10^9N/m^2, the second term in equation 4.29 is O(10^{-3}) or in the order of magnitude of 1 mm. Or a relative systematic error of 10^{-4}, as such being insignificant to the relative variation in the density of sewage (circa $6*10^{-3}$).

([37]) Long-term indicates a period of several weeks to several months.

is by far the most important and is easily compensated for. Still, under practical conditions, the echo principle is not as robust as the pressure method; for several reasons;

- Floating objects or foam layers can seriously disturb the measurements by absorption of the sound pulse or by influencing the angle of reflection.
- In practical cases sound reflections due to the geometry of the space (normally a manhole) in which the device is located may occur, leading to inaccuracy.

Another drawback is that the minimum distance that can be measured is about 0.5 m, implying that flooding cannot be recorded. The overall achievable accuracy (95% confidence interval) for commercially available sensors is estimated by Uhl (1993) to be circa 0.5%-1.0% of the measuring range, under favourable conditions. The long-term stability was estimated to be 0.5%-0.9% of the measuring range. Furthermore the resolution depends on the frequency of the sound applied. At high frequencies (1MHz), the resolution is circa 0.3-0.4 mm and for a resolution of 2 mm at least a frequency of 170 KHz is needed (Michalski & Berger (1984)).

4.4.3 Flow measurement.

4.4.3.1 General

The method of choice to measure discharge depends strongly on the location at which the discharge is to be known in an urban drainage system. In this context the following situations are briefly discussed:

- Discharge over special constructions (i.e. weirs or sluices).
- Discharges in conduits.

Accurate discharge measurement in pressure mains is mostly relatively easy to implement and is therefore not discussed in detail. Techniques like magneto-induction, Doppler-shift or vertex principles have been developed to a high degree of perfection, guaranteeing a relative accuracy of 0.5% or even better. This level of accuracy is only achieved when a fully developed velocity profile is present and no air pockets are present in the measuring section. Detailed information on measuring principles and achievable accuracy levels are found in literature (see e.g. ISO (1983[a])). In general, an accurate discharge measurement is relatively expensive when compared to water level measurements. This is especially so when constraints are put on allowable changes that may be made in the urban drainage system under study.

4.4.3.2 Weirs

As discussed in Chapter 2 the general relation between water level and discharge over a weir has the following form $Q(h) = \alpha h^{\beta}$. If the values for α and β are known, an accurate water level measurement is used to calculate the discharge. For special (standardised) constructions the values for both coefficients are known and can be used without prior calibration. However, even when a well-calibrated weir is applied the accuracy of the discharge calculation depends on the stability and accuracy of the level measurement.

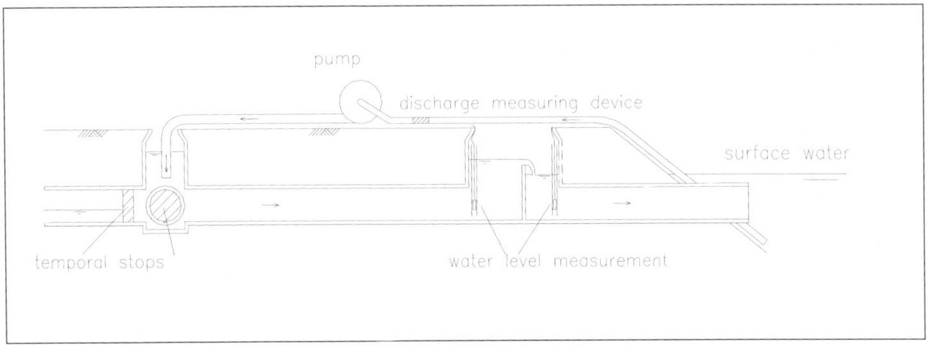

Figure 88: Experimental set-up for Q-h calibration of a weir (after STOWA (1996[b])).

When calculating the discharge using a measured value for h, the uncertainty in the discharge obtained is calculated using the general formula for error accumulation according to Gauss (see e.g. Taylor (1982)):

$$\left.\begin{aligned}\sigma_q^2 = & \left(\left.\frac{\partial f}{\partial h}\right|_{\substack{h=\mu_h\\ \alpha=\mu_\alpha\\ \beta=\mu_\beta}}\right)^2 \sigma_h^2 + \left(\left.\frac{\partial f}{\partial \alpha}\right|_{\substack{\alpha=\mu_\alpha\\ h=\mu_h\\ \beta=\mu_\beta}}\right)^2 \sigma_\alpha^2 + \left(\left.\frac{\partial f}{\partial \beta}\right|_{\substack{\beta=\mu_\beta\\ \alpha=\mu_\alpha\\ h=\mu_h}}\right)^2 \sigma_\beta^2 + \\ & + 2\left(\left.\frac{\partial f}{\partial \alpha}\right|_{\substack{h=\mu_h\\ \alpha=\mu_\alpha\\ \beta=\mu_\beta}}\right)\left(\left.\frac{\partial f}{\partial \beta}\right|_{\substack{h=\mu_h\\ \alpha=\mu_\alpha\\ \beta=\mu_\beta}}\right)\rho_{\alpha,\beta}\end{aligned}\right\} \qquad \text{(eq. 4. 32)}$$

Equation 4.32 is only correct when there is no covariance between h,α and β. In practice the values for the coefficients are obtained from a data fit using n data pairs (Q_i,h_i), this means that there exists a covariance between α and β. During operation, the measured values for h are independent from α and β, implying that any correction for these interdependencies can be left out. When applying the Least Squares Method (see also Chapter 5) with the assumption that the measured data (i.e. Q and h) have a normally distributed uncertainty, the variance-covariance matrix is calculated from:

$$\underline{\mathrm{cov}} = \sigma_r^2 \left(\underline{J}^T \underline{J}\right)^{-1} \qquad \text{(eq. 4. 33)}$$

In which σ_r is the standard deviation of the residuals estimated by $\sigma_r = \sqrt{\dfrac{\sum_{i=1}^{i=n} (r_i - \mu_r)^2}{n-1}}$ and $r_i = Q_i - \alpha_i^\beta$, the Jacobean is defined as:

$$J = \begin{bmatrix} \frac{\partial r_i}{\partial \alpha} & \frac{\partial r_i}{\partial \beta} \\ \cdot & \cdot \\ \cdot & \cdot \\ \cdot & \cdot \\ \frac{\partial r_n}{\partial \alpha} & \frac{\partial r_n}{\partial \beta} \end{bmatrix}$$

In this manner the measuring uncertainties in Q and h in the calibration experiment are directly translated into a variance-covariance structure for the coefficients α and β. The elements in the Jacobean are: $\frac{\partial r_i}{\partial \alpha} = -h_i^{\beta}$ and $\frac{\partial r_i}{\partial \beta} = -\alpha h_i^{\beta} \ln(h_i)$.

The standard deviation in Q when using a calibrated relation is easily derived in this case:

$$\sigma_Q^2 = \left(\alpha\beta h^{(\beta-1)}\right)^2 \sigma_h^2 + \left(h^{\beta}\right)^2 \sigma_\alpha^2 + \left(\alpha h^{\beta} \ln(h)\right)^2 \sigma_\alpha^2 + 2\left(h^{\beta}\alpha h^{\beta} \ln(h)\right)^2 \rho_{\alpha\beta} \qquad \text{(eq. 4. 34)}$$

As an example, the results of an experiment in Delft at 18/1/2000 that followed the procedure as described in this section are discussed. The overflow was prepared to obtain a favourable geometry; i.e. a sharp crested weir, made out of thin sheet, was installed and set exactly at level. The measuring results are shown in Table 36.

pair	H (mm)	Q(m³/h)	pair	H(mm)	Q(m³/h)	Pair	H (mm)	Q(m³/h)
1	21	45	10	132	600	19	115	500
2	31	73	11	145	712	20	95	400
3	32	80	12	167	800	21	80	300
4	36	100	13	186	900	22	58	189
5	50	152	14	180	905	23	35	101
6	59	202	15	170	800	24	30	74
7	79	301	16	145	700	25	21	47
8	100	403	17	145	700			
9	115	501	18	130	600			

Table 36: Measuring results calibration experiment Delft.

During the experiment, problems were encountered when the discharge was > 700 m³/h. The effects on the accuracy of the Q-h relationship obtained are shown in Figure 89 and in Table 37 the values found for the parameters and the variances and the covariance are shown.

	Range for h	α	β	σ_α^2	σ_β^2	$\rho_{\alpha\beta}$
Q-h_1	21-145 (mm)	0.6153	1.4145	0.0016	$1.8509*10^{-4}$	$-5.4539*10^{-4}$
Q-h_2	145-186 (mm)	4.1704	1.0296	4.8637	0.0107	-0.2277
Q-h_3	21-186 (mm)	0.9542	1.3183	0.0130	$5.7667*10^{-4}$	-0.0270

Table 37: Fitted Q-h relations.

The effect of the experimental problems is directly seen, the parameters in relation Q-h_1 result in significantly smaller values for variance-covariance than relation Q-h_2 that holds for the upper range of the measurements in which problems were encountered. When fitting a relation for the whole measuring range (relation Q-h_3), the effect of the inaccuracies translates back into significantly larger values for variance-covariance. Therefore, it must be concluded that for practical applications the relation Q-h_1 should be used only, limiting the usable data to the lower range (h_{max}=145 mm.)

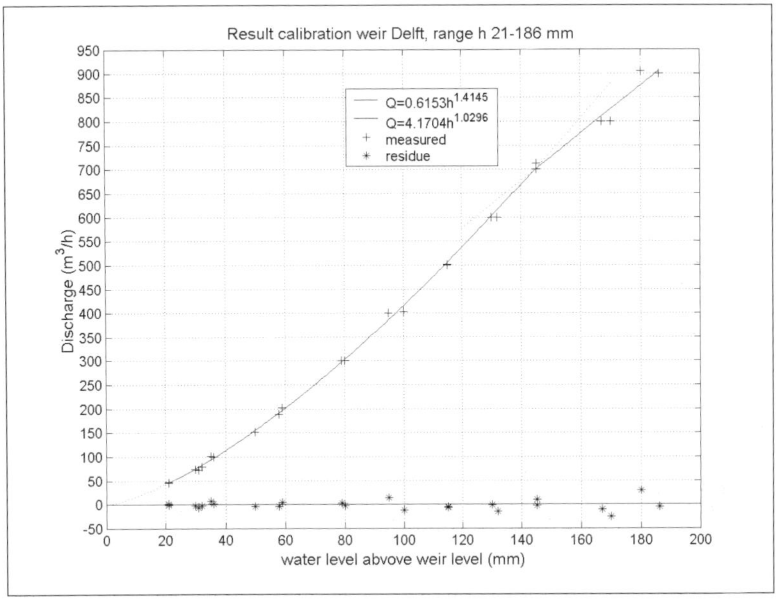

Figure 89: Q-H relations obtained from the calibration experiment at Delft January 18 2000.

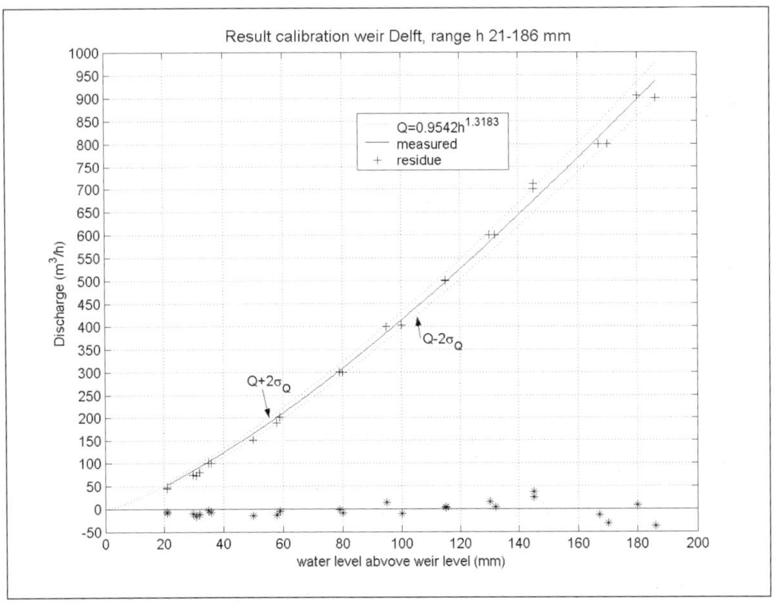

Figure 90: Q-H relations obtained from the calibration experiment at Delft January 18 2000.

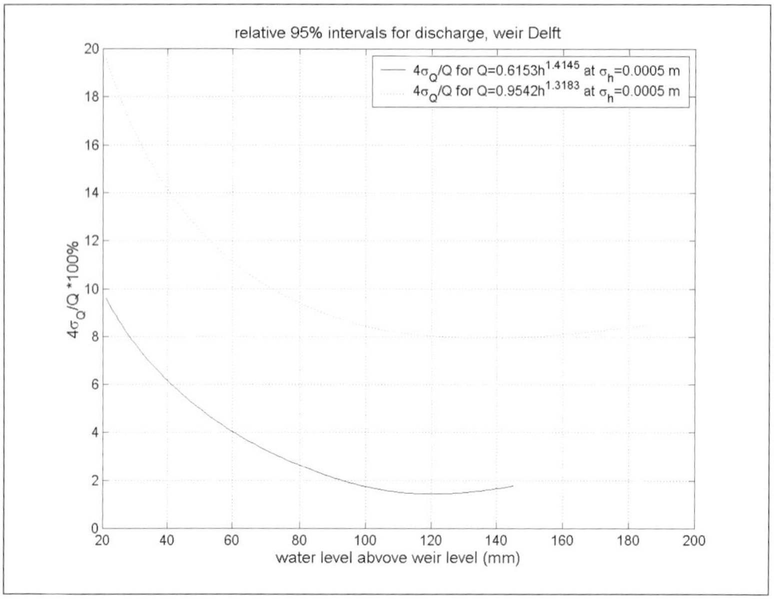

Figure 91: 95% confidence range relative to the measured discharge for a calibrated weir.

The accuracy of the discharge calculated using a fitted relation is largely influenced by the magnitudes of the variance-covariance in the parameters of the relation. Therefore, it is more relevant to perform the calibration experiment with great care and precision than to heavily invest in extraordinary accuracy in level measurement during the measuring period after the weir calibration; accurate operational measurements cannot compensate for an inaccurate calibration.

4.4.3.3 Conduits

4.4.3.3.1 General

The flow in a conduit can be measured using several principles;

- Velocity measurement combined with a level measurement.
- When the relation between discharge and difference between up- and downstream water level is known the discharge can be obtained using two water level measurements.
- Using a constriction in the conduit (e.g. a venturi channel or a crump-weir) can be considered.
- Flow-through measurement (e.g. using a magneto-induction principle).

Which method is applied depends on the desired accuracy, the budget and the question whether or not a backwater effect due to the measuring devices is to be accepted.
In discharge measurement in conduits in urban drainage systems it is difficult to obtain high levels of measuring accuracy (as will be seen in the next sections). Therefore, it should only be done when there is a clear purpose for knowing such a discharge. When compared to water level measurements, discharge measurement is more expensive while resulting in information of relatively inferior quality.

4.4.3.3.2 Velocity/water level.

This principle is based on the fact that $Q = \int_A udA$, the cross sectional area A depends on the water level and the, known, geometry of the channel under study (for circular channels this relationship is found in annexe I). The main problem in this method is to obtain an accurate value for $\int_A udA$. In most cases the following formula is applied:

$$Q = \gamma(h) * A(h) * \hat{u} \qquad \text{(eq. 4. 35)}$$

In which:

A(h)	cross-sectional area	(m^2)
h	water depth	(m)
$\hat{u}$	measured velocity	(m/s)
γ(h)	correction coefficient	(-)
Q	discharge	(m^3/s)

In most cases either a Doppler effect measurement or a magneto inductive method is used. In either case the velocity in only a part of the cross sectional area is measured. This implies that a correction on the measured value for the velocity has to be introduced. This correction depends on the water level, the shape of the profile and the apparatus applied. For circular conduits being surcharged, it is fairly easy to give an accurate value for the correction factor because the wetted area is known at a large accuracy. In such a case an accurate discharge measurement is possible. This is not the case when channels are only partially filled; the velocity profile depends on the water level and the geometric profile and is commonly not exactly known. Therefore, only an in-situ calibration of the measuring device as a function of water level will result in accurate results. Without in-situ calibration, an accuracy of about 20% is achievable. In case of pressurised conduits, this accuracy is about equal to the accuracy of the velocity measurement (about 1- 2 % of the measured value).
The accuracies that can be obtained under practical conditions have been investigated by Watt & Jefferies (1996). Their main conclusion was that the accuracy of flow sensors based on ultrasonic Doppler-shift measurement and flow depth measurement is between 17%-49% depending on practical conditions. One of the most important practical problems mentioned is the temporary absence of suspended solids in the sewage resulting in a serious under-estimation of the velocity. The problem of low concentrations of suspended solids is most likely encountered under storm conditions; this is typically the mode in which measurements contain important information. Thenard et al (1999) report similar estimates on the accuracy of the Doppler-shift measuring principle, they furthermore report systematic errors depending on flow conditions (i.e. DWF versus storm conditions).

4.4.3.3.3 Venturi's

A venturi functions on the principle that in a constriction in a flow field a pressure drop (or in the case of a free surface a local drop in water level) occurs. Combining this with the (known) geometrical properties an accurate value for the momentary discharge can be obtained. (Theoretical backgrounds are described in annexe V). When a venturi works within its design range the flow enters the construction in a sub-critical manner and becomes critical in the narrowest part of the venturi (this implies Fr=1). Using this information opens the possibility to calculate the discharge directly from a single upstream water level measurement.
A venturi has several favourable properties:

- A high degree of accuracy can be obtained.
- It can be used for free-surface flow as well as for surcharged flow.
- It is relatively cheap (circa half the costs of Doppler-shift devices).

A major drawback is that incorporating a venturi does result in a backwater effect that can impose an unwanted influence upon the system under study. Furthermore, it must be realised that a venturi has an upper and a lower limit to its measuring range.
If reversal of flow direction is expected to occur, a symmetric construction using two level gauges is a minimum demand. When flow measurements are to be done when no transition between sub-critical and super-critical flow occurs, then at least three measuring points are needed (i.e. upstream, in the 'throat' and at the downstream end). In literature, several developments with respect to venturi discharge measurement in urban drainage networks are reported. Diskin (1963) reports the properties of the so-called 'Diskin' venturi. Due to its shape this device can only be used when no reversal of stream direction will occur. Wenzel (1975) developed a venturi that is able to measure under free surface as well as under surcharged condition and is able to handle stream reversal. Hassapis (1991) even commercialised a discharge measuring system for application in urban drainage.
As stated before, inserting a venturi in an existing system will in most cases be impossible because the backwater effect is not acceptable. In newly built systems however, a venturi can be incorporated in the system, especially when the designer considers using some kind of sluice construction. The achievable measuring accuracy using a venturi in flat sewer systems (the predominant situation in the Netherlands) has, to the knowledge of the author, not been investigated under practical conditions. In order to get an indication a limited laboratory investigation was done. The experimental setup, the results and theoretical backgrounds are described in annexe V. The diameter of the pipe was 189 mm and only a limited range of discharges and water levels was investigated. Therefore, no general conclusions with respect to obtainable accuracy can be draw. As can be seen, the measuring results match the theory satisfactorily (see Figure 92 and Figure 93). At depths below circa 0.04 m backwater effects tend to disturb the water level at the upstream measuring point, indicating the lower limit of the measuring range. Furthermore, it is clear that at small inclination angles of the conduit small changes in this factor have a noticeable influence. Since it is hard to measure the inclination level in a conduit exactly, application in practice cannot go without in situ calibration, indicating a similar problem is encountered when compared with regular weirs used as measuring devices.
In Figure 94 the measuring accuracy (under laboratory conditions) of the Q-h relationship is shown. As can be seen the accuracy of the water level measurement is about equally important as the inherent accuracy obtained in the Q-h relation; an accuracy of about 10 % of the nominal value is obtainable. It is unlikely that under practical conditions (i.e. in a real sewer) this level of accuracy can be obtained.

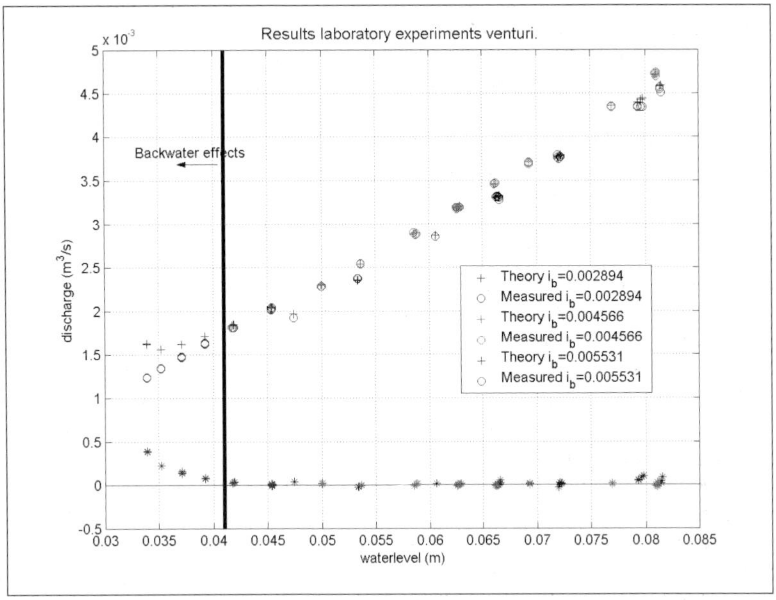

Figure 92: Experimental results obtained in laboratory experiment using a venturi in a circular shaped conduit.

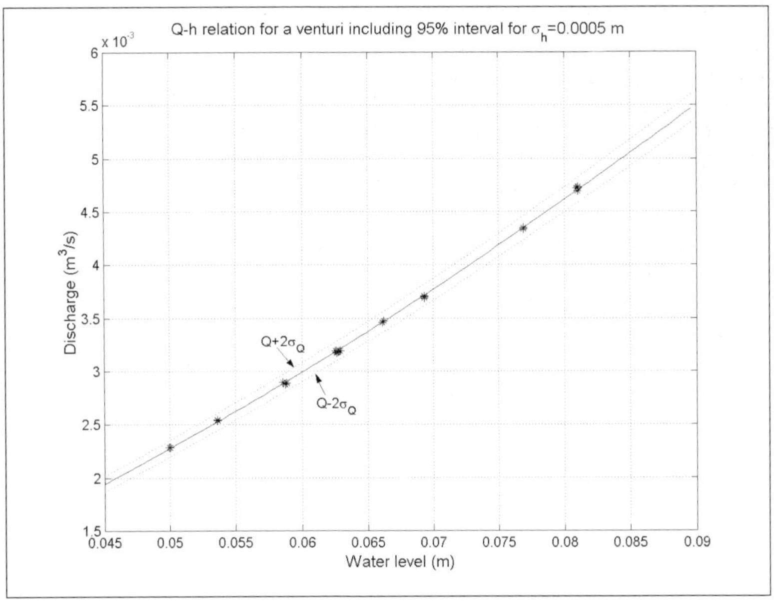

Figure 93: 95% accuracy interval for the laboratory experiments.

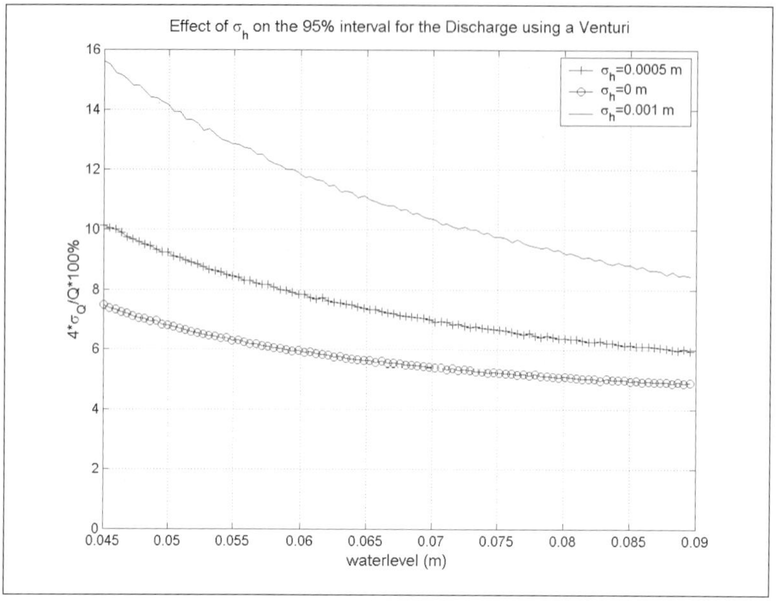

Figure 94: Accuracy obtainable by a venturi.

4.4.4 Rain measurement

4.4.4.1 General

Rain shows a large variation in space and in time. Therefore, it is important to have a monitoring network with a sufficient spatial density and a relative highly monitoring frequency in time. The aspects of spatial density will only be discussed shortly here, while the frequency in time is mostly between 1 and 5 minutes which has proven to be adequate for all practical applications. An important observation with respect to rain gauges is that the accuracy that can be obtained in practice depends to a large extent on the location at which the device is situated. Under unfavourable conditions, the systematic errors induced by location overwhelm the inherent inaccuracy of the apparatus employed.
Although several principles are applied for rain measurement (see e.g. KNMI (1983)) only the inherent accuracy level of the predominant type of rain gauges applied in urban drainage field research is shortly discussed here; namely the tipping bucket rain gauge. (Another system enjoying increasing popularity in the field of urban drainage is the rain gauge based on weighing).

4.4.4.2 Tipping bucket

In Figure 95 the principle of the tipping bucket is shown. The rain is caught in a bucket with a precisely known catch area (in most cases 200 cm^2 or 400 cm^2). The rain passes a filter and flows via a syphon into the bucket assembly mounted on an axle allowing it to rotate. When the bucket is filled to a certain level the equilibrium changes and the bucket assembly tips over, every tip being recorded. When the volume held by the bucket at the moment of tipping

is known, it is possible to calculate the averaged rain intensity by counting the tips in a well-defined time span.
Normally this time span is taken to be 5 minutes.
So, the measuring principle is relatively simple, but it has some inherent disadvantages:

- Regular cleaning of the filter is a necessity, especially during autumn or spring the filter quickly becomes blocked by debris like leafs or blossom.
- At low rain intensities the number of recorded tips in a given time span may be too small to obtain accurate results.
- At high rain intensities the number of counts is limited due to inertia of the moving parts and/or frictional forces induced by rotation. This implies that a substantial part of the rain is not registered, namely during the time-interval in which the bucket-assembly tips over.

Therefore, every rain gauge based on this principle has a lower rain intensity limit and an upper limit at which the bucket cannot respond quickly enough to the water stream being poured on it. In between these two limits the effect of inertia and friction increases and should be compensated for, deviations up to circa 5% of the observed measurements are normally present at high rain intensities (i.e. > 50 l/s.ha). Normally this is corrected using device specific software.

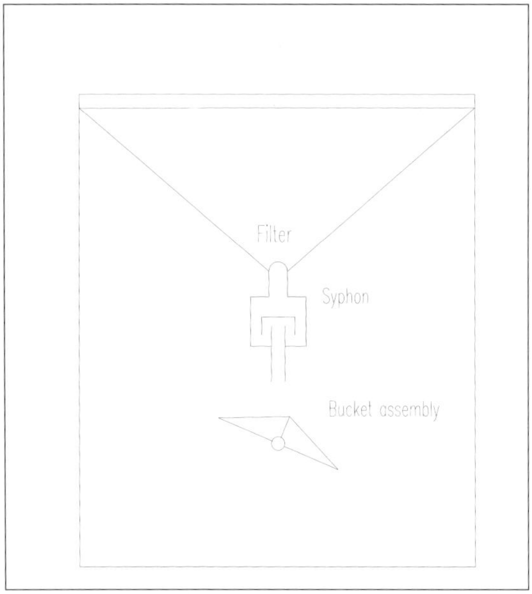

Figure 95: Schematic presentation of the tipping bucket rain gauge.

Furthermore, the presence of a syphon influences the accuracy of the measurement. The syphon is applied to obtain a rain-intensity independent momentum on the bucket assembly. An extensive research into the effects of the syphon has been done by Overgaard et al (1998). Their conclusion was that the effect of the syphon is minimal if the volume of the syphon is much smaller than the bucket volume or if the ratio between these two volumes is an integer value. In the first case, the resolution is nearly equal to the bucket volume (i.e. for most devices equal to 0.2 mm rainfall), in the latter case it is equal to the volume of the syphon. In order to avoid 'false double tips' or 'lack of tips', the optimal syphon volume should be circa half the volume of the bucket.

4.4.4.3 Effects of location on rain intensities.

The most preferable measuring setup for a rain gauge is to have the edge of the rain gauge at ground level, located in such a position that the influence of wind and ricocheting raindrops is eliminated. A possible setup that meets these ends is to place the rain gauge in an artificial terrain depression in such a manner that the upper edge of the gauge is at ground level. (See also Figure 96). This type of measuring setup is widely applied for meteorological purposes and is probably the best practically obtainable. As such, it is used as a reference for other measuring setups. It is noted that in the urban environment the spatial variation in rain intensities will be different from that measured in the open field due to differences in temperature and the influence of buildings on wind patterns (see e.g. Buishand & Velds (1980)). When designing a measuring setup for a field research for urban drainage, the ideal setup as discussed, is hardly ever obtainable. This is due to the following reasons:

- The investments are considered too large, since the duration of the measurements is mostly limited.
- The constant threat of vandalism (see Figure 96).
- In the urban environment it is hardly ever possible to find a suitable location at ground level.

Therefore, normally the rain gauge is mounted on a pole at a safe level above the ground at a location where wind influences are expected to be moderate or it is placed on a flat roof in the research area. Several researches (e.g. NWRW (1989)) have shown that a rain gauge mounted on a pole at a level of 3 m above ground level shows a deviation of about -6 % compared to a rain gauge at ground level. It is questionable whether this effect is to be corrected when processing measuring data; after all the rain is caught at least for a substantial part on roofs where wind effects are also present. Another important aspect is the geographical position of rain gauges in a research area. For practical reasons at least 2 rain gauges should be applied, so as to avoid the risk of having no data at all when a device does not function correctly. Another reason for implementing more than one rain gauge is because of the spatial variability of rain intensity.
Without going into detail, the following known results with respect to accuracy and reliability are shortly presented. Lei & Schilling (1993) state that

> *"2 rain gauges per catchment is a threshold number that is necessary for a reliable estimation of the statistical performance characteristics of a sewer system".*

From their research it was concluded that the accuracy level changes dramatically when stepping from 1 to 2 gauges, while adding any more gauges leads to a marginal increase of the accuracy with respect to rain volume over the catchment. Furthermore, they state that the catchment size does not seem to influence this threshold number.
This statement is supported, though weakly, by observations that claim that rain has fractal properties in time and space, see Mandelbrot (1984) and Lovejoy & Mandelbrot (1985).
The supposed fractal character of rain intensity in the space-time domain implies that the variability of the rain intensity is invariant for scaling in space and time.

Figure 96: Application of a rain gauge as gravel container.

More recent research into statistical modelling of the variability of rain intensities in space and time (see Willems (1998) and Willems & Berlamont (1998)), has shown that rain intensity in rain cells can be modelled as a moving bivariate Gaussian distribution. Rain cells are the smallest structures recognised in a rain field, the rain is concentrated in the cells. Rain cells cover an area in the order of magnitude of 1-10 km^2. Clusters of rain cells (or so-called small mesoscale areas) cover areas of 10^2 to 10^3 km^2. For most urban catchments only the scale of individual cells or clusters of rain cells are of importance; in these cells the spatial correlation of the rain intensities is high. Based on these observations, one rain gauge per km^2 is a defendable choice, though variation is present on even smaller scales. Especially on the small time scales of interest for urban drainage these variations can be considerable (De Bruin (1973) reports differences of approximately 30 mm per day over distances of some kilometres.)

The main conclusion with respect to design of a monitoring network of rain gauges in urban drainage is that it is not simple to obtain a reliable estimate for the accuracy of measured rain intensities valid for a catchment using a limited number of rain gauges. For practical purposes, one rain gauge per km^2 and at least two rain gauges per catchment seems a satisfying criterion.

4.5 Conclusions

The main conclusions with respect to measurements and their accuracy in urban drainage are:

- When designing a measuring network for an urban drainage system the temporal sampling distance can be determined using classical concepts as applied in designing networks for e.g. groundwater flow. The sampling intervals to be applied in a given case depend on the system under study, the location of the gauging point in the system and on the characteristics of the storm event. The latter conclusion is of importance since it precludes any method for obtaining a 100% safe prior estimate for the sampling interval. However, it was made plausible that, applying the standard Dutch design storms, the risk of making an incorrect estimate (i.e. a too large sampling interval) is acceptable.

- Using simple techniques based on simplified process descriptions along with some basic characteristics of the system under study (water-level height relation, total weir length and pumping capacity) a first indication for sampling intervals can be obtained. A further, more refined value for the sampling interval is obtained by applying time domain analysis and frequency domain analysis. These techniques allow for a design based on the characteristics of the drainage system at the gauging locations as well as on the accuracy of the measuring apparatus to be used.
- Discharge measurement in conduits or special constructions like existing weirs allows only a limited accuracy under practical conditions. For weirs, even after calibration, 95% intervals can be as large as circa 10-20 % of the nominal discharge.
- Measurement of water levels is possible with high precision, 95% intervals of 0.6% of the measuring range are possible under practical conditions.
- Application of a venturi in a circular conduit leads, under laboratory circumstances, to high accuracy levels, while the results match the results obtained from basic hydraulic theory accurately. Practical application in flat system, however is prohibited mainly by the backwater effect being an inherent consequence of a venturi. In existing systems application interferes with the hydraulic behaviour of the system under study and, more seriously, may even cause flooding.
- The accuracy that can be obtained with respect to rain intensities are dominated by the measuring setup rather than by the device applied. The accuracy obtained for rain intensities on small time and spatial scales is hard to quantify. For calibration purposes, this is an inconvenient conclusion since the inaccuracy of these measurements will translate back into the quality of the calibration results (see Chapters 5 and 6).

When a comparison is made between the expected accuracy levels in modelling (see Chapter 3) and the obtainable accuracy with measurements the following conclusions are drawn:

- Water level measurement is to be preferred (when compared to flow measurement) in measuring programmes aimed at quantifying the processes in an urban drainage system because:
    - Water level measurement is simpler to implement without interfering with the processes to be measured.
    - Water level measurement is much cheaper per gauging location.
    - Water level measurement can be done in a precise manner.
- Flow measurement can be done with an accuracy varying between 10 and 50 % of the nominal value, depending on the location at which it is measured and the measuring method applied. These accuracies are in the same order of magnitude as the inherent inaccuracies of a non-calibrated model. Therefore, flow measurement should not be applied for the purpose of calibration of a hydrodynamic model.

CHAPTER5: CALIBRATION

5.1 Introduction

5.1.1 General

As was explained in Chapters 3 and 4 information on the hydraulic behaviour of an urban drainage system can be obtained either by using a model or by studying information obtained from field measurements.
Both hold inherent discrepancies when compared to reality, they both merely are methods to obtain insight in the behaviour of such a system. It is stated that 'reality' can not be described in every detail, not by a model, nor by measuring data. Therefore, a practising engineer has to work with an incomplete or at least a more or less 'vague' image of the reality on which to decide on reconstruction or modification.
Models are used in engineering practice to judge designs or modification in design with respect to their functioning. In many cases large investments are based upon the results of models and the interpretation of these results by engineers. In the Chapters 2 and 3 in this thesis, it was outlined that hydrodynamic models in urban drainage, as used in practice nowadays, are based on several simplifications in process description and suffer from potential data errors. This is especially so with respect to geometry and structure of the system and the type and amount of the catchment area involved. Furthermore, several model parameters have to be introduced like hydraulic loss-coefficients, run-off coefficients that are seldom accurately known for the specific case in hand. The result of a good calibration process is: combining the benefits of a model with the strong point of information by measuring data, that is, combining relatively quick answers to 'what if' designs questions, and a detailed image in time and space of the hydraulic behaviour at one hand with a 'precise' image of the systems behaviour, though limited in space and time related to only one geometrical situation. The calibration of a model is the process in which a model is adapted in such a manner that it reproduces measured behaviour a well as possible. Validation, calibration and verification are terms closely related to each other and are easily mixed up; the terminology is subject to discussion. For instance, Oreskes et al (1994) argue that 'verification' and validation' of numerical models is inherently impossible unless closed systems like formal logic are addressed, models of natural systems are seldom closed. Their main problem with the use of the terms verification and validation is that the suggestion of 'truth' is raised, while any numerical model relies on assumptions and simplifications (see chapter 2 of this thesis). For clarity, while ignoring the linguistic discussion, the following definitions are applied in this thesis:
Validation is defined as:

> *To verify if the model reproduces the processes as observed in the real world. This implies that in qualitative terms the reproduction of the model response to a given input is in line with observed behaviour in the real world. E.g., a model describing the behaviour of a body on which a constant force works should show that this body accelerates in the direction in which this force works. If the model result is not in line with this observed reality the modelling of the process is wrong.*

Calibration is defined as:

> *The process in which a set of model parameters is defined which, combined with a validated process model, reproduces a specific measured situation as good as possible. This implies that in the calibration process the model-input is adjusted in*

such a manner that the model result 'fits' as closely as possible to a specific observed reality. It should be noted that the model input consists of many different types of parameters (see Chapters 2 and 3).

Verification is defined as:

The process of testing if the set of model parameters obtained from a calibration results in a correct reproduction by the model of a measured reality that was not taken into account in the calibration process. In fact, the result of the verification may show that the portability of a set of model parameters is limited, indicating either that the calibration process needs improvement or that the model applied does not contain all relevant processes.

It is clear that validation, calibration and verification are interconnected in an iterative process aimed at refining existing models. The process of calibration and verification results in new insights to make alterations in the description of processes in the models applied. Models used in urban drainage are based upon generally accepted descriptions of hydraulic processes (see e.g. Chapter 2) and process descriptions of the run-off process that are still subject of discussion. In this chapter, it is assumed that the models applied are validated; it is assumed that the De Saint-Venant equations are a correct description of the hydraulic process in an urban drainage system. Furthermore, the run-off models applied are also assumed to be verified. Therefore, the focus is on calibration; the verification of a calibrated model is discussed in Chapter 6 in which a case study of a calibration process is described. The objective is to present and demonstrate methods for calibration that can be applied by practising engineers using simple means.

5.1.2 Calibration of models in urban drainage

Ahyerre et al (1998) mention the lack of calibrated models as one of the three ([38]) main difficulties in the further development of storm water quality modelling. As was argued in Chapter 1, a first prerequisite for the further development of storm water quality modelling is the availability of a calibrated hydrodynamic model.

Calibration of hydrodynamic models is only scarcely applied in practice; this is due to several reasons:

- Calibration is regarded a time consuming, and therefore expensive job, with uncertain results.
- Field measuring campaigns are regarded as expensive.
- A lack of practical accessible methods to calibrate and objectively judge the results of a calibration process.

With respect to the first two arguments related to costs, the element of 'calculated risk' is involved. One should find a balance between a cheap model study with a risk of a less correct functioning design as a consequence, and a more costly model study resulting in a more secure tailor made design. In simple, small designs it may well be that the benefits of calibration do not counterweight the expenses. For optimisation studies (redesigning existing systems), in which the expected effects of modifications are normally relatively small, a sound calibration is to be performed. This is even more so when the effects of modification fall within the expected uncertainty range of the model results (see Chapter 3). The need for calibration of models is stressed by several authors (e.g. Price & Catterson (1997)). The calibration of a hydrodynamic model for urban drainage is no trivial task; when setting out to

([38]) The other problems identified are: a sparse knowledge of the processes involved in water quality modelling and the uncertainty and lack of data.

do so several problems arise which have been summarised as follows by, Harremoës & Madsen (1998):

- In many cases information needed proves to be missing from the data set.
- The fact that several parameter combinations may lead to an equally good fit between measuring data and model results.
- The problem of falsification.

The first point put forward is correct, but this can be investigated in advance by making an analysis of the information available (see Chapter 4 and Section 5.3). The second point is in fact known as ill-posedness, meaning that either the parameters chosen for calibration are wrong given a certain data set or the modelling concept is wrong given the parameters and the available data set. The last point put forward is an important observation; it takes only one set of measuring data that does not fit with model results based on 'calibrated' parameters to falsify the whole calibration. Therefore, it is never certain that at some moment a storm event will occur resulting in a falsification. This implies that any claim for a model to be calibrated/verified is hard to state with great certainty and extrapolation of the results is doubtful. The real benefit of a calibrated model is that in the calibration process several sources of uncertainty in data and/or process models are quantified or at least detected, in this manner enabling the modeller to enhance the model applied. Several authors report experiences with the calibration of (sub) models used in urban drainage; these will be discussed briefly.

Amongst others, Babovic & Wu (1994), Liong et al (1995), Veltri & Pecora (1999) and Rauch & Harremoës (1999) describe the application of evolutionary algorithms for calibrating hydrodynamic models. Main problem with such an approach is the absence of the possibility to judge the calibration results in terms of statistical properties. This does not imply that evolutionary algorithms don't find application in the calibration of hydrodynamic models, their benefits in identifying local minima in the goal function being recognised (see Section 5.10.4). A calibration strategy for SWWM was proposed by Ibrahim & Liong (1992), this strategy aimed at fitting of peak flows, which is a limited objective ignoring the evolution of processes prior to and posterior to the peak flow. In the strategy they propose the quality of the obtained model parameters is judged by the value of the objective function only, which is not regarded as adequate for an in-depth analysis (see Sections 5.2 and 5.3).

Ahyerre et al (1998) present the results of the SWWM model (Huber (1981)), for modelling TSS (total suspended solids). They conclude that various parameter vectors result in equally 'good' results. In conclusion, the amount of measuring data available was far too little to obtain a correct calibration. Cagiao et al (1998) report the calibration of a small urban catchment in Santiago de Compostela using SWWM and EXTRAN. Some of their results are shown in Figure 97. In their calibration the imperviousness was taken as a parameter, in its essence this parameter is not a process parameter but should be a hard value in the database. The tendency to 'calibrate' geometrical or structural parameters is seen in many other reports also (see. e.g. Foundation RIONED et al (1998) and Delleur (1998)). Lei et al (1999) report a parameter optimisation procedure based on Monte Carlo simulation. This method is straightforward but has serious drawbacks when compared to other methods of parameter optimisation (this issue is dealt with in Section 5.5.4.2). The application of expert systems as calibration tool has been put forward by Baffaut & Delleur (1989, 1990) and Griffin et al (1994). A nice example of the attitude many urban drainagists have on calibration is the following statement (Fuchs et al (1998)):

> *'Calibration is an optimisation or trial and error procedure in which a set of model parameters is obtained that minimise the difference between the measured and simulated values of the quantities being modelled....The possibility exists that the*

procedure could produce different sets of model parameters. In this case another optimisation procedure is needed to result in a single set of parameters.'

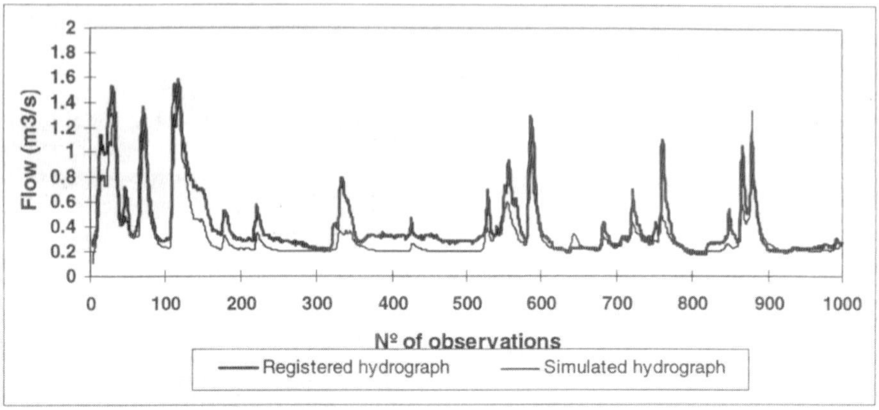

Figure 97 : Calibration results reported by Cagiao et al. (1998) (reprinted by courtesy).

The main tendencies in the literature on calibration of models in urban drainage are:

- There is little practical experience.
- The procedures focus on parameter optimisation only.
- Trial and error procedures are applied on a large scale.
- Calibration is performed in research projects mainly.
- The quality of the results obtained in terms of statistical properties and error analysis is hardly discussed.

In the remainder of this chapter the following issues are discussed:

- The calibration process.
- Definition of optimisation objectives.
- Parameter optimisation techniques.
- Statistical analysis of the calibration results.
- Error analysis.
- A priori judgement of measuring setups for calibration of models by the concept of auto calibration.

As was stated earlier, in this thesis the aim is to present methods and applications that can be used and implemented by any experienced engineer familiar with hydrodynamic models, using relatively simple means.

Calibration is often, incorrectly, referred to as parameter optimisation (Fuchs et al(1998)). Parameter optimisation is only one facet, albeit an important one, in the calibration process. The definitions according to Hemker (1996) are adopted. The first aspect is the choice of a modelling concept. In day-to-day practice, the number of alternatives is often limited because in most cases only one model (software-tool) is available. Within these models however, some freedom in processes to take into account is possible. The possibilities are:

- A deterministic model based on known physical process descriptions (see Chapter 2 of this thesis).
- Stochastic models.

- Conceptual models ([39]).

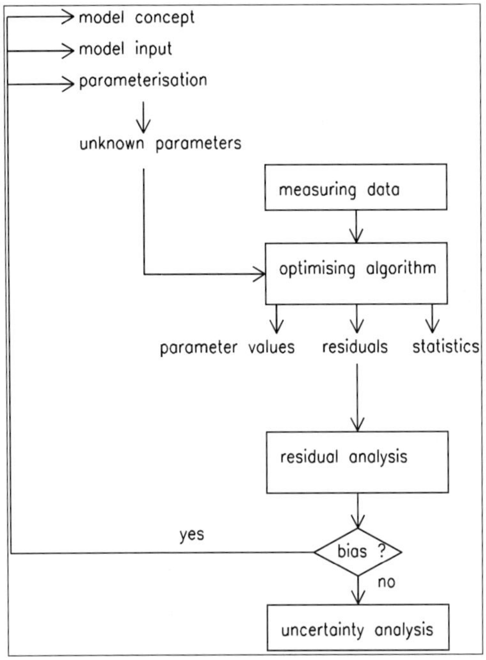

Figure 98: Definitions after Hemker.

A second important step is to define the model input; this aspect is discussed in Chapter 3 of this thesis. As has been pointed out, the accuracy of the database used and the extent of removal of errors in this database are crucial with respect to the achievable accuracy of the calculation results. As will become clear in this chapter, database errors are an important source of bias when calibrating a model. In order to achieve a good calibration the quality of the basic information must meet high standards. After choosing the modelling concept and preparing the database and other model input, the parameters on which the calibration is based must be chosen (parameterisation). When realising the intrinsic amount of the unknown parameters in even a relatively small drainage network it is immediately clear that choices have to made with respect to which parameters should be taken into account.

The model parameters are distinguished into several groups as defined by Ummels & Clemens (1998): Run-off parameters, Catchment area, Hydraulic parameters, Geometry parameters (e.g. profile measures) and Structure of the drainage system. The only parameters introduced in the model that should be incorporated in the parameter optimisation process are

([39]) The nomenclature for this type of models as advocated by e.g. Ruan (1999) is somewhat misleading. Since every model is based on some concept of thought ranging from physical process via mathematical description into eventually a tool (in most cases a software product, see Section 2.1) the term 'conceptual' is to be understood in its context. The term 'conceptual' is used to characterise a set of models that do not use physical processes as a starting point. These models merely try to find and identify relations between input (rain intensity) and output of 'real world' systems' (i.e. water levels or flows). These models have some favourable characteristics, for instance a low demand on computational capacity.

so-called process related parameters. Geometrical and structural data also play a role in the calibration process but are to be regarded as fixed values in the parameter optimisation process. The total number of process related parameters can be overwhelming in practice; the total number of parameters can be estimated to be:

$$N = N_m + N_c + N_w + 3N_L + N_s \qquad \text{(eq. 5. 1)}$$

In which:

N_c	number of conduits
N_L	number of area types
N_m	number of manholes
N_s	number of special constructions
N_w	number of weirs

An example is a small system having 123 manholes, 142 conduits, 3 weirs and distinguishing 3 area types, leading theoretically to a number of 277 parameters([40]) to be introduced in the model. As will be seen later (Section 5.11) it is impossible to introduce all these parameters in the optimising process because the number of measuring data should be one or two orders of magnitude larger than the number of parameters to be optimised. So, choices must be made in order to limit the number of parameters to be optimised. One way to do this is clustering, meaning that e.g. Nikuradses roughness is supposed to be equal for groups of conduits (distinguished by e.g. material or age). The same can be done with respect to hydraulic loss coefficients in manholes. Another number of parameters can be eliminated from the optimising process when their accuracy and influence on the calculation results is studied. In doing so, the focus is on parameters regarded as essential to the goal of the calibration project under hand. Consequently, a reduction to circa 10 parameters should be aimed at, given the normally available amount of measuring data.
An important observation is that the type of parameters taken into account in the calibration should be process parameters. Adjustment in the calibration process of parameters that describe the geometry and/or the structure of the drainage system or catchment is to be regarded as bad practice since it precludes beforehand the possibility of verification. Especially adjustment of the catchment area is frequently applied in calibration (see e.g. Foundation RIONED et al (1998)) and is referred to as 'seasonal variation'. In fact it would be more logical to accept that some processes in the run-off model (like infiltration rates) are poorly described as a function of time when studying time scales in the order of months
The measures of height and the structure of the system are assumed to be exactly known. It is well understood this is never totally so. Eventual errors in the structure of the system should become clear when analysing the results of the parameter optimisation process. Since e.g. the hydraulic roughness depends on the materials used and the age of the system it is postulated that the roughness in the model can be represented using one single value for all pipes. This implies that when obtaining a parameter value for the hydraulic roughness this can not be a physical meaningful parameter in the sense that it represents *the* hydraulic roughness of the pipes. The parameter value obtained is corrupted and merely represents a mean resistance parameter. As stated earlier (Chapter 3), in flat systems the calculation results are relatively insensitive to the hydraulic roughness, therefore it is accepted a priori to obtain a corrupted physically meaningless value for the hydraulic roughness.

([40]) Here it is assumed that every conduit has its individual roughness value, every manhole has its unique local loss coefficient, the runoff is described by three parameters per surface type (infiltration, routing constant and depression storage, and the weirs are characterised by an overflow coefficient. In this case this leads to a number of 277 parameters.

The choice of the parameters should be based on the model on one hand; on the other hand, the information held by the measuring data is to be considered also. Furthermore, it should be noticed that during a calibration process the modeller is often tempted to add more and more parameters in the optimisation process. In general, the quality of the calibration deteriorates when adding more parameters, it should be tried to define the minimal number of parameters needed to get a good description of the processes involved. The problem of obtaining a correct parameterisation is difficult to solve, since it largely depends on the model applied, the measuring data available and the optimisation method used (see also Section 5.11).
In the parameter optimising process, the goal is to find a set of parameter values that are most likely to be correct given a model and a set of measuring data. Extending this notion to the schema of Hemker implies that the model concept should also be incorporated in the optimising process. This would mean that the known set of applicable process models for a given problem should be used in the calibration process. This manner of operation is called 'Bayesian' (see Bernardo & Smith (1994)) and is in most cases hard, if not impossible, to implement to its full extent in practice because of the time consuming nature of such a process.

5.1.3 The 'spin-off' when calibrating a model

Calibrating a model reveals the discrepancies between model and observed 'reality'. Apart from obtaining a well-defined fit between model and reality, the process followed while calibrating, results in identifying the causes of the differences found. Sources of systematic errors are found in either:

- The processes incorporated in the model (in fact only after verification).
- The software.
- The database (structure, geometry, catchment area).
- The measuring data.

Identifying these systematic errors is hard to catch in general rules, and largely depends on the experience of the engineer performing the study. As was explained in Chapter 3, a database error of a few centimetres in any measure of height will pass unnoticed in the process of database verification. If a systematic discrepancy between model and measuring data shows up (e.g. the measured water level drops below the lowest invert level in a manhole), this points at some errors in the databases or it may point at some systematic error in the measuring data. In such an event, a further scrutiny of both information sources is necessary.

5.2 Optimisation goals

5.2.1 General

When conducting 'parameter optimisation', a first question to be answered is what the exact objective of the optimisation is; in this respect the following objectives are distinguished:

- Obtaining a 'reasonable fit between model results and measuring data'.
- A minimum deviation between model results and measuring data.
- Obtaining a parameter set that optimally fits the model with the measuring data and a quantification of the uncertainties in these parameter values.

The first objective formulated is the one mostly applied in engineering practice nowadays, and is mostly unwittingly used due to fact that a 'trial and error' optimisation method is

applied. The latter two objectives are more useful from a scientific point of view, since they are open for a mathematical formulation of the objective and allow the use of algorithms to reach the objective formulated.
Several optimisation objectives have been reported in literature (see Delleur (1998)):

- Minimisation norms based on the difference of a series of measured and modelled processes (y_i and $\hat{y}_i$)

$$\min\left[\sum (y_i - \hat{y}_i)^2\right] \qquad \text{(eq. 5. 2)}$$

$$\min\left[\sum |y_i - \hat{y}_i|\right] \qquad \text{(eq. 5. 3)}$$

$$\min\left[\sum (\ln y_i - \ln \hat{y}_i)^2\right] \qquad \text{(eq. 5. 4)}$$

In which y_i is a measured quantity and $\hat{y}_i$ is the corresponding model result

- Norm based on spilled volume:

$$\min\left[\frac{1}{N}\sum \frac{V_m - V_p}{V_m}\right] \qquad \text{(eq. 5. 5)}$$

In which V_m is a measured volume passing at a certain location in a given period and V_m is the corresponding volume resulting from the model.

- Norm based on reproduction of peak values (either in discharge or in water level):

$$\min\left[\frac{1}{N}\sum \frac{P_m - P_p}{P_m}\right] \qquad \text{(eq. 5. 6)}$$

In which P_m is a measured maximum of some quantity (either water level or discharge) at a certain location over a given period and P_m is the corresponding result obtained by the model.

- Norm based on time shift between model and measured values

$$\min\left[\frac{1}{N}\sum (T_m - T_p)\right] \qquad \text{(eq. 5. 7)}$$

In which T_m is the moment at which a (local) maximum for some quantity is measured, and T_p is the corresponding result obtained by the model.

- Weighing between two aspects in one norm

$$\min\left[\frac{1}{N}\left[a\frac{\sqrt{\sum (m-p)}}{m}\right] + b\left|\frac{V_m - V_p}{V_m}\right|\right] \qquad \text{(eq. 5. 8)}$$

In which a and b are weighing factors.

- The classical least squares norm (Yang & Parent (1996))

$$\min\left[\sqrt{\frac{1}{N}\sum (Q_m - Q_p)^2}\right] \qquad \text{(eq. 5. 9)}$$

When studying the definition of these norms it is clear that some of them are unbounded. Especially the norms defined in eq. 5.5, 5.6 and 5.7 pose no lower limit to their numerical value. This immediately implies that no sound minimisation technique can be employed. Some of these norms are defined in conjunction with the parameter optimisation method applied. This is especially so for the peak reproduction (eq. 5.6) and for the norm based on time shifts (eq. 5.7). These norms do not have to result in an overall fit between model and measurement, implying that they are relatively easy to apply in a trial and error calibration (see Section 5.6.2). The choice of the parameter in the norm (water level, discharge or volume) is based on several considerations:

- The possibility to obtain an accurate measured value (see Chapter 4).
- A relation with some regulation.

Using the norm based on volume or discharge (eq. 5.5 and eq. 5.9) most likely originates from regulations limiting the allowable volumes or discharges spilled into the receiving surface water. It must be remembered that obtaining accurate measurements of flow (and therefore of volume) is difficult (see Chapter 4). Choices for a norm and the parameter used in this norm translate back into requirements on the measuring data and the monitoring network. Application of a weighed norm (see e.g. eq. 5.8) has the disadvantage that some subjective choice has to be made for the values of the weights (a and b in eq. 5.8) being applied on the 'sub norms', making such a norm open for subjective discussions on the choices made. In the next section, a method to derive a norm based on Maximum Likelihood Estimates is described. Contrary to the norms described in the present subsection, this method is directly derived from mathematical principles using some basic assumptions that are open for verification during the calibration process.

5.2.2 Maximum Likelihood Estimates

The parameter optimisation technique, known as Maximum Likelihood Estimates (from now on referred to as MLE) was originally developed by Daniel Bernoulli (1700-1782) (van Gelder (2000)), and is described in great detail by Carrera & Neumann (1986[a,b,c]) and Olsthoorn (1998). A short description is presented here. The purpose is to get estimated parameter values based on hydraulic head measurements obtained at some points in a drainage system at discrete time intervals and (not necessarily) prior estimates for some parameters. These prior estimates for parameters are obtained by separate measurements. MLE is based on minimising residuals defined by:

$$\underline{e_x} = \underline{\hat{X}} - \underline{\tilde{X}} \qquad \text{(eq. 5. 10)}$$

in which $\underline{\hat{X}}$ is a vector containing the measured values of parameter x (or measured heads) and $\underline{\tilde{X}}$ is the vector containing an improved estimate for parameter x. The residuals reflect both measurement as well as model and software errors (like unnoticed database errors, numerical errors or software bugs); these errors are not identified separately. The 'real' value of a certain parameter x remains unknown; in this sense parameter optimisations is a kind of 'trading off' between model and measured reality. Based on the central limit theorem it is assumed that prior errors (head as well as prior estimates for parameters) are Gaussian ([41])

([41]) In fact choosing the Gaussian distribution is often chosen implicitly, because it essentially leads to a generalised least square method. Assuming an exponential distribution leads to a method of first power (the difference between model and measurement), whereas an extreme value distribution leads to higher powers thus adding more weight to the extremes in the observed differences between model and measuring data.

with a mean value of zero and a variance of σ^2. This hypothesis needs a posterior check when the optimisation process is completed. Any known covariance in the error structure is represented in a covariance matrix:

$$\left.\begin{aligned} \underline{\underline{C_h}} &= \sigma_h^2 \underline{\underline{V_h}} \\ \underline{\underline{C_{p_i}}} &= \sigma_{p_i}^2 \underline{\underline{V_{p_i}}} \end{aligned}\right\} \qquad \text{(eq. 5. 11)}$$

The global covariance matrix of all prior errors is block diagonal:

$$\underline{\underline{C_z}} = \begin{bmatrix} \underline{\underline{C_h}} & 0 & 0 & 0 & 0 \\ 0 & \underline{\underline{C_{p_1}}} & 0 & 0 & 0 \\ 0 & 0 & \underline{\underline{C_{p_i}}} & 0 & 0 \\ 0 & 0 & 0 & \underline{\underline{C_{p_{I-1}}}} & 0 \\ 0 & 0 & 0 & 0 & \underline{\underline{C_{p_I}}} \end{bmatrix} \qquad \text{(eq. 5. 12)}$$

It is assumed that no correlation exists between distinct groups of parameters, implying that prior estimates can not be based on the same measured head values used in the further optimisation process. The likelihood of a certain set of parameters $\beta(\underline{p},\underline{\psi})$ ($\underline{p}$ denoting the vector of parameter values and $\underline{\psi}$ the vector containing statistical properties of these values given a set of measuring data z^* ($L(\beta|z^*)$)([42]) can be expressed as the probability of obtaining the result z^* if β were correct. Therefore, the joint probability density functions for the model parameters and their statistical properties that maximise the probability of obtaining the measuring data are to be found. To this end the means and variances of the probability density functions of the model parameters are to be optimised. By assuming these functions to be Gaussian, the likelihood function to be optimised is:

$$L(\beta|z^*) = f(z^*|\beta) = (2\pi|\underline{\underline{C_z}}|)^{-0.5} \exp(-\frac{1}{2}\underline{e_z^T}\,\underline{\underline{C_z^{-1}}}\,\underline{e_z}) \qquad \text{(eq. 5. 13)}$$

For practical purposes the negative logarithm of the likelihood function ($S=-2\ln(L(\beta|z^*))$) is used for the actual optimisation:

$$\left.\begin{aligned} S = -2\ln\{L\{\beta \mid z^*\}\} = \frac{F_h}{\sigma_h^2} + \sum_{i=1}^{i=N}\frac{F_{p_i}}{\sigma_{p_i}^2} + N_h \ln|\underline{\underline{V_h}}| + N_h \ln(\sigma_h^2) + \\ + \sum_{i=1}^{i=N}\{N_{p_i} \ln(\sigma_{p_i}^2)\} + N\ln(2\pi) \end{aligned}\right\} \qquad \text{(eq. 5. 14)}$$

Omitting constants in the object function S, the actual function to be minimised is:

$$S = \frac{F_h}{\sigma_h^2} + \sum_{i=1}^{i=N}\frac{F_{p_i}}{\sigma_{p_i}^2} + N_h \ln(\sigma_h^2) + \sum_{i=1}^{i=N}\{N_{p_i} \ln(\sigma_{p_i}^2)\} \qquad \text{(eq. 5. 15)}$$

Or, in terms of residuals:

([42]) Notations applied: β is a vector containing the parameter values, z^* is a vector containing all measuring data, water levels, discharges and parameter values obtained by separate measurements.

$$S = \frac{1}{\sigma_h^2} \underline{e_h} \underline{\underline{V_h^{-1}}} \underline{e_h^T} + \sum_{i=1}^{i=N} \frac{1}{\sigma_{p_i}^2} \underline{e_{p_i}} \underline{\underline{V_{p_i}^{-1}}} \underline{e_{p_i}^T} + N_h \ln(\sigma_h^2) + \sum_{i=1}^{i=N} \{N_{p_i} \ln(\sigma_{p_i}^2)\} \qquad \text{(eq. 5. 16)}$$

Multiplying by σ_h^2 yields:

$$S' = S\sigma_h^2 = \underline{e_h} \underline{\underline{V_h^{-1}}} \underline{e_h^T} + \sum_{i=1}^{i=N} \lambda_{p_i} \underline{e_{p_i}} \underline{\underline{V_{p_i}^{-1}}} \underline{e_{p_i}^T} + \sigma_h^2 N_h \ln(\sigma_h^2) + \sigma_h^2 \sum_{i=1}^{i=N} \{N_{p_i} \ln(\sigma_{p_i}^2)\} \qquad \text{(eq. 5. 17)}$$

In which:

$$\lambda_{p_i} = \frac{\sigma_h^2}{\sigma_{p_i}^2} \qquad \text{(eq. 5. 18)}$$

λ_p may be interpreted as weighing-factors for the prior information (e.g. when for a certain parameter λ_p=1, the prior information has an equal influence on the result as the head measurements). In the special case in which the values for the variances are known, the method reduces to the well known least-square method (see Section 5.2.1), in practical applications this will seldom be the case.

5.3 Uniqueness, identifiability and stability

The correctness of the parameters found after optimisation is judged in terms of uniqueness and identifiability. The uniqueness of the parameters, meaning that the acquired vector $\underline{p}$ is the only combination leading to the given set of measuring data using the model applied, can be corrupted by two mechanisms. These mechanisms are obtaining a local minimum and instability (i.e. the object function being nearly flat in the region of the minimum). The latter is detected by slow convergence of the optimising routine and random stops. The risk of the existence of local minima in the goal function is reduced effectively if prior estimates are available for every parameter in the optimisation process (Carrera & Neuman 1986[b]). In practical situations this will seldom be the case, therefore several optimisation runs should be made using different starting vectors $\underline{p}$. Identifiability means that one set of parameters leads to just one possible result. This implies that identifiability is a necessary (but insufficient) condition for uniqueness. The information needed to judge the identifiability of the parameters is available in the Jacobean matrix (see Section 5.6.3.2 also). If the Jacobean is rank deficit, or nearly so, this implies that no orthogonal basis for the parameter space can be defined. In other words, the dimensionality of the optimisation problem is chosen too large given the available information. Unfortunately, this can be quantified only posteriorly, since the Jacobean matrix only contains trustworthy information for the minimum of the object function. In practical situations the rank of the Jacobean is difficult to calculate, an efficient manner to judge whether a matrix is (nearly) rank deficit is by applying Singular Value Decomposition (SVD). SVD is a technique for decomposing non-square matrices according to:

$$\underline{\underline{J}} = \underline{\underline{U}}\,\underline{\underline{\Omega}}\,\underline{\underline{V}}^T \qquad \text{(eq. 5. 19)}$$

In which the matrix $\underline{\underline{U}}$ is a NxN matrix containing the normalised orthogonal left hand vectors, $\underline{\underline{V}}$ is an MxM matrix containing the normalised orthogonal right hand vectors and $\underline{\underline{\Omega}}$ is a NxM matrix with the M singular values on its diagonal (at least in the first M rows), all other elements in this matrix being zero. N is the number of observations and M is the number of parameters in the optimisation and $\underline{\underline{U}}\,\underline{\underline{U}}^T = \underline{\underline{V}}\,\underline{\underline{V}}^T = \underline{\underline{I}}$.

The singular value are the square roots of the eigen values of $\underline{\underline{J}}^T\underline{\underline{J}}$ whereas the colums of $\underline{\underline{U}}$ are the eigenvectors of $\underline{\underline{J}}^T\underline{\underline{J}}$ and the colums of $\underline{\underline{V}}$ are the eigenvectors of $\underline{\underline{J}}\underline{\underline{J}}^T$.
Essentially SVD can be used for an overall sensitivity analysis for a model when the Jacobean matrix is defined as:

$$\underline{\underline{J}} = \frac{d\underline{h}}{d\underline{p}} = \left[\frac{\partial \underline{h}}{\partial p_1}, \ldots, \frac{\partial \underline{h}}{\partial p_M}\right] \qquad \text{(eq. 5. 20)}$$

If the definition of the Jacobean is used (namely based on the *difference* between observed and modelled results, either water levels or flows, see Section 5.6.3.2; eq. 5.35) a measure for the sensitivity of the object function for each parameter is obtained. In this manner, the SVD decomposition gives valuable information on the identifiability of the chosen parameters *given* the model applied and *given* the set of observations. If the matrix is (nearly) rank deficit then one or more singular values are (nearly) zero. The effect of this on the parameter is easily seen when writing (see also Olsthoorn (1998), p. 224):

$$\underline{\underline{J}} = \frac{d\underline{e}_h}{d\underline{p}} \Rightarrow \underline{\underline{J}}d\underline{p} = d\underline{e}_h \Rightarrow d\underline{p} = \underline{\underline{V}}\underline{\underline{\Omega}}^{-1}\underline{\underline{U}}^T\underline{e}_h \qquad \text{(eq. 5. 21)}$$

Since the matrices $\underline{\underline{V}}$, $\underline{\underline{U}}$ and the vector $\underline{e}_h$ have bounded norms the individual elements of $\underline{dp}$ tend to grow to infinity when the corresponding singular value tends to zero:

$$dp_i = \frac{C_i}{\omega_i} \Rightarrow \lim_{\omega_i \to 0} \frac{C_i}{\omega_i} = dp_i = \infty \qquad \text{(eq. 5. 22)}$$

This implies that, regardless of the residual $\underline{e}_h$ (the difference between observed and modelled water level or flow), the parameter value can vary within (theoretically) unlimited value bounds. Or, in reverse, a large change in the residual (caused by e.g. a measuring error) only causes a small difference in the parameter value for which the singular value tends to zero. Therefore, when the SVD of the Jacobean reveals one or more singular values (almost) equal to zero, the corresponding parameters are not identifiable. A further implication is that proceeding with a gradient based search method easily leads to instability of the algorithm applied. Because of the large possible variation of one or more of the parameter values the algorithm will easily become caught in locations in the search space that are wide apart. In other words, identifiabilty and instability are related to each other. In the remainder of this thesis the eigen vectors of the covariance matrix are studied (as will be explained hereafter), the relation between the singular values of the Jacobean matrix and the eigen values of the covariance matrix is easily demonstrated:
First, the following, well known decomposition using eigen vectors and eigen values is used:

$$\underline{\underline{J}}\underline{\underline{J}}^T = \underline{\underline{E}}\underline{\underline{\Lambda}}\underline{\underline{E}}^{-1} \qquad \text{(eq. 5. 23)}$$

In which $\underline{\underline{E}}$ is an NxN matrix of which the columns are the eigen vectors of $\underline{\underline{J}}\,\underline{\underline{J}}^T$ and $\underline{\underline{\Lambda}}$ is a diagonal matrix with the eigen values as diagonal elements.
From eq. 5.19 follows:

$$\underline{\underline{J}}\underline{\underline{J}}^T = \underline{\underline{U\Omega V}}^T\left[\underline{\underline{U\Omega V}}^T\right]^T = \underline{\underline{U\Omega V}}^T\underline{\underline{V\Omega}}^T\underline{\underline{U}}^T = \underline{\underline{U\Omega}}^2\underline{\underline{U}}^T \qquad \text{(eq. 5. 24)}$$

Using eq. 5.21 and 5.22 gives:

$$\left.\begin{aligned}
&\underline{\underline{E\Lambda E}}^{-1} = \underline{\underline{U\Omega}}^2\underline{\underline{U}}^T \Rightarrow \underline{\underline{E}} = \underline{\underline{U\Omega}}^2\underline{\underline{U}}^T\left[\underline{\underline{\Lambda E}}^{-1}\right]^{-1} \Rightarrow \\
&\Rightarrow \underline{\underline{U}}^{-1}\underline{\underline{E}} = \underline{\underline{\Omega}}^2\underline{\underline{U}}^T\underline{\underline{E\Lambda}}^{-1} \Rightarrow \left[\underline{\underline{\Omega}}^2\underline{\underline{U}}^T\right]^{-1}\underline{\underline{U}}^{-1}\underline{\underline{E}} = \underline{\underline{E\Lambda}}^{-1} \Rightarrow \\
&\Rightarrow \left[\underline{\underline{U}}^T\right]^{-1}\underline{\underline{\Omega}}^{-2}\underline{\underline{U}}^{-1}\underline{\underline{E}} = \underline{\underline{E\Lambda}}^{-1} \Rightarrow \underline{\underline{\Omega}}^{-2}\underline{\underline{U}}^{-1}\underline{\underline{E}} = \underline{\underline{U}}^T\underline{\underline{E\Lambda}}^{-1} \Rightarrow \\
&\underline{\underline{\Omega}}^{-2}\underline{\underline{E}} = \underline{\underline{U}}^T\underline{\underline{E}}\left[\underline{\underline{U}}^{-1}\underline{\underline{E}}\right]^{-1} = \underline{\underline{U}}^T\underline{\underline{EE}}^{-1}\underline{\underline{U}} = \underline{\underline{I}}
\end{aligned}\right\} \qquad \text{(eq. 5. 25)}$$

Equation 5.25 implies that for an individual eigen value and the corresponding singular value the following relation applies:

$$e_i = O_i^2 \qquad \text{(eq. 5. 26)}$$

In which e_i is the ith eigen value and O_i is the corresponding singular value. So, obviously it is senseless to perform a singular value as well as an eigen value decomposition, since the two are related by equations 5.25 and 5.26.
The reason for performing SDV on the Jacobean additional to a eigen value decomposition is because numerical effects. In practical cases all matrices involved originate from numerical processes and are numerically inverted, multiplied etc. As was shown by Patel, Laub & Van Dooren (1993) (page 10-11), the rank indicated by eigenvalue decomposition of a matrix $\underline{\underline{B}}=\underline{\underline{A}}^T\underline{\underline{A}}$ may be wrong due to small numerical errors introduced in the multiplication process. The effect of losing essential information is avoided when determining the rank of the matrix A by singular value decomposition.

The covariance matrix is defined by (see Section 5.6.3.2 also):

$$\underline{\underline{\mathrm{cov}}} = \left(\frac{1}{\sigma_h^2}\underline{\underline{J_h^T C_h^{-1} J_h}} + \sum_{i=1}^{i=N}\underline{\underline{C_{p_i}^{-1}}}\right)^{-1} \qquad \text{(eq. 5. 27)}$$

When no prior information is available, the equation 5.27 reduces to:

$$\underline{\underline{\mathrm{cov}}} = \sigma_h^2\left(\underline{\underline{J_h^T C_h^{-1} J_h}}\right)^{-1} \qquad \text{(eq. 5. 28)}$$

This clarifies the effect of adding prior information (the term $\sum_{i=1}^{i=N}\underline{\underline{C_{p_i}^{-1}}}$ in eq. 5.27); it enhances the diagonal terms in the covariance matrix, and therefore it leads to larger singular values. This in turn leads, as was explained, to an increase of identifiability and stability of the optimisation process. This also implies that care should be taken in using prior information. For instance, when adding prior information on a parameter for which the additional observations do not hold any substantial information, the parameter optimisation process seems to work well, but in fact it does only confirm the prior information added to it. Basically such a parameter should be taken as a fixed value, thus decreasing the dimensionality of the optimisation process.
The diagonal terms of the covariance matrix give an indication of the quality of the parameter estimates and should therefore be as small as possible. When parameter estimates are correlated, the corresponding off-diagonal terms give an indication of this correlation. The

diagonal terms are used to calculate a 95% confidence interval for the individual parameter values, this is done by defining the standard deviation factor as $stdfac_{corr} = 10^{\sqrt{cov_{ii}}}$ ([43]).
In which cov_{ii} is the diagonal term of the parameters covariance matrix, the 95% interval is then defined by: $p*stdfac^{-2} < p < p*stdfac^{2}$. Therefore the smaller the individual elements of the covariance matrix are, the better the results of the optimisation.
A far better quantification of the correlation between parameter estimates is obtained when studying the eigen-values and eigen-vectors of the matrix. While the coefficients of the corresponding eigen-vector reveal the components of the linear combination of parameter estimates, a large eigen-value indicates a large uncertainty of this combination of parameters. Using the eigen values and vectors a 95% confidence interval for each optimised parameter combination is computed using $stdfac_{eigen} = 10^{\sqrt{\xi_i}}$, in which ξ_i is the eigen value for parameter i. In cases where no significant covariance terms are present, $stdfac_{corr}$ and $stdfac_{eigen}$ are nearly identical. In Figure 99 the relation between these two intervals is shown for a 2 dimensional case. In the case of covariance between parameters, the intervals found for the two parameters are projections of the ellipse with axes along the two eigen vectors found. The accuracy intervals found using the eigen values are valid only for a linear combination of the eigen vectors. If covariance is absent, the eigen vectors define exactly the same parameter space as is defined by the original parameter axes.

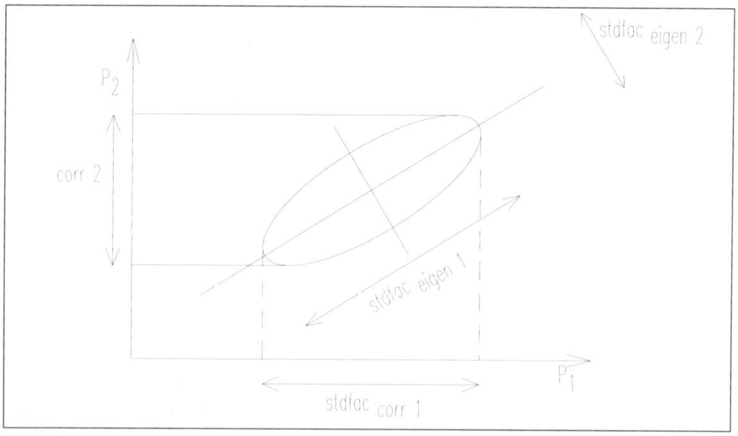

Figure 99: Relation between stdfac $_{eigen}$ and Stdfac $_{corr}$.

5.4 analysis of the residuals

Referring to Figure 98, the residuals resulting from the optimising process are to be analysed. Since one of the basic assumptions introduced in MLE is a Gaussian distribution of the variables, the residuals should be checked on this point. Furthermore the residuals should be randomly distributed in space and time implying that auto-correlation of the residues on one gauging point must be absent. Furthermore, the cross-correlation between residues at distinct gauging points must be absent also. If this is not the case, the cause of it is to be investigated. At this point in the calibration process, the experience and the expertise of the modeller is crucial. Bias is a measure for the existence of systematic errors, as long as bias is significant the results of the optimising process can not be trusted.

([43]) The power relation originates from the application of ^{10}log transformed values in the calibration process.

Bias is defined as:

$$Bias = \frac{1}{T_s}\int_0^{T_s}(\{\frac{1}{n_w}\sum e^2{}_i\}_w - \sigma^2_{e_w})dt \qquad \text{(eq. 5. 29)}$$

$\sigma^2_{e,w}$ is the variance of the residues, the first term in the integrand is the mean squared error, or MSE for short. The subscript w denotes a shifting time window defined by a time interval $[t_0-T, t_0]$, in which T is a characteristic time scale for the system.
Bias should be minimal over every sub-domain of the measuring values, this implies that $(\frac{1}{n_w}\sum e^2{}_i{}_w - \sigma^2_{e_w})$ should be studied over the whole time domain.
Furthermore bias should also be virtually absent in the space domain, which implies that the residues from different measuring locations should shown no correlation in the time domain. In this case, a sub-domain in time must be considered also, since it is possible for a certain time shift to exist between distinct measuring locations.
A measure for an acceptable amount of bias is e.g.

$$|Bias| < \alpha\sigma^2_{e_w}. \qquad \text{(eq. 5. 30)}$$

In which α is a small number (e.g. in the order of magnitude of 10^{-1}). Removal of bias is difficult to catch in general rules, a wrong measuring setup, non-synchronicity of measuring gauges, zero-drift of measuring devices, data errors in the model or model errors in general may be the underlying problem. Recognising and removing them is by far the most time-consuming activity. In this respect, the value of automated parameter-optimisation is highlighted; time spent otherwise in trial and error optimisation may now be used to enhance the quality of the calibration as a whole.

5.5 Shape of the object function

The shape of the object function is an important subject with respect to optimisation. Several minimisation methods demand a purely convex form in order to guarantee finding a minimum. As will be seen in the sections on minimisation methods, when only one minimum exists and the object function is convex in the whole parameter domain, finding that minimum is a relatively trivial task. When however, several local minima exist or the gradient of the goal function with respect to one or more parameters is small serious problems may arise with respect to convergence. In Figure 100 three cases of the shape of the object function are shown for a one-parameter model.

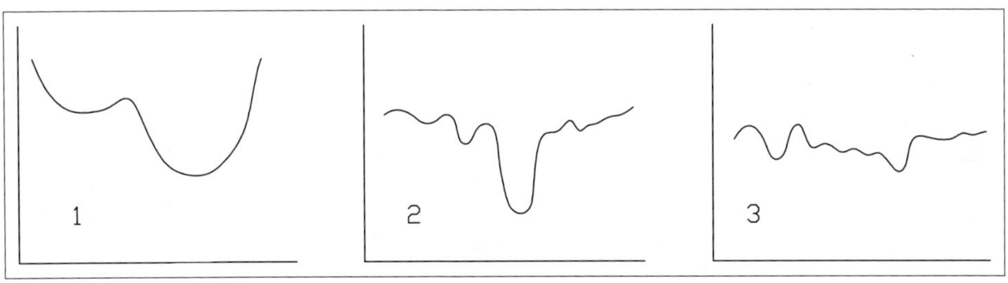

Figure 100: Three possible cases for an object function.

In the first case two minima are present, depending on the initial guess, a gradient based optimisation routine will find either one of the two; using a limited number of starting vectors easily reveals the absolute minimum. In case 2 however, this method would probably ask for a large number of starting vectors, whereas in case 3 the method is practically impossible due to the calculation efforts involved and the time needed for analysing the results obtained.
A poorly defined object function may be due to several causes:

- Poor parameterisation.
- The fact that measuring errors overwhelm the influence of one or more parameters.
- The existence of 'subsystems' in the problem under study having their 'own' minima.

The latter case is suspected to exist in parameter optimisation for urban drainage systems, especially when several catchments are linked to each other by means of e.g. internal overflows or pumping stations.
In practical cases it is not always possible to predict whether a poor shape of the object function will occur or to identify the cause of it. Therefore, the strategy applied for minimisation has to incorporate the handling of function shapes in category 2 and 3.
In those cases it is practical to obtain a more or less clear picture of the shape of the object function prior to starting a precise optimisation routine.
With respect to possible strategies the following methods are distinguished:

- Methods based on generating detailed information with respect to the MLE (and therefore being able to produce information on the derivatives of the object function with respect to the parameters used in the calibration).
- Methods that do not use derivatives (also known as DUD's).
- Methods to handle object functions with several local minima, in order to obtain information on the shape of the object function.

Therefore, some hybrid strategy has to be followed when solving more complicated problems. In Section 5.6 several methods and algorithms to this end are discussed.

5.6 Parameter optimising methods

5.6.1 General

Because the goal function S' as well as the heads are non-linear with respect to the parameters an iterative process is to be used for the actual minimisation. A large body of literature exists on solving algorithms (see e.g. Rustagi (1994), Fletcher (1987) or Luenberger (1973)). In the next sections some methods are briefly described. On the subject of choosing the 'best' methods for a given case several studies have been performed (see e.g. Neuman (1980)). It seems to be virtually impossible to identify the optimal method for a given optimisation problem a priori. In most cases applying several methods in succession (i.e. switching methods during the optimisation process) enhances the convergence of the process as a whole. The main problem is that convergence of several methods depends on the local shape of the goal function. This implies that the performance of a given method depends on the shape of the goal function and on the initial parameter vector chosen. So, for choosing the optimal method the object function should be known in some detail. The main reason for applying optimisation algorithms is to avoid the calculation effort needed to obtain exact knowledge of the goal function. This reasoning results in a situation in which it is hard to obtain a guarantee that any search algorithm will identify the global minimum. For practical purposes, the task is therefore to try to obtain as much information as possible on the object function within the available possibilities (limited by calculation power, deadlines and budgets).

5.6.2 Trial and error

The method of 'trial and error' is probably the one most applied method in urban drainage; an experienced engineer tries to achieve a 'reasonable' fit between his model results and the measuring data acquired. At its best, some kind of strategy is applied with respect to parameterisation and event choice in relation to it. By experience the author finds this method time consuming and hard to report, except for simple cases. This observation is confirmed in literature (see e.g. Foundation RIONED (1992)).

The fundamental drawbacks of trial and error methods are summarised by Cooley & Naff (1990):

- *'No guarantees are present that the best set of parameters is obtained.*
- *It is hard to tell whether this best set has been obtained.*
- *A judgement of the parameterisation can not be given.*
- *The quality of the calibration can not be judged.'*

Apart from these drawbacks, the irreproducibility of this 'calibration process' must be mentioned. Considering the difficulties with trial and error methods, it is concluded that they should not be used in practice, though they still are in use even in recent researches (see e.g. Foundation RIONED et al (1998)).

5.6.3 Methods aimed at obtaining detailed information of the minimum

5.6.3.1 General

In the next subsections, two simple and straightforward gradient search based algorithms are described. There is no doubt that more efficient algorithms exist, though as stated in Section 5.5.1 it is hard to tell a priori what algorithm is optimal for a given case. Therefore, the main condition applied throughout this thesis, a high accessibility by the practising engineer, is taken as a lead in the algorithms described.

5.6.3.2 Levenberg-Marquart method

A straightforward and easy to implement method for minimisation of the object function is the Levenberg-Marquart method (Marquart (1963)).
Assuming σ_h and λ_i constant, the goal function to be minimised is:

$$S' = \underline{e_h}\,\underline{\underline{V_h^{-1}}}\,\underline{e_h^T} + \sum_{i=1}^{i=N} \lambda_{p_i}\,\underline{e_{p_i}}\,\underline{\underline{V_{p_i}^{-1}}}\,\underline{e_{p_i}^T} \qquad \text{(eq. 5. 31)}$$

The minimisation of this goal function is based on a multi-dimensional Gauss-Newton relation:

$$\frac{\partial^2 S'}{\partial \underline{p}^2}\,\underline{\Delta p} = -\frac{\partial S'}{\partial \underline{p}} \qquad \text{(eq. 5. 32)}$$

In which:

$$\frac{\partial S'}{\partial \underline{p}} = 2\underline{\underline{V_h^{-1}}}\,\underline{\underline{J_h^T}}\,\underline{e_h} - \sum_{i=1}^{i=N} 2\lambda_{p_i}\,\underline{\underline{V_{p_i}^{-1}}}\,\underline{e_{p_i}} \qquad \text{(eq. 5. 33)}$$

and:

$$\frac{\partial^2 S'}{\partial \underline{p}^2} = 2\left(\underline{\underline{J_h}}^T \underline{\underline{V_h^{-1}}} \underline{\underline{J_h}} + \underline{\underline{B}}\right) + \sum_{i=1}^{i=N} 2\lambda_{p_i} \underline{\underline{V_{p_i}^{-1}}} \qquad \text{(eq. 5. 34)}$$

$\underline{\underline{J_h}}$ is the Jacobean:

$$\underline{\underline{J_h}} = \begin{bmatrix} \frac{\partial e_{h_1}}{\partial p_1} & \cdot & \cdot & \cdot & \frac{\partial e_{h_1}}{\partial p_n} \\ \cdot & \cdot & \cdot & \cdot & \cdot \\ \cdot & \cdot & \cdot & \cdot & \cdot \\ \cdot & \cdot & \cdot & \cdot & \cdot \\ \frac{\partial e_{h_m}}{\partial p_1} & \cdot & \cdot & \cdot & \frac{\partial e_{h_m}}{\partial p_n} \end{bmatrix} \qquad \text{(eq. 5. 35)}$$

$\underline{\underline{B}}$ contains second derivatives, which become virtually zero at the optimum, so they can be neglected in the numerical process (see e.g. Jenning & McKeown (1992)). It should be noted, however, that when the optimising process has not approached the minimum value of the goal function yet, the resulting parameter values can not be trusted since $\underline{\underline{B}}$ will deviate from $\underline{\underline{0}}$. Here N is the number of parameters is the optimisation process and M is the number of head measurements. In practice the measurements of different parameters $\underline{p}$ and the parameter values themselves posses a certain degree of correlation.

This leads to the following expression for the Gauss-Newton method:

$$\underline{\Delta p} = \left(\underline{\underline{J_h^T V_h^{-1} J_h}} + \sum_{i=1}^{i=N} \lambda_{p_i} \underline{\underline{V_{p_i}^{-1}}} \right)^{-1} \left(-\underline{\underline{J_h^T V_h^{-1}}} \underline{e_h} + \sum_{i=1}^{i=N} \lambda_{p_i} \underline{\underline{V_{p_i}^{-1}}} \underline{e_{p_i}} \right) \qquad \text{(eq. 5. 36)}$$

In practice, the matrix $\left(\underline{\underline{J_h^T V_h^{-1} J_h}} + \sum_{i=1}^{i=N} \lambda_{p_i} \underline{\underline{V_{p_i}^{-1}}} \right)$ may become almost singular which inhibits the use of the straightforward Gauss-Newton method. This can be overcome by adding a positive definite diagonal matrix, resulting in:

$$\underline{\Delta p} = \left(\underline{\underline{J_h^T V_h^{-1} J_h}} + \sum_{i=1}^{i=N} \lambda_{p_i} \underline{\underline{V_{p_i}^{-1}}} + \mu \underline{\underline{I}} \right)^{-1} \left(-\underline{\underline{J_h^T V_h^{-1}}} \underline{e_h} + \sum_{i=1}^{i=N} \lambda_{p_i} \underline{\underline{V_{p_i}^{-1}}} \underline{e_{p_i}} \right) \qquad \text{(eq. 5. 37)}$$

In which μ is a small positive number (e.g. 0.1). This so-called Levenberg-Marquart method is solved using a perturbation algorithm like the one described by e.g. Olsthoorn (1995). Taking μ at a relatively large value, Levenberg-Marquart is in fact a 'steepest decent' algorithm, which is favourable when being far away from the minimum. As the process is in the direct vicinity of the minimum, μ is decreased. The Levenberg-Marquart algorithm mimics ‘steepest decent’ when far from the minimum and mimics Gauss-Newton when near the minimum. In practice, the optimising algorithm uses the log-transformed parameter values, this enhances the shape of the goal function and decreases effects of round-off errors. The stop criterion applied is:

$$\left| \frac{\Delta S'}{S'} \right| < \varepsilon \qquad \text{(eq. 5. 38)}$$

In which ε is any convenient small real number (e.g. 10^{-5}). So far, the proposed algorithm is able to optimise model parameters if there is an a priori value for λ_i.
This method is extended so as to make it possible to optimise the goal function also in cases when the variances are unknown (i.e. unknown values for λ_i). This done is by application of a stage-wise optimisation using different starting values for λ_i. Initially all values are set to 1 and the results of the first parameter optimisation is used to give estimates for the variances:

$$\left.\begin{aligned}\sigma_h^2 &= \left[\underline{e}_h V_h^{-1} \underline{e}_h{}^T\right]\frac{1}{M}\\ \sigma_{p_i}^2 &= \left[\underline{e}_{p_i} \underline{V_{p_i}^{-1}} \underline{e}_{p_i}{}^T\right]\frac{1}{N_i}\end{aligned}\right\} \qquad \text{(eq. 5. 39)}$$

Using these new estimates the goal function is optimised again until a given stop criterion is reached:

$$\left|\frac{\Delta\lambda_i}{\lambda_i}\right| < \varepsilon \text{ for all } \lambda. \qquad \text{(eq. 5. 40)}$$

5.6.3.3 Estimating the Jacobean

In the Levenberg-Marquart method outlined in the previous section, the Jacobean is essential (as it is for many parameter optimisation algorithms, see e.g. Rustagi (1994)). In order to obtain this matrix a finite difference approximation is used, applying the following procedure for each iteration:

- Run the hydrodynamic model using the vector $\underline{p}_i(p_1, ...,p_n)$ resulting in a vector of heads $\underline{h}_i(h_1,..........,h_m)$.
- Then run the model N times, using the vectors $\underline{p}_{ij}(p_i,...p_j+dp,....,p_n)$, resulting in vectors $\underline{h}_{ij}(h_{1j},..........,h_{mj})$.
- The jth row of the Jacobean is obtained by $\dfrac{(h_{ij}-h_i)}{dp}$

This implies that for an N-dimensional parameter space the model has to be run N+1 times. For large catchments, this imposes a practical limitation on the dimensionality of the parameter space. By experience, it takes about 40 iterations for a well-parameterised problem to converge. Considering the acceptable waiting time for a calibration run to be in the order of magnitude of a few days, a search space of 10 parameters is regarded as the practical limit. Another drawback of the method proposed is the approximate character of the Jacobean matrix obtained. The approximation is due to two reasons:

- A finite difference method results by definition (see Chapter 2 also) in an approximation of the 'real' derivative.
- The fact that the numerical accuracy is limited by the accuracy of data passing through the database used by the hydrodynamic model, see Section 5.7.

The latter drawback can effectively be removed when applying an adjoint model (see Chavent (1971)). Without going into detail on this technique, the interested reader is referred to the literature, an adjoint model can be understood as a model 'running in reversed time direction'. In other words, it reconstructs how an actual state (set of water levels or flows) developed from history. In the process the derivatives (i.e. the Jacobean) can be extracted directly from the numerical heart of the adjoint model. This technique has the advantage that the Jacobean becomes available without extra model runs. Adjoint models find wide application in

parameter optimisation, data assimilation and sensitivity analyses of a wide variety of models (see e.g. Cacucci (1981)), Ten Brummelhuis et al (1993) and Mouthaan et al (1994),
From literature, see e.g. Griewank (1989), the adjoint method is known to be economical, in terms of calculation time, for large parameter spaces (i.e. > 20-30 parameters), since the calculation effort is independent of the number of parameters taken into account.
Main drawback of the adjoint modelling technique related to urban drainage is that, to the knowledge of the author, no adjoint model code has been developed so far in this field of engineering. Since building an adjoint code asks for the availability of the source code of the original program (Giering & Kaminski (1989)), ordinary users of commercial software can not, or with great difficulty only, built an adjoint model. When however, model calibration is going to be applied on a larger scale an investigation into the possibilities of adjoint models in urban drainage is to be hoped for.

5.6.3.4 Nelder-Mead algorithm

The Nelder-Mead algorithm (also known as simplex method, Nelder & Mead (1965)) is a beautiful minimisation (or optimisation) method because its functioning is intuitively understood. The algorithm comprises a limited number of moves as shown in Figure 101.

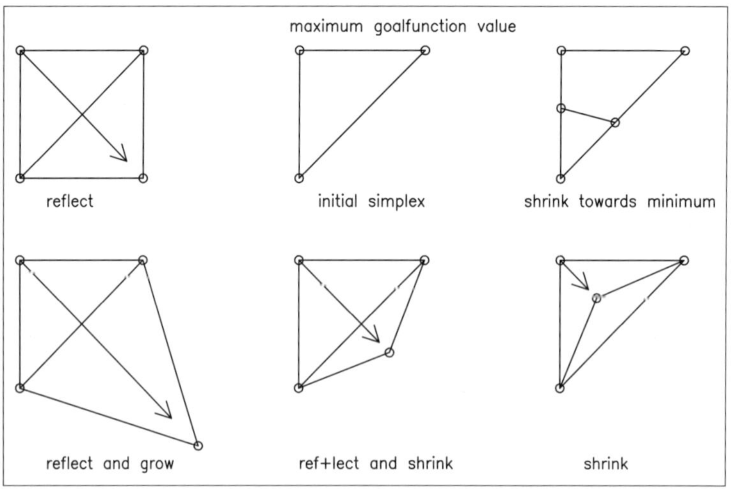

Figure 101: The Nelder-Mead algorithm in a 2-dimensional parameter space.

The algorithm is initiated by choosing a simplex (i.e. a geometrical Figure having one vertex more than dimensions; a triangle in 2D, a tetraedon in 3D etc.).
The goal function f for each of the points in the simplex is calculated and ordered in such a manner that:

$$f(x_1) > f(x_2) > \ldots\ldots > f(x_{D+1}) \qquad \text{(eq. 5. 41)}$$

In which D is the dimensionality of the goal function.
The next step is to calculate the mean value of the vertices except for the one with the highest goal function values. (This is for minimisation, for maximisation the one with the smallest value is left out):

$$x_{mean} = \frac{1}{D}\sum_{i=2}^{i=D+1} x_i \qquad \text{(eq. 5. 42)}$$

A new vector is calculated by 'reflecting' x_1 in x_{mean}:

$$x_{1,new} = 2x_{mean} - x_1 \qquad \text{(eq. 5. 43)}$$

If the goal function for the new vector meets the following constraint:

$$f(x_{1,new}) < f(x_{D+1}) \qquad \text{(eq. 5. 44)}$$

Then the effect of 'reflect and grow' is tried:

$$x_{1,new} = 3x_{mean} - 2x_1 \qquad \text{(eq. 5. 45)}$$

If this leads to an even smaller value for the goal function, this vector is accepted, if not, the result from just reflecting is accepted, and the now worst point in the new simplex is treated.

If after reflecting the new goal function value is still the worst point, this implies the moves made are too large (overshooting the minimum). In this case, the move 'reflect and shrink' is made:

$$x_{1,new} = \frac{3}{2}x_{mean} - \frac{1}{2}x_1 \qquad \text{(eq. 5. 46)}$$

If after this move $f(x_{1,new}) < f(x_2)$, the simplex is updated and the now worst point is updated. If after 'reflect and shrink' $f(x_{1,new}) > f(x_2)$, the move 'shrink' is applied:

$$x_{1,new} = \frac{1}{2}(x_{mean} + \frac{1}{2}x_1) \qquad \text{(eq. 5. 47)}$$

If after shrinking the new vector is still the worst point this implies the measure of the simplex is too large, therefore the simplex is shrunk towards the best point:

$$x_{i,new} = \frac{1}{2}(x_i + x_{D+1}) \qquad \text{(eq. 5. 48)}$$

Then, a new evaluation starts along the sequence described. A stopping criterion is reached if the algorithm keeps shrinking towards the same best point in succession (repetition of eq. 5.48 at the same 'best' point').

5.6.4 Methods to obtain general information on the shape of the object function

5.6.4.1 General

5.6.4.2 Monte Carlo methods

The well known Monte Carlo method (or 'Tour du Wino', see Farlow (1982), section 43) is a method that can be applied for calibration. It is used in urban drainage mainly for sensitivity analysis (see Ummels & Clemens (1998) and Clemens & von der Heide (1999) and Section 3.8 of this thesis). A large number of runs using some hydrodynamic model are made using slightly different starting values for the parameters on which the calibration is done. The

results of every run are compared with the measuring data, after making 'n' runs the combination of parameters having the best 'fit' is chosen.
The method has the advantage of being simple to implement, even for inexperienced users in everyday practice. The, what is called, undirected Monte Carlo method can be used for optimising models with a maximum of five parameters simultaneously. It was shown by Olsthoorn (1998)([44]) that the method fails to be reliable when using a larger number of parameters.
This proof is best illustrated in the 2-dimensional case, the minimum of the goal function is present within a circle with radius r at (p_1,p_2). The Monte Carlo approach will pick randomly parameter combinations positioned in the square with boundaries (p_1,p_2), (p_1+r,p_2), (p_1,p_2+r),(p_1+r,p_2+r). The chance of 'hitting the circle' is proportional to the quotient of the area of the circle and the square

$$p = \frac{\pi r^2}{4r^2} = \frac{1}{4}\pi \qquad \text{(eq. 5. 49)}$$

For higher dimensional cases hyper spheres and hyper cubes evolve, resulting in:

$$p = \frac{V_{globe}}{V_{cube}} = \frac{\pi^{d/2}}{d * 2^{d-1}\,\Gamma(d/2)} \qquad \text{(eq. 5. 50)}$$

In which:

d	Dimensionality of the problem	(-)
Γ(.)	The Gamma function	

For d=5 the value of p becomes circa 0.16, this is a poor result as compared to 0.79 for d=2. Therefore, Monte Carlo methods are less likely to be successful in higher dimensional parameter spaces.
Still, Lei et al (1999), advocate Monte Carlo calibration for hydrodynamic models.
An advantage of the Monte Carlo method, though, is that problems with convergence to an optimum, often encountered using automated optimisation process are absent. In low dimensional problems the Monte Carlo method may be considered, eventually in conjunction with gradient search methods.
The Monte Carlo method can be used in a more intelligent manner; a first investigation of the shape of the goal function being achieved by the undirected method. In this manner the area in which the optimum may be found is limited and a better directed search method can be applied. One way to do this is constructing a meta-model; a hyper plane described by some analytical function that fits the goal function as closely as possible. Such a goal function is more easily accessible for obtaining the global minimum. Main drawback of this approach is that, depending on the shape of the object function, the fitted hyper plane may be only a rough approximation resulting in a poor identification of the global optimum of the real object function.

5.6.4.3 Fuzzy logic and expert systems

The optimisation of the goal function can be treated as a control problem. In real time control problems the use of fuzzy controllers is under development (see e.g. Monsma (1999)). Expert systems have been applied for parameter optimisation problems (see e.g. Baffaut & Delleur (1989 and 1990)). The advantage of these methods is that in a relatively small number of

([44]) A similar proof is found in Gershenfield (1997).

calculations a first estimate is obtained, that can be used as the starting point for the optimisation methods outlined in the previous sections.
Expert systems and fuzzy logic applications render the use of linguistic reasoning in computer programs feasible (Kasabov (1996)). Basically, the reasoning of an expert is translated into a formal reasoning method. This implies that the strategies an expert would use to calibrate a model are transcribed in a computer code.
An expert would, when calibrating a hydrodynamic model in urban drainage, for example, apply rules like:

- *If the calculated level in the DWF period is slightly lower than the measurement then the DWF should be increased a little.*
- *If there is rain and the moment at which a raise in water level occurs in the calculation is earlier than in the measurements, then the surface storage should be increased.*

If the reasoning of the expert is formalised and the linguistic qualification like 'slightly lower than', 'much earlier', 'later' or 'about correct ' are made accessible for quantification, the calibration by 'trial and error' can be implemented in a computer program. In this manner one of the main drawbacks of 'trial and error' is avoided, namely the problem of non-repeatability. Since the results of these technologies depend on the acquisition of much practical experience for building either a good knowledge and rule based (for expert systems) or for the correct implementation of membership functions (fuzzy logic) the portability from one calibration process to another is questionable. Furthermore, applying these techniques no information on the statistical properties of the parameter set obtained becomes available without further analysis.

5.6.4.4 Genetic algorithms

The term 'genetic algorithm' and its underlying principles were introduced in the mid 1970's by Holland (1975). Genetic algorithms, or evolutionary strategies, have found application in a wide field of optimisation problems (see e.g. Quagliarella et al (1998) and Willemsen (2000)), ranging from complicated control problems to the design of airfoils. In this thesis, the ability of GA's to effectively solve multi-dimensional optimisation problems is of main interest.
The basic thought in GA is to mimic the optimisation process as is postulated to exist in living nature; 'survival of the fittest'.
The terms and specific jargon used when working with genetic algorithms are therefore somewhat peculiar comprising terms like 'father, mother, mutate, mate or crossover'. For convenience these terms are adopted, while it is stated that genetic algorithms (GA's, for short) are methods using a kind of statistical trial and error optimisation process.
First, the basic principles of genetic algorithms will briefly be discussed and demonstrated using some examples for the optimisation of a multidimensional function in N-dimensions having a large number of local minima.

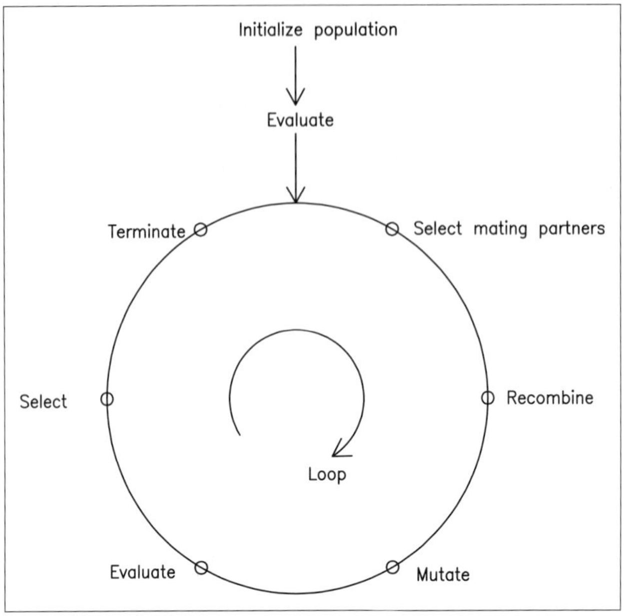

Figure 102: Basic principles for a genetic algorithm.

In Figure 102 the principles used in GA's are shown. The process is initialised by defining a set of parameter combinations called 'individuals' ([45]), for each individual the value of the goal function is evaluated. The goal function is a measure for the 'fitness' and is used in the selection process. Depending on the fitness of each individual, it creates offspring or 'children' by mating with another relatively 'fit' individual. The most 'unfit' individuals are removed from the process and 'die'. During the mating process 'mutations' occurs, meaning that some random changes are introduced in the parameter values composing a 'child'. In addition, the crossover effect can be introduced in the process.
After selection, mating, dying, mutation and crossover, a new population is created for which the 'fitness' is calculated again, initialising a new generation. Eventually the process leads to a set of individuals adapted to the 'environment' defined by the goal function and restrictions put on the allowable parameter space. The effect introduced by mutations and crossover avoids the algorithm to be caught in local minima. This specific feature of GA's makes them an interesting alternative for the optimisation methods described in the preceding sections.
A formal notation for GA's is defined by Rudolph (1994) and is known as the $(\mu,\kappa,\lambda,\rho)$ evolutionary strategy ES:

$$(\mu,\kappa,\lambda,\rho)ES := (P^{(0)},\mu,\kappa,\lambda,rec,p_r,p_m,\rho,\gamma,mut,sel,\zeta,\varepsilon,t) \quad \text{(eq. 5. 51)}$$

In which:

$P^{(0)}$	Start population
rec	Recombinant operator
mut	Mutation operator
sel	Selection operator

([45]) The number of parameters m is equal in each individual,

t	Termination criterion
p_m	Mutation probability
p_r	Recombination probability
ρ	Number of ancestors for one individual
ε	Required accuracy
κ	Upper limit for the life span (at least 1 cycle)
μ	Number of parents for one individual
γ	Number of crossover locations
ζ	Number of tournament participators
λ	Number of children per couple of parents

In this thesis a simplified version, which incorporates the essentials as described by Rudolph (1994) is applied; $(\mu,\rho,\kappa,\lambda)=(2,1,1,1)$, reducing the GA to:

$$ES := (P^{(0)}, rec, p_r, p_m, \gamma, mut, sel, \zeta, ter) \quad \text{(eq. 5. 52)}$$

The initial population $P^{(0)}$ is a set of N vectors (individuals) consisting out of M elements (parameter values). For each element in every vector a random value is picked in the interval $[pmin_i, pmax_i]$ with i=1,...,M. In practice, the parameter intervals are defined by the allowable limits in the software used to calculate the goal function (e.g. in HYDROWORKS® the interval for the linear reservoir constants is in the interval [0,999], any value not being a member of this interval inevitably leads to a 'program crash').

The selection operator

Formally, the *sel* operator is defined as:

$$sel(P^{(i)}) \rightarrow P_i \quad \text{(eq. 5. 53)}$$

In which $P^{(i)}$ is the ith population of individuals and P_i is a set of probabilities for the population to create offspring.
The selection of parents allowed for creating offspring can be effected in various manners.
From the fitness values of each parameter combination, a probability distribution is calculated. For each individual, the probability of generating offspring is calculated. This probability is based on the difference between the objective function values S_i and the minimum objective function value S_{min} in the generation of which an individual is a member. The probability P_i for each individual i to create offspring is given by:

$$P_i(j: j=1,...,N) = \left[\frac{(S_i - S_{\min})}{\sum_{i=1}^{N}(S_i - S_{\min})}\right]^{-1} \Bigg/ \sum_{i=1}^{N}\left\{\left[\frac{(S_i - S_{\min})}{\sum_{i=1}^{N}(S_i - S_{\min})}\right]^{-1}\right\} \quad \text{(eq. 5. 54)}$$

S_{min} is defined as $2S_1$-S_2 to prevent a 100% domination of the S_1 parameters, which provides the best objective function value.
Thus, individuals resulting in function values close to the minimum receive high probability values for being allowed to create offspring, and vice versa.
There are various other methods possible to transform fitness values into a probability value. Another method applied successfully is simply deleting the 'unfittest' portion

individuals, and using the 'fittest' portion to propagate in parameter generation, applying equal chances. In this case the set P_i might look like $\{a_1,......a_{n/2},0_{n/2+1},...0_n\}$, with $a_i \in [0,1]$.

The termination operator

In the algorithm applied, since only one lifecycle for each individual is allowed, the termination operator holds two sub-operators. Suppose the *sel* operator is based on the set $P_i=\{a_1,......a_{n/2},0_{n/2+1},...0_n\}$, then the ter_1 operator is the conjugate of the *sel* operator;

$$ter_1(P^{(i)}) \rightarrow PT^{(i)}; \Rightarrow PP^{(i)} \cup PT^{(i)} = P^{(i)} \qquad \text{(eq. 5. 55)}$$

In which $PP^{(i)}$ is the subset of individuals having a non-zero probability to be selected to create offspring. The second termination operator defines the subset of individuals from generation i that are allowed to be member of the next generation.

$$ter_2(PP^{(i)}) \rightarrow PNG^{(i)}; \qquad \text{(eq. 5. 56)}$$

In the algorithm used, this operator is simple; $PNG^{(i)} = \{0\}$, although other possibilities have been applied successfully; see e.g. Schefel (1995). In some cases allowing $PNG^{(i)}$ to hold individuals introducing some form of 'ancestral memory' in the next generation is beneficial when the objective is to obtain an accurate quantification of the optimum. In the case considered here the objective is merely to find out whether or not local minima are present and if so, to globally identify their position in parameter space.

The recombination operator

The recombination operator defines how the offspring is generated from the individuals $P^{(i)}$ in the previous generation:

$$rec(P^{(i)}) \rightarrow PC^{(i+1)} \qquad \text{(eq. 5. 57)}$$

In the algorithm applied (since $PNG^{(i)} = \{0\}$) this reduces to:

$$rec(P^{(i)}) \rightarrow P^{(i+1)} \qquad \text{(eq. 5. 58)}$$

The *rec* operator is a general recombination-crossover technique, and therefore simply indicated by 'rec'. A new individual is composed from a 'mother' and a 'father', both selected from $P^{(i)}$ by random number generation imposing the set of probabilities $P^{(i)}$ by which the selection chance is given. The 'child' (the new vector) is defined as the randomly weighed average of the parents ([46]):

$$\mathrm{rec}[\mathrm{P}(i,j)] := \mathrm{RAN} * \mathrm{P}(i_{\mathrm{father}}, j) + (1 - \mathrm{RAN}) * \mathrm{P}(i_{\mathrm{mother}}, j) \qquad \text{(eq. 5. 59)}$$

In which RAN $\in [0,1]$. By recombination-crossover, a population of i=1,...,n new individuals, each containing j=1,...,m parameters are generated, together they form $P^{(i+1)}$ (the next generation).

([46]) The value for RAN is picked in the interval [0,1] using a uniform distribution.

The mutation operator

Any GA will result in a process of endlessly copying the 'best' individuals without generating new, better, individuals, unless a mutation operator is applied. After creating a new generation, mutation is imposed at the level of parameter values (individual vector elements) with a selected probability. Two types of mutation are applied; relative and absolute mutation, being imposed in succession. The mutation operator is defined as

$$mut(P^{(i+1)}) := mut_{abs}(P^{(i+1)}) \circ mut_{rel}(P^{(i+1)}) \qquad \text{(eq. 5. 60)}$$

In which mut_{abs} is the 'absolute mutation sub-operator' and mut_{rel} is the 'relative mutation sub-operator'.

Relative mutation is defined as:

$$\text{mut}_{\text{rel}}[\text{P(i, j)}] := (1 + (((\text{RAN} - 0.5) * 2 * \text{MAXMUT})/100)) * \text{P(i, j)} \qquad \text{(eq. 5. 61)}$$

The mutation rate is randomly chosen in the range [0,MAXMUT], given by program input 'MAXMUT' (%).
The 'absolute mutation sub-operator' is applied as:

$$\text{mut}_{\text{abs}}[\text{P(i, j)}] := \text{P(i, j)} + (\text{RAN} - 0.5) * (\text{PMAX(j)} - \text{PMIN(j)}) \qquad \text{(eq. 5. 62)}$$

allowing a mutation dependent on the range of the parameter value. Absolute mutation provides the possibility of moving to a completely different part of the search space of the objective function. Especially when dealing with small parameter values this is essential to escape from local function minima. In case the 'father' and 'mother' are identical vectors the probability for absolute mutation is separately identified by $MAXMUT_{ident}$.

The mutation rate is calculated by a random number generator. Four mutation parameters should be provided as program input:

- Probability of absolute mutation (%).
- Probability of relative mutation (%).
- Maximum mutation rate (%).
- Probability of absolute mutation in case of identical parents (%).

The latter mutation type is added to avoid the generation of exact copies.

In order to illustrate the effect of GA's with respect to functions with several local minima, the Rastingin's function is used:

$$F(x) = \sum_{i=1}^{i=D} (x_i^2 - 10\cos(2\pi x_i) + 10) \qquad \text{(eq. 5. 63)}$$

In which D is the dimensionality of the function. In Figure 103 a contour plot of this function is shown for D=2 in the domain [(-5,5),(-5,5)].

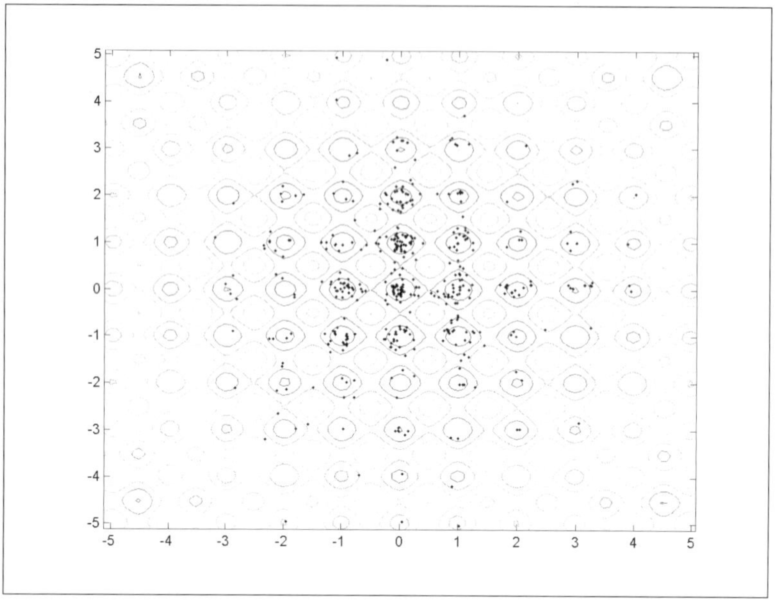

Figure 103: Contour plot of Rastrigin's function and results obtained applying the GA.

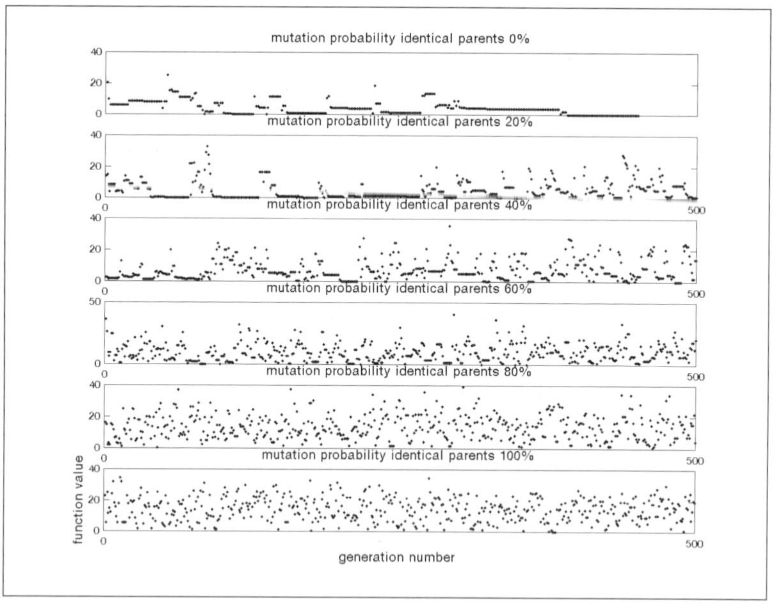

Figure 104: GA results of Rastrigin's function (D=2); effect of mutation chance.

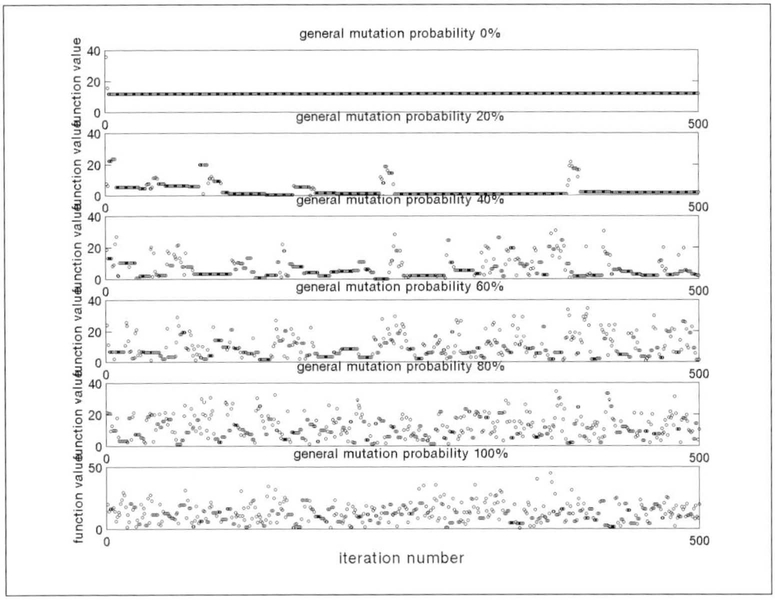

Figure 105: GA results of Rastrigin's function (D=2); effect of mutation chance.

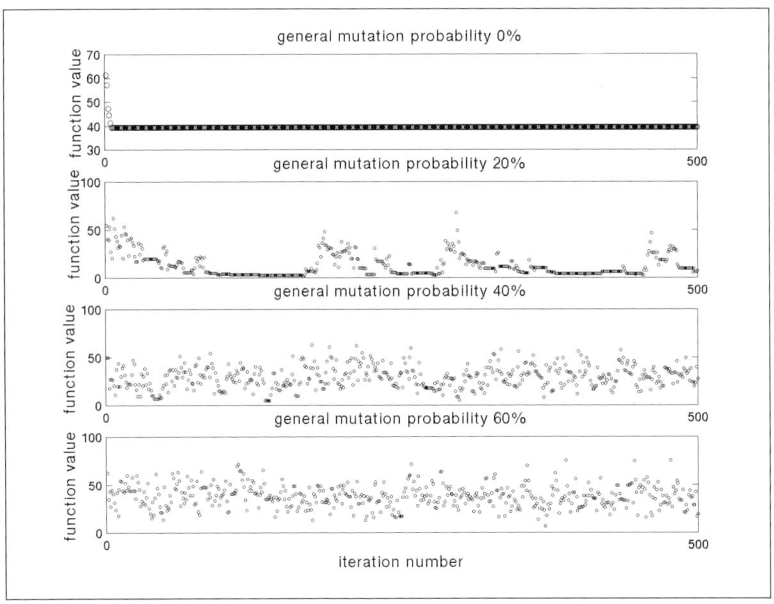

Figure 106: GA results of Rastrigin's function (D=5); effect of mutation chance.

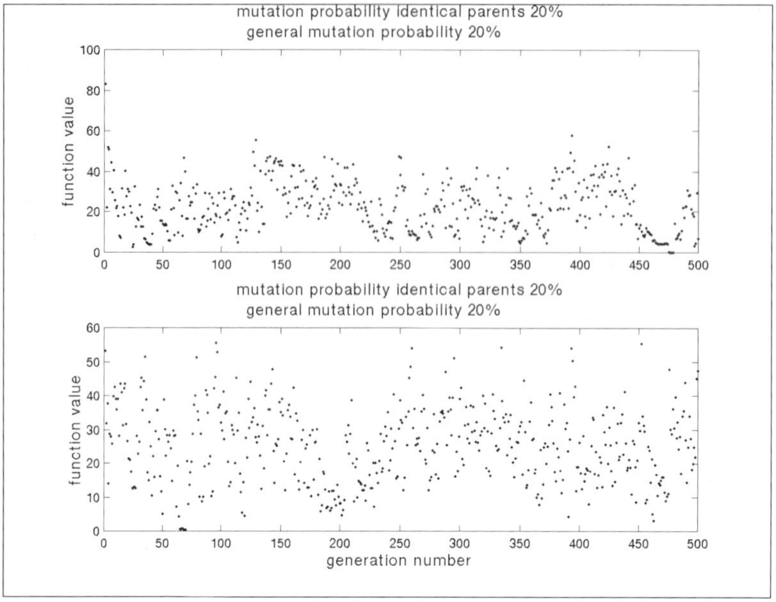

Figure 107: GA results for two runs with identical settings. As can be seen the two processes are different, this is because of the random nature of the algorithm applied.

Test runs on Rastrigin's function have been performed for the 2 and 5 dimensional case (D=2 and D=5). The respective populations applied consist of 5 and 10 individuals. The results are presented in graphs, showing the minimum function value of the population as a function of the iteration number. The mutation probabilities were varied in order to investigate their influence on the GA's behaviour. In this respect the focus is on the ability to trace local optima, rather than exactly identifying the global optimum. Only in the 2-dimensional case the results can be shown by contour plot (Figure 103). Local minima are located at all data points for which (X_1,X_2) are whole numbers, e.g. (0,1), (1,2). The contour plot (Figure 103) clearly shows that most calculated function values are in the direct vicinity of local minima, surrounding the global minimum at (X_1,X_2) = (0,0). The global minimum R=0 is found, in this case, after 75 iteration loops. This means that 5*75 = 375 function evaluations have been done([47]).Figure 104 shows the global effect of an increasing mutation chance in case of identical 'parents' (PIP). The general mutation chance (PRM and PAM) is set at 20%. High mutation chances (i.e. > 40%) lead to a diffuse field of function values, showing a tendency towards the minimum with irregular cycles over the generations. The exact minimum 0,00 at (0,0) is not found, however, in some cases the values found are quite close to this minimum. Furthermore, several other nearby local minima are identified. At a lower mutation probability (i.e. < 20%) numerous exact copies are generated, due to dominating parent values. Figure105 shows the results of an increasing general mutation chance. PIP remains 0%. Thus, the first graph shows a run without any mutation, ending up in a process of endlessly copying the same value, which is definitely not the global minimum. When

([47]) By calculating the function systematically on a partition of 0.1, an amount of almost 10500 function values should be calculated to be sure of the global minimum. A partition of 0.01 results in a number over one million data points to be calculated. Apart from this, the number of data points increases rapidly with the dimensionality of the function.

increasing PRM and PAM, a larger part of the search space is covered resulting in a wider variation of function values. More local minima are found as a result. Figure 106 shows the results of GA-runs using Rastrigin's function (D=5) with a population of 10 individuals. The number of local peaks is over 160.000 (11^5) in the interval $[-5.12..5.12]_{i=1..5}$. From these graphs it is concluded that the probability to generate exact copies decreases with an increasing number of parameters. Small mutation probabilities have a relatively strong impact on variation of function values. The exact value of the global minimum is less likely to be found. Figure 107 shows the results of two runs with identical settings (D=5, (PAM, PRM, PIP)=20%). In the first run, the global minimum is found after about 475 iterations. In the second run the global minimum is found after about 60 iterations. This particular behaviour of GA's prohibits a strict definition of a termination criterion. It is stressed that every individual GA run with *identical* settings will result in a sequence of different generations. This is the result of different randomly picked individuals constituting the initial population and the fact that the whole process is directed by random number generation. From the test runs on Rastrigin's function the following conclusions are drawn:

- The 'optimal' mutation settings give a balance between covering a great part of the total range of the objective function, and identification of the global minimum.
- The 'optimal' mutation settings depend on the number of parameters to be optimised and the objective function itself.
- Identical parents generate copies of function values in the next generation, thus providing no additional information unless mutation is added to the GA. On the other hand, generation of a limited number of copies strengthens the search for (local) optima.
- The 'fitness' values (transformed into a probability of propagation) attributed to the parameter vectors, provide a satisfactory selection procedure. There are various other methods however, to achieve this.

SOFTWARE

In every minimisation method the goal function has to be evaluated, when optimising a model this implies that this model should be forced into a kind of 'master-slave' construction with the optimising algorithm. This means that the optimising algorithm decides which parameters are used in successive model runs. This is schematically shown in Figure 108 **.** Potentially the optimising algorithm can be incorporated in a software tool that also contains the hydrodynamic model, but this thesis separate software tools for optimisation, model evaluation, loop control and I/O were used. The minimisation methods outlined in Section 5.6 are implemented in a computer program. The actual optimisation program is totally independent from the hydrodynamic model used; the only link between them is a special I/O format. The optimisation routine can be used for whatever type of computational model. Actually the code was debugged and tested on a simple numerical model for the dispersion of salt in a continuous medium.

The program CALSEW has an output in which the following information is made available:

- The value of the object function and the actual parameters, per iteration.
- Per iteration, a file in which the statistics of the optimisation are shown.

The user can decide on which optimisation algorithm to put into effect, in this manner during the optimisation process it is easy to switch from one algorithm to another.

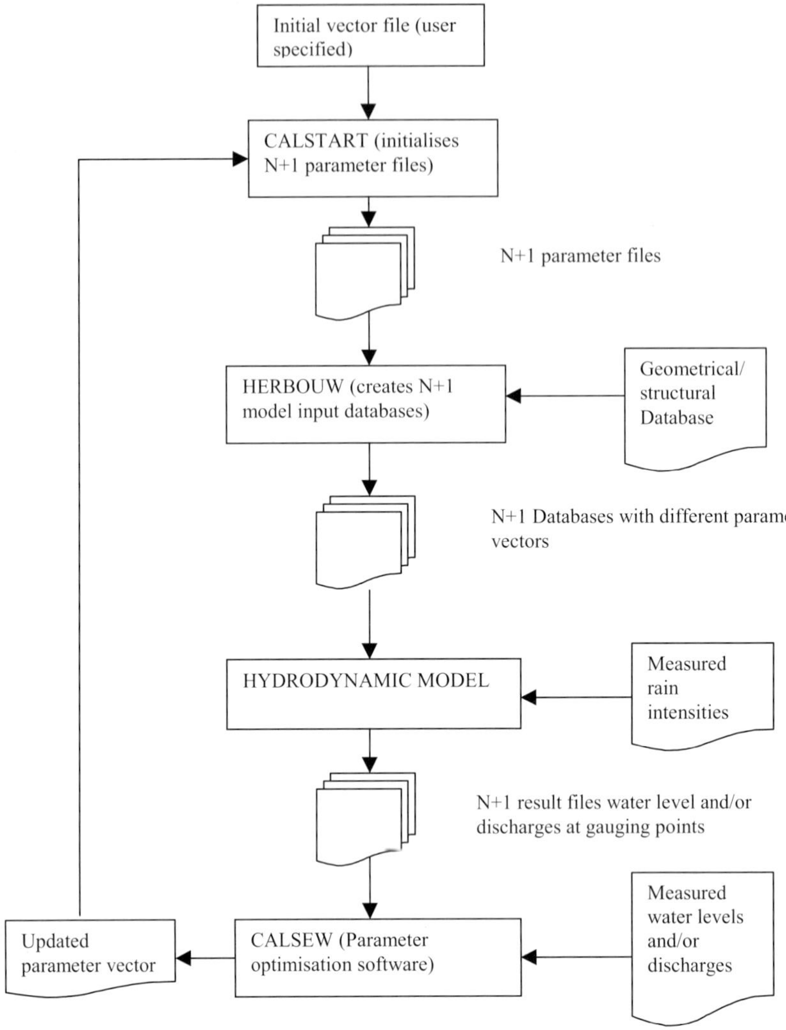

Figure 108 : Relations between the software tools applied in calibration

The program HERBOUW translates the updated parameter values into databases used by the hydrodynamic program. The program CALSTART initialises the optimisation run, the user defines the number of iterations and the starting vector from which the calibration starts and a termination criterion. All programs are written in FORTRAN 90 code to assure a high degree of portability. Because a large number of matrix operations and ^{10}log transformations are used in the algorithms, the use of double precision calculus is a necessity to avoid accuracy loss and the occurrence of matrix singularity during the calculation process. A practical problem to be dealt with is the occurrence of truncation errors between optimising algorithm and hydrodynamic program. The optimising algorithm determines in each iteration a new model parameter vector $\underline{p}_i$ up to say 10 decimals, the database format used by the hydrodynamic model can handle only up to 2 or 3 decimals. This implies that the actual parameter vector

$\underline{p}_{i,actual}$ differs slightly from the initial vector $\underline{p}_i$. This is of importance with respect to the determination of the Jacobean, the step size vector $\underline{dp}_i$ used in the finite difference formula is not exactly equal to δ_i but is determined by

$$\underline{dp}_i = \underline{p}_{i-1} - \underline{p}_i \qquad \text{(eq. 5. 64)}$$

The overall management in order to get the several programs working in the right mode was done using a macro-recorder ([48]). The further analysis and presentation in graphs has been done using general-purpose software like spreadsheets and MATLAB®.

5.8 Artificial examples applying the Levenberg-Marquart method

5.8.1 General

In this section a number of artificial examples is given using different minimisation methods. The examples are used to show the several aspects associated with the MLE theory and the strategies followed in a calibration process. In addition, the use of artificial simulations for the preparation of measuring campaigns is demonstrated. In order to get artificial measuring data the hydrodynamic model was run with known parameters, the calculation results from a selection of objects were stored (these objects represent the artificial gauging points in the system). In order to avoid singularity in matrix calculations and making these artificial measuring data more realistic, white noise was added to them. This was done using a random generator (Press et al (1992)).

5.8.2 Artificial example 'DIAMOND'

5.8.2.1 General

This simple network 'DIAMOND' is one of the standard networks that are supplied with HydroWorks®. In its essence it is a hydraulic analogy of the famous electrical circuit used to measure electrical resistance (the well-known bridge of Wheatstone). In Figure 109 the network is shown. The 'measuring data' were obtained by making one run with known parameters for the overflow coefficient C and the hydraulic roughness k_n. The roughness is chosen because it is known that the model is relatively insensitive for variation in this parameter, therefore the use of this parameter is regarded as a good test case.
Using the DIAMOND three initial parameter combinations are applied in the optimisation process. In these runs perfect 'measuring' data are used, σ_h was assumed to be known, in fact a generalised least squares method was applied:

$$S'' = \underline{e_h}\,\underline{\underline{V_h^{-1}}}\,\underline{e_h^T} \qquad \text{(eq. 5. 65)}$$

([48]) Actually, the Windows Marco scheduler 5.0 supplied by MJT Net Ltd. was used, obtainable at www.MJTnet.com.

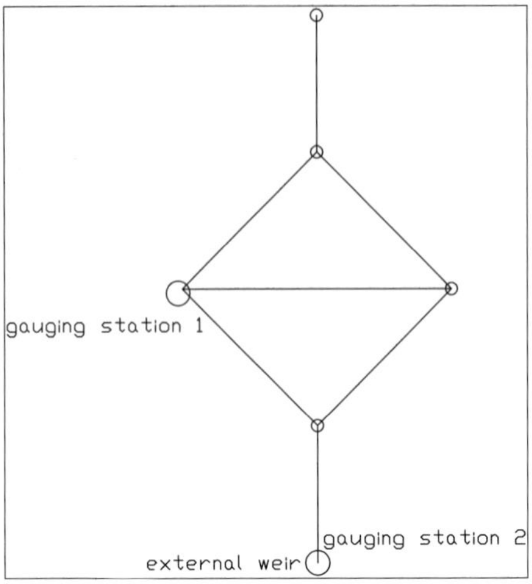

Figure 109: Lay-out 'DIAMOND'.

Three different initial vectors $p_i(C,k_n)$: $p_0(0.5,5.0)$, $p_1(5.0,15.0)$ and $p_2(0.2,15.0)$ are used. The 'real' values are C=2.0 $m^{0.5}/s$ and k_n=10.0 mm. The results of the optimisation are summarised in Table 38.

	End value C	End value k_n	Iterations
Run 1 (p_0)	2.00	9.99	335
Run 2 (p_1)	2.00	10.22	470
Run 3 (p_2)	2.00	10.00	231

Table 38: Result parameter optimisation for the 'DIAMOND', runs 1, 2 and 3.

Using the same vectors p_0 and introducing a 'measuring error' with $\sigma=5*10^{-3}$ m and mean value 0, run 4 through 6 are made. The value for σ_h was assumed to be unknown and was initially set to a value of 0.01 m. the results are summarised in Table 39.

	End value C	End value k_n	Iterations
Run 4 (p_0)	2.01	6.35-8.37	133
Run 5 (p_1)	2.02	10.23	207
Run 6 (p_2)	2.02	10.21	178

Table 39 : Result parameter optimisation for the 'DIAMOND', runs 4,5 and 6.

In runs 7,8 and 9 prior information was added for both parameters, the MLE-method was here implemented in its full extent, meaning that the variances for the heads as well as for the prior estimates were taken into account in the optimisation process. The results are shown in Table 40.

	End value C	End value k_n	Prior estimate C	Prior estimate k_n	Optimum value for λ_C	Optimum value for λ_{kn}
Run 7 (p_2)	2.02	9.99	1.50	9.99	$3.34*10^{-4}$	$5.68*10^{10}$
Run 8 (p_2)	2.00	9.13	1.95	8.00	$3.55*10^{-2}$	$1.73*10^{-3}$
Run 9 (p_2)	2.03	9.59	1.20	11.00	$1.11*10^{-4}$	$1.62*10^{-3}$

Table 40: Result parameter optimisation for the 'DIAMOND', runs 7, 8 and 9.

5.8.2.2 Analysis results using the 'DIAMOND'

Run 1,2 and 3 are used to validate the implementation of the Levenberg-Marquart algorithm, judged by its capability to reveal the 'true' parameter values. After 130 iterations the exact values are found to within 2 decimals. The value for the hydraulic roughness takes far more iterations to be estimated correctly, i.e. C has reached its correct value in about 10 iterations, but it takes at least 130 iterations to reach a good approximation for k_n. This is perfectly understandable since the calculated water levels are relatively insensitive with respect to the value of k_n at the velocities occurring ($v < 0.5$ m/s). The starting vector in run 4 proved to lead to instability in the optimisation routine, several trails were made resulting in correct values for C but completely different values for k_n. Furthermore, it can be seen that the 'measuring error' leads to a slight deviation in the parameter estimates when compared with the results of run 1,2 and 3.

Gauge	MSE	VAR	BIAS	BIAS/VAR
1	$2.73 * 10^{-5}$	$2.74 * 10^{-5}$	$1.57 * 10^{-7}$	0.00375
2	$3.24 * 10^{-5}$	$3.20 * 10^{-5}$	$-1.40 * 10^{-6}$	-0.0437
Overall	$2.99 * 10^{-5}$	$2.98 * 10^{-5}$	$-6.19 * 10^{-7}$	-0.0210

Table 41: residual analysis 'DIAMOND', run 6.

	C	K_n	$Stdfact_{corr}$
C	1,0000	0,0216	1,0110
k_n	0,0216	1,0000	1,0757

Table 42 Correlation matrix 'DIAMOND', run 6.

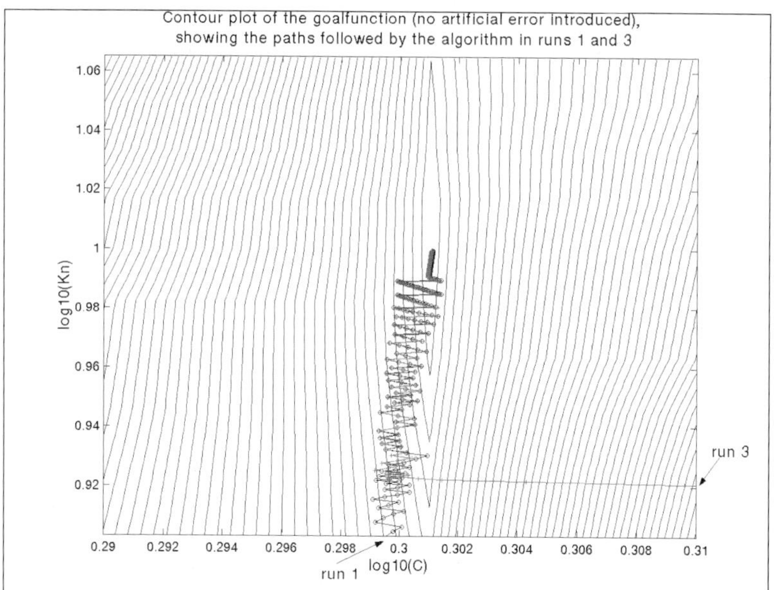

Figure 110: Contour plot of the ideal goal function. As can be seen, the algorithm tends to show wiggles close to the minimum. These can effectively be suppressed by imposing a filtering technique, e.g. a Moving-Average filter during the optimisation.

	Eigen value	$Stdfac_{eigen}$	Singular value	Relative condition number	Prior estimate
C	$2{,}7099 * 10^{-5}$	1,0120	192,096	1,00	-
k_n	$1{,}022 * 10^{-3}$	1,0764	31,278	0,16	-

Table 43: Analysis eigen values run 6

The results for run 5 and 6 are identical; therefore only the analysis for run 6 is presented. The eigen vectors and eigen values reveal that there exists certain interdependency (however small) between the parameter estimates. This is easily explained from a hydraulic point of view: in a certain sub domain of the measuring data, both parameters have an influence on the measured levels simultaneously. (This is illustrated in Figure 111 in which the elements of the Jacobean are shown).

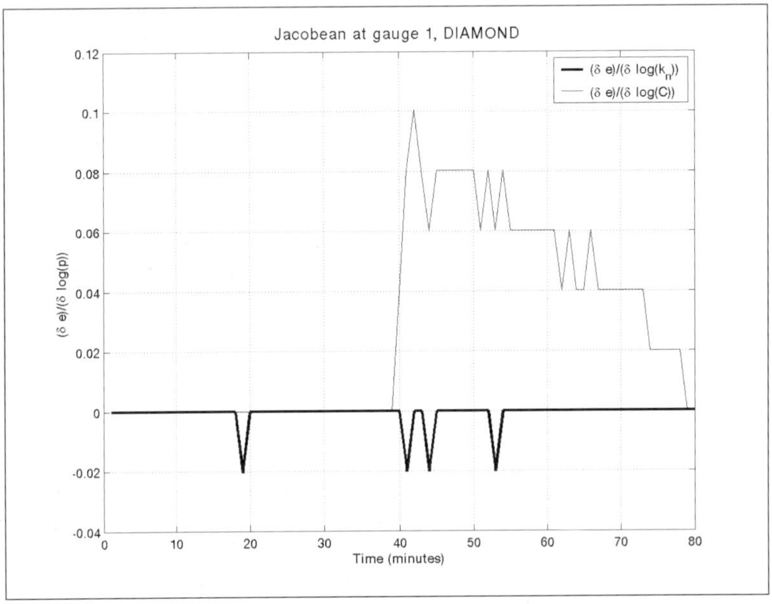

Figure 111: Jacobean gauge 1, the DIAMOND.

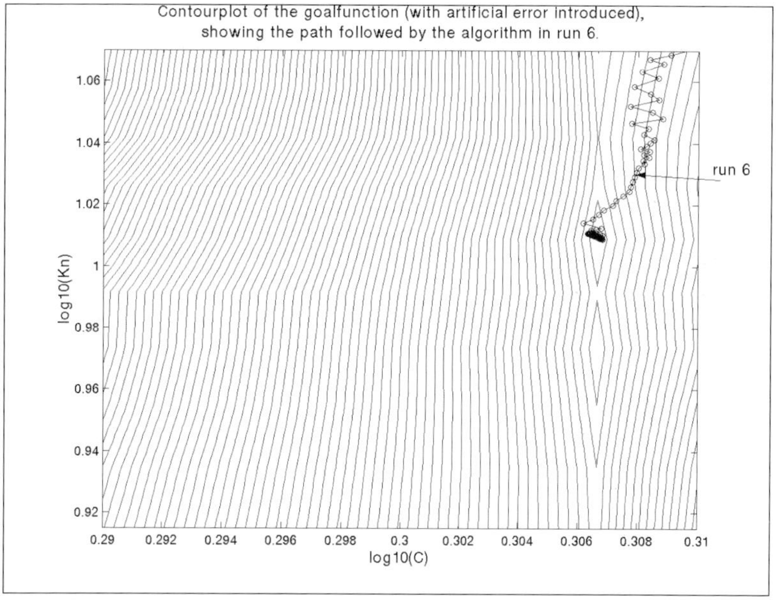

Figure 112: Contour plot of the goal function with the addition of an artificial 'measuring' error. As can be seen, measuring errors result in the occurrence of extra local minima.

Run 7 ,8 and 9

In the runs 7,8 and 9 prior knowledge was added to the optimising process.
The effects can be seen especially with respect to the results for k_n. In run 7 in which the prior knowledge matches the 'true' value for k_n almost perfectly, the singular value increased tremendously when compared to run 5 and 6 (compare Tables 7 and 11). The correlation between the parameter estimates is reduced to almost zero (see Table 14) meaning that the identifiabilty of the two parameters is perfect. In runs 8 and 9 the prior estimates did not match the 'true' values, the influence of the prior information on the results is reduced (compared to run 7) quantified by the values for λ_C and λ_{kn} (see Table 4).
This fact can be used to track discrepancies in the measuring data. If the prior knowledge does not match the information gained from head measurements either the head measurements are not correct or the prior information is not correct. In Figure 113 the probability distribution of the residuals is shown, it complies to a fair extent with the assumption to be Gaussian.

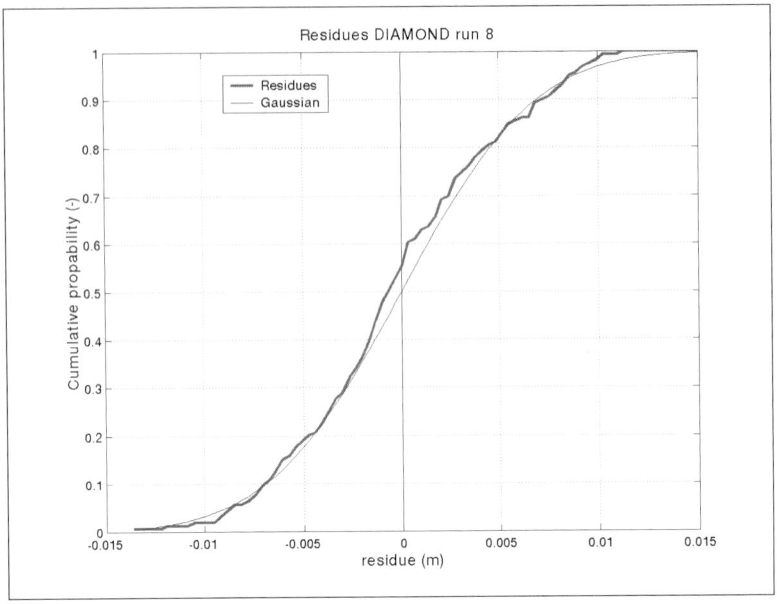

Figure 113: Probability distribution for the residues in run 8, as can be seen the distribution is close to the theoretical Gaussian distribution.

Figure 148 shows MSE, VAR and BIAS as a function of time in a shifting window of 300 s. The value of BIAS is over the whole time domain significantly smaller than VAR and MSE, implying no systematic errors are present.

Gauge	MSE	VAR	BIAS	BIAS/VAR
1	$2.79*10^{-5}$	$2.82*10^{-5}$	$-7.17*10^{-8}$	-0.00254
2	$3.12*10^{-5}$	$3.16*10^{-5}$	$-1.52*10^{-6}$	-0.04804
Overall	$2.96*10^{-5}$	$2.97*10^{-5}$	$-1.59*10^{-6}$	-0.05336

Table 44: residual analysis DIAMOND run 7

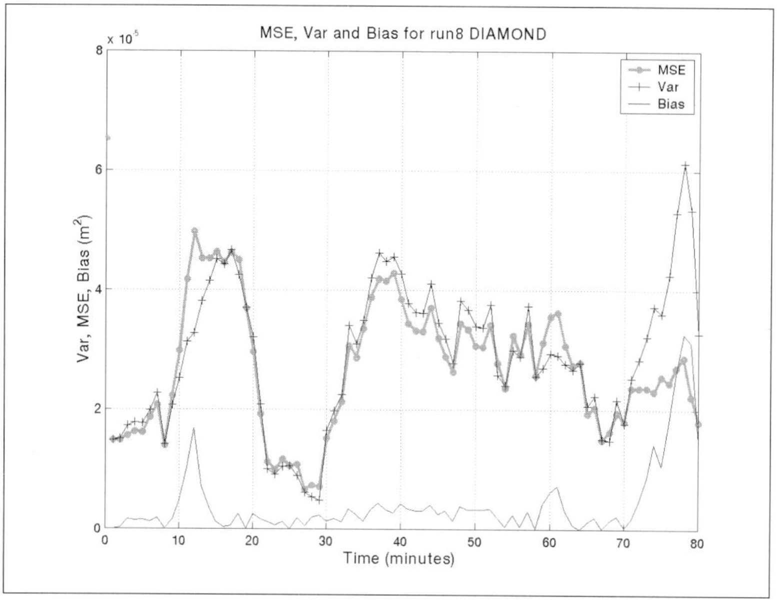

Figure 114: Mean squared error (MSE) variation (VAR) and bias in a shifting time window.

Gauge	MSE	VAR	BIAS	BIAS/VAR
1	$2.73*10^{-5}$	$2.74*10^{-5}$	$-3.41*10^{-8}$	-0.00124
2	$3.09*10^{-5}$	$3.13*10^{-5}$	$-1.79*10^{-6}$	-0.05713
Overall	$2.91*10^{-5}$	$2.92*10^{-5}$	$-1.82*10^{-6}$	-0.06236

Table 45: residual analysis DIAMOND run 8.

Gauge	MSE	VAR	BIAS	BIAS/VAR
1	$2.73*10^{-5}$	$2.74*10^{-5}$	$-3.41*10^{-8}$	-0.00124
2	$3.09*10^{-5}$	$3.13*10^{-5}$	$-1.79*10^{-6}$	-0.05713
Overall	$2.91*10^{-5}$	$2.92*10^{-5}$	$-1.82*10^{-6}$	-0.06236

Table 46: residual analysis DIAMOND run 9.

	Eigen value	$Stdfac_{eigen}$	Singular value	Relative condition number	Prior estimate
C	$2,674*10^{-5}$	1.0120	193,361	$2,004*10^{-4}$	1.50
k_n	$1,074*10^{-12}$	1.0000024	$9,64857*10^{5}$	1.00	9.99

Table 47: Analysis eigen values run 7.

	Eigen value	Stdfac	Singular value	Relative condition number	Prior estimate
C	$2,183*10^{-4}$	1,0108	214,02	1,00	1,95
k_n	$7,15*10^{-5}$	1,0635	37,398	0,17	8,00

Table 48: Analysis eigen values run 8.

	Eigen value	$Stdfac_{eigen}$	Singular value	Relative condition number	Prior estimate
C	$2,175*10^{-5}$		214,403	1,00	1,20
k_n	$8,786*10^{-4}$		33,735	0,157	11,00

Table 49: Analysis eigen values run 9.

	C	K_n	$Stdfact_{corr}$
C	1.000	$5.964*10^{-15}$	1.0120
k_n	$5.964*10^{-15}$	1.000	1.0000

Table 50: Correlation matrix, run 7.

	C	K_n	$Stdfact_{corr}$
C	1.00000	0.00777	1,0109
k_n	0.00777	1.00000	1,0635

Table 51 Correlation matrix, run 8

	C	K_n	$Stdfact_{corr}$
C	1.00000	0.01842	1,0109
k_n	0.01842	1.00000	1,0706

Table 52 Correlation matrix, run 9

5.8.3 Artificial example 'De Hoven'

5.8.3.1 General

In Figure 115 the second urban drainage network is shown, it contains three external weirs and one pumping station. The results obtained at the 'gauging points' are used as 'measuring data' for the calibration.

- In the example, it is assumed that the three overflow coefficients are to be calibrated.
- All other information, like geometry, runoff etc, is assumed to be exactly known.

In order to decluster the measuring data in the time dimension a characteristic time T =300 s is used. A measuring error with mean value 0 and σ_h=0.005 m was added to the measuring data. Four measuring set-ups were simulated using 'De Hoven'

- 1: gauges 1,2,3 and 4
- 2: gauges 1,2,3, and 4 introducing a synchronicity error at station 2
- 3: gauges 1,3,4 and 5
- 4: gauges 6 and 7

The 'real' values were C_1=0.50, C_1=0.60, C_1=0.70, three starting vectors are used: (2.00,2.00,2.00), (0.50,0.50,0.50) and (5.00,0.50,5.00).

Parameter	'real' values	Sit 1	sit 2	sit 3	sit 4
C_1	0.50	0.50	0.46	0.50	0.44-0.61
C_2	0.60	0.59	0.79	0.59	1.00-0.29
C_3	0.70	0.70	0.69	0.70	0.67-0.72

Table 53: results for the different initial situations defined for 'De Hoven'.

In situation 2 a synthetic synchronicity error was introduced; the 'measuring data' for gauging point 2 were shifted 120 s forward in time. In situation 3 the gauge 2 was cancelled and replaced by gauge 5 at some distance of the weir construction. In situation 4, only two gauges are present. As can be seen from the results in situation 3 the 'true' values for the overflow coefficients were obtained implying that moving one gauging station does not significantly reduce the information gained from the measurements. When looking at the results from situation 4 it becomes clear that in this case the information gained from the measurements is not good enough to reach unique (correct) results, this implies the goal function has several local minima. In Figure 116 the paths followed by the optimisation algorithm are shown for situation 1.

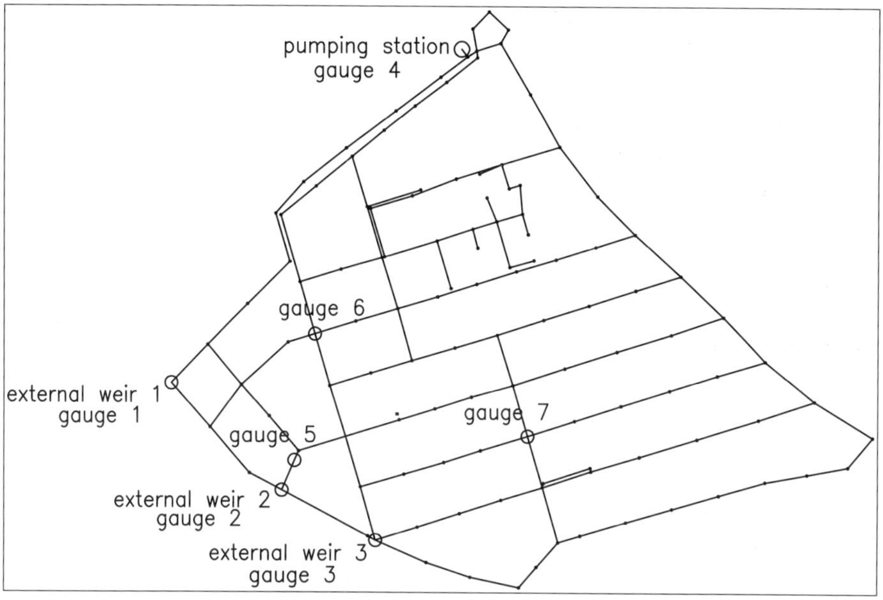

Figure 115: Catchment 'De Hoven'.

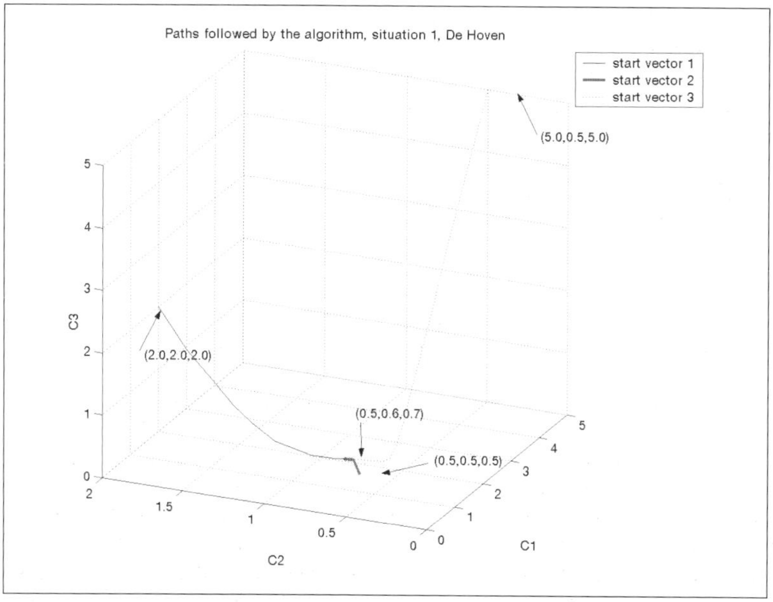

Figure 116: Paths followed by the Levenberg-Marquart algorithm for situation 1, De Hoven.

5.8.3.2 Analysis results using 'De Hoven'

The results in situation 1 using the three starting vectors were identical. In Figure 117 the probability distribution of the residuals is shown, as can be seen it matches the normal distribution to a fair extent. In Table 54 the mean squared error (MSE), the variance (VAR) and the bias (BIAS) is reported for every gauging station.

Gauge	MSE	VAR	BIAS	BIAS/VAR
1	$2.92*10^{-5}$	$2.94*10^{-5}$	$-1.33*10^{-6}$	-0.045
2	$2.57*10^{-5}$	$2.77*10^{-5}$	$-1.80*10^{-6}$	-0.065
3	$2.19*10^{-5}$	$2.17*10^{-5}$	$1.95*10^{-7}$	0.009
4	$1.87*10^{-5}$	$1.81*10^{-5}$	$8.85*10^{-8}$	0.005
Overall	$2.43*10^{-5}$	$2.43*10^{-5}$	$-2.85*10^{-6}$	-0.117

Table 54: Residual analysis 'De Hoven' situation 1.

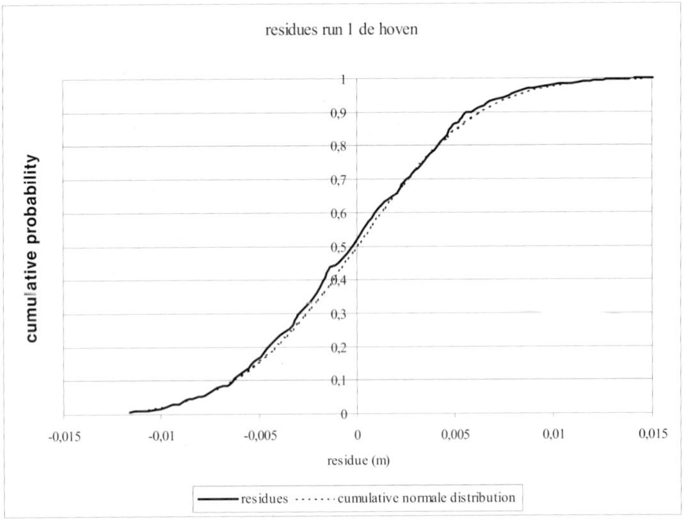

Figure 117: Residues for run 1 'de Hoven'.

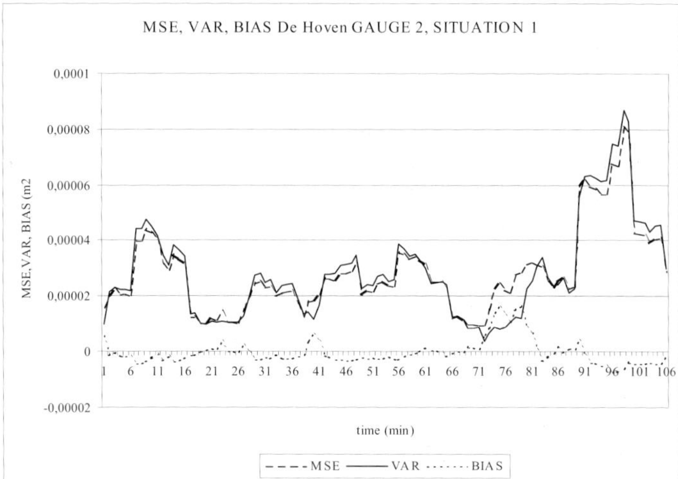

Figure 118: Bias at gauge 2, situation 1, bias is limited and in this artificial case only due to numerical processes applied.

As can be seen in Table 55 there is a relatively strong negative correlation between the parameter values. This is perfectly understandable from a hydraulic point of view; if one of the overflow coefficients is increased, the others must be decreased in order to avoid violation of the mass-balance during the overflow period. C_1, C_2 and C_2, C_3 have a correlation of -0.447 and -0.459 respectively, while the correlation between C_1 and C_3 is -0.195. Apparently, the correlation is stronger between weirs that are located closely together. This implies that implementing an a priori guess for the correlation between measured heads in the space dimension in order to decluster this information will enhance the results (i.e. this will result in a smaller correlation between different parameter values.)

	C_1	C_2	C_3	$Stdfac_{corr}$
C_1	1.000	-0.447	-0.195	1.00784
C_2	-0.447	1.000	-0.459	1.01792
C_3	-0.195	-0.459	1.000	1.00626

Table 55: Correlation matrix 'De Hoven' situation 1.

	C_1	C_2	C_3	Eigen value	Singular value	$Stdfac_{eigen}$
C_1	1.0000	-0.2187	0.6496	$1.1487*10^{-5}$	295.052	1.00783
C_2	0.1071	1.0000	0.3057	$6.3605*10^{-5}$	125.387	1.01853
C_3	-0.6824	-0.1635	1.0000	$3.2251*10^{-6}$	554.578	1.00416

Table 56: Eigen vectors situation 1 'De Hoven'.

In situation 2, a systematic error was introduced at gauging station 2. The results for the different starting vectors were identical. The effect of non-synchronicity becomes apparent when analysing the residuals, especially when looking at the bias at gauging station 2. In Figure 119 the probability distribution of the residuals is shown, it is clear it deviates substantially from Gaussian. In Figure 120 the bias as a function of time at gauge 2 is shown, a significant maximum is found. This highlights the need for a close inspection of the residues, when looking at the overall figures in situation 2, no large discrepancies become clear at first sight. The effect on the parameter estimates is seen in all overflow coefficients, however the effect on overflow coefficient 2 is the largest (see Table 53).

Gauge	MSE	VAR	BIAS	BIAS/VAR
1	$3.36*10^{-5}$	$3.22*10^{-5}$	$4.96*10^{-6}$	0.154
2	$3.34*10^{-3}$	$3.17*10^{-3}$	$1.69*10^{-3}$	0.535
3	$2.71*10^{-5}$	$2.71*10^{-5}$	$6.79*10^{-6}$	0.250
4	$1.98*10^{-5}$	$2.02*10^{-5}$	$2.68*10^{-6}$	0.133
Overall	$2.44*10^{-5}$	$2.45*10^{-5}$	$1.71*10^{-3}$	69.89

Table 57: Residual analysis De Hoven situation 2.

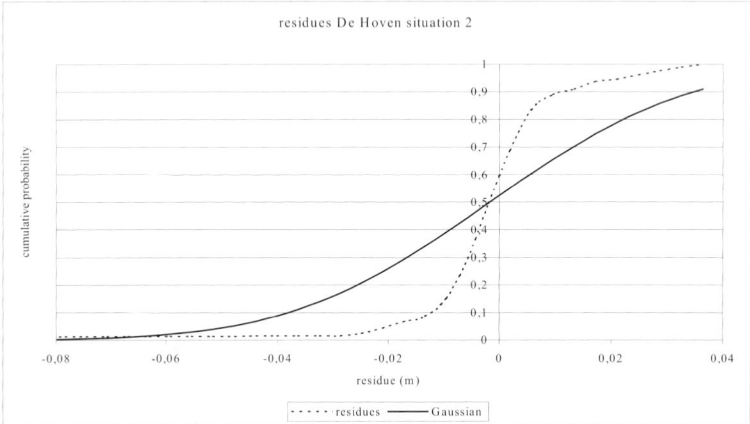

Figure 119: Probability distribution situation 2, as can be seen a serious deviation exists between the obtained distribution and the Theoretical Gaussian distribution. Along with the bias this is a sign some kind of systematic error is present (i.c. a time shift).

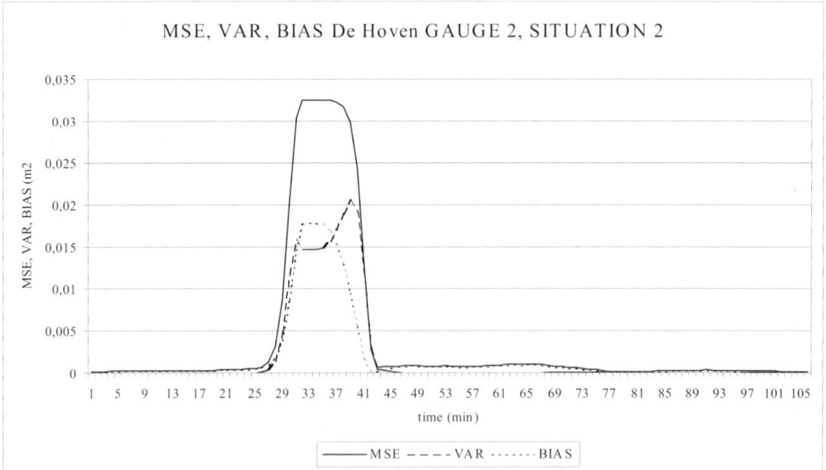

Figure 120: Bias at gauge 2 (see also Table 57).

The measuring set-up in situation 3 is judged tot be practically identical to situation 1. (Compare Tables 16 and 20, 17 and 21 and 18 and 22.) This implies that if for some practical reason no measuring devices could be installed in the overflow construction no. 2, gauge station 5 is a good alternative.

Table 58: Residual analysis De Hoven situation 3

Gauge	MSE	VAR	BIAS	BIAS/VAR
1	$2.557*10^{-5}$	$2.579*10^{-5}$	$-9.210*10^{-7}$	-0.0357
5	$2.543*10^{-5}$	$2.508*10^{-5}$	$7.020*10^{-7}$	0.02799
3	$1.925*10^{-5}$	$1.894*10^{-5}$	$7.901*10^{-7}$	0.04172
4	$2.789*10^{-5}$	$2.812*10^{-5}$	$-1.341*10^{-7}$	0.00478
Overall	$2.454*10^{-5}$	$2.449*10^{-5}$	$4.370*10^{-7}$	0.01784

Table 59: Correlation matrix 'De Hoven' situation 3

	C_1	C_2	C_3	$Stdfac_{corr}$
C_1	1.000	-0.493	-0.174	1.008
C_2	-0.493	1.000	-0.448	1.019
C_3	-0.174	-0.448	1.000	1.008

Table 60: Eigen vectors situation 3 'de Hoven'

	C_1	C_2	C_3	Eigen value	Singular value	$Stdfac_{eigen}$
C_1	1.000	-0.218	-0.891	$4.402*10^{-6}$	476.633	1.00484
C_2	0.403	1.000	0.014	$7.527*10^{-5}$	115.261	1.02017
C_3	0.885	-0.209	1.000	$1.475*10^{-5}$	260.336	1.00888

In situation 4 however, it becomes clear that the information gained from only two gauging stations at a relatively large distance from the overflow constructions is not good enough to obtain correct results. This is revealed when using different starting vectors leading to different solutions. Probably the goal function has several minima, when studying the residuals no aberrations are found (see Table 61). This demonstrates that in order to judge the quality of the results all aspects are to be taken into account:

- Uniqueness of the results by exploring the goal function for local minima using several starting vectors.
- Identifiability: by inspecting the correlation matrix and the properties of the eigen vectors and values of this matrix.
- A check for the statistical properties of the residues: a posterior check whether or not the initial assumptions are justified (i.e. the residues showing a Gaussian distribution).

Gauge	MSE	VAR	BIAS	BIAS/VAR
6	$2.107*10^{-5}$	$2.091*10^{-5}$	$1.089*10^{-6}$	0.0521
7	$2.017*10^{-5}$	$3.033*10^{-5}$	$-1.194*10^{-6}$	-0.0394
Overall	$2.562*10^{-5}$	$2.553*10^{-5}$	$-1.040*10^{-7}$	-0.0041

Table 61: Residual analysis 'De Hoven' situation 4, two gauging points at distance of the external weirs.

5.8.3.3 Example of an auto-calibration

In this section the concept of 'auto-calibration' is explained and demonstrated by means of a historical storm event as also discussed in Chapter 6. The catchment is 'De Hoven' having gauging points 1,2 and 3 only (this is in line with the actual situation that evolved in the case study as discussed in chapter 6.) In order to judge the possibility for such a measuring setup to obtain enough information for a high dimensional parameter optimisation the following procedure was followed:

- Perform a model calculation using known model parameters.
- Extract the calculated water levels for the gauging point from the output.
- Add white noise to the results, so as to mimic 'measured' results.
- Start the parameter optimisation process using some optimisation algorithm with an initial parameter vector that deviates from the known parameter vector.
- Judge the ability of the algorithm to reproduce the known parameter vector.
- Judge the statistical aspects of the results obtained.

If the conclusion from this procedure is that problems occur with either identifiability, uniqueness or stability, the measuring setup has to be rejected for the parameterisation chosen. In other words, the measuring setup does not produce enough information to obtain a good calibration result. It must be remembered that this is no guarantee that in the fully practical case things will work out correctly; in the auto calibration no systematic errors are present with respect to data base or process modelling.

parameter	symbol	'real' value	initial 1	result 1	initial 2	result 2
DWF	N 1	45	65	46.0	40	46.0
routing 1	F 1	100	50	74	70	86
infiltration 1	I 1	0.5	1.0	0.49	0.3	0.47
storage 1	B 1	1.5	1.0	1.59	2.0	1.55
routing 2	F 2	200	100	144	150	239
infiltration 2	I 2	0.1	0.2	0.13	0.2	0.16
storage 2	B 2	0.2	0.4	0.39	0.3	0.29

Table 62: Results auto-calibration.

In any case; if the auto-calibration does not work, it will certainly not work in practice. To this end, the storm event measured on September 2 1998 was used. This event did not lead to an external overflow situation. The parameters to be calibrated are the (constant) infiltration rate, the surface storage and the linear reservoir constant for two types of surfaces and the amount of DWF. The results of two initial vectors are presented in Table 62.
As is immediately seen, the results for N1, I1,B1, I2 and B2 result in values relatively close to the 'real values'. The results for the routing constant, however deviate substantially from the 'real' values.
From the ranking of the parameters, it is clear that the parameter values associated with surface type 2 are less identifiable than those associated with surface type 1. The eigen vector with the highest ranking is determined by the parameters N1 and I1 (DWF and infiltration rate for surface type 1 respectively). As can be seen in Figure 122, the eigen vector has a negative component in N1 direction and a positive component in I1 direction. This implies that the residues are sensitive when the DWF is decreased and at the same time the infiltration rate is increased (both parameter changes result in a decreases of the total inflow). On the other hand, when the parameter variation is done along a vector orthogonal to the eigen vector the total inflow does not change, resulting in hardly any effect on the residuals. Therefore, it must be concluded that without further information it is impossible to identify the DWF and the infiltration rate for surface type 1 separately.

parameter	$Stdfac_{corr}$	$Stdfac_{eigen}$	Singular value	relative condition number	rank
N1	1.0172	1.0182	127.76	0.090	2
F 1	1.1739	1.1749	14.29	0.010	7
I 1	1.0105	1.0106	1420.48	1.000	1
B 1	1.0376	1.0316	73.99	0.052	3
I 2	1.0527	1.0424	55.45	0.039	4
B 2	1.0588	1.0607	39.06	0.027	5
F 2	1.0602	1.0679	35.04	0.025	6

Table 63: 95% intervals for the parameters in the auto-calibration. The ranking of the parameters is done based on the eigen vectors, the largest component of eigen vector i indicates which parameter is predominant.

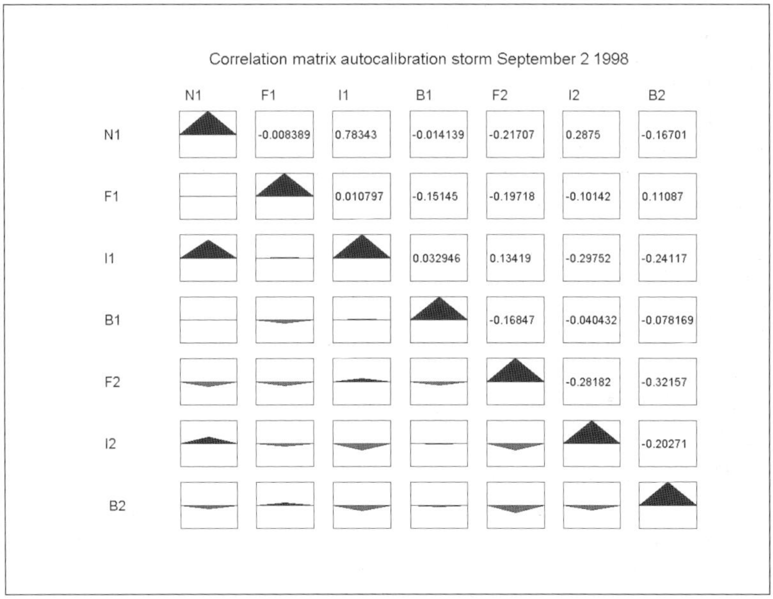

Figure 121: Correlation matrix autocalibration. As can be seen the two parameters N1 (DWF) and the infiltration rate for surface type 1 show a large positive correlation. This positive correlation is easily understood; a decrease of the infiltration rate results in an increase of run-off, in order to keep the residues small this is compensated by an increased value for the DWF. The other correlations are insignificant.

This is an important observation, especially when, for some reason, the DWF has to be known with a high accuracy. In such a case, a separate measurement is needed. On the other hand, the eigen vector with singular value 127.0832 forces an opposite line of reasoning. This is explained by the fact that the DWF is working on the system during the whole time domain taken into account, whereas the infiltration is only of importance in the period in which rain occurs. Imposing larger values for DWF and I1 in the model will therefore result in an increased bias in the residues. This becomes apparent during the period after the rain has stopped when the system is being emptied gradually by the pumping station. Due to the incorrect ratio between DWF and pumping capacity, the system is emptied too slowly, resulting in a systematic error between measurements and model results.

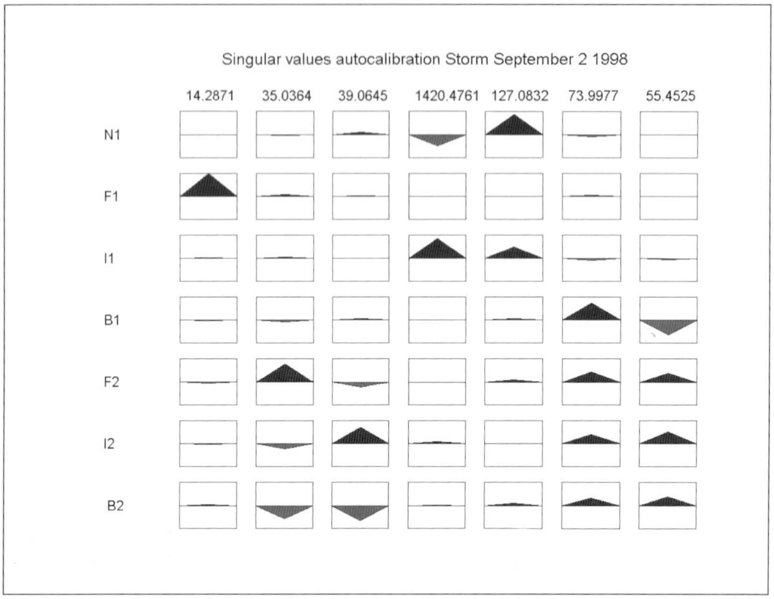

Figure 122: Eigen vectors and singular values for the auto calibration using storm on September 2 1998. As can be seen, the eigen vector with the largest singular value shows that the parameters N1 (DWF) and I1 (infiltration rate for surface type 1) work in opposite directions. This is perfectly logical from the physical process underlying these parameters influence. This effect seems to be somewhat reduced by the eigen vector with singular value 127.0832. Furthermore, the eigen value with the smallest singular value (14.2871) is almost entirely dominated by parameter F1. This implies that N1 and I1 cannot be identified separately and that the value for F1 can only be determined with a large inaccuracy range. Generally, the parameters associated with surface type 2 are less identifiable than those referring to surface type 1 are.

As can be seen, the residuals show no correlation in time or in place, and bias is at an acceptable low level (see Figure 123 and Figure 124), implying that the residuals comply with the demand put on them. Yet, it is still doubtful if the measuring setup is capable of obtaining enough information since the values found using two different starting vectors are for some parameters different (like the routing constants). This doubt is further strengthened when studying the eigen vectors (see Figure 122). It may be possible that some local minima exist in the object function, which has not been investigated in this case.

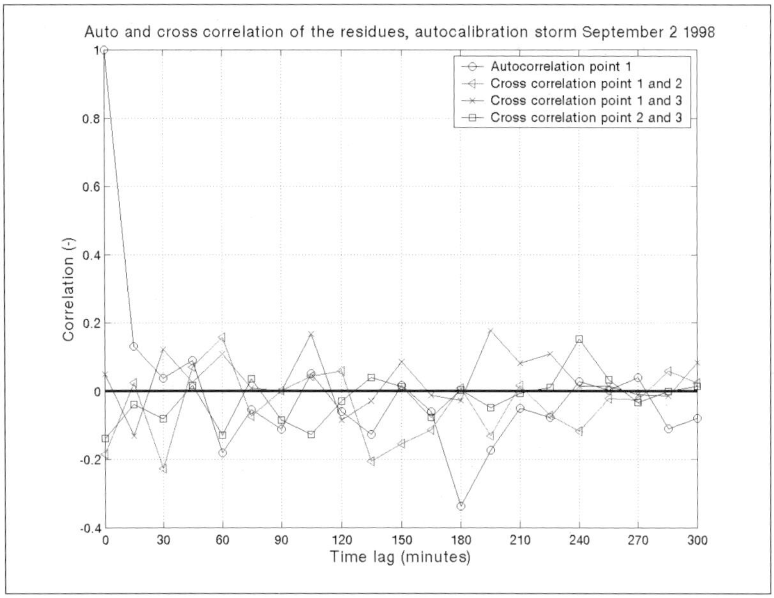

Figure 123: Auto-correlation function for gauging point 1 and the cross correlation function for the residues. As can be seen there is no correlation either in time or in space, indicating the residues are white noise.

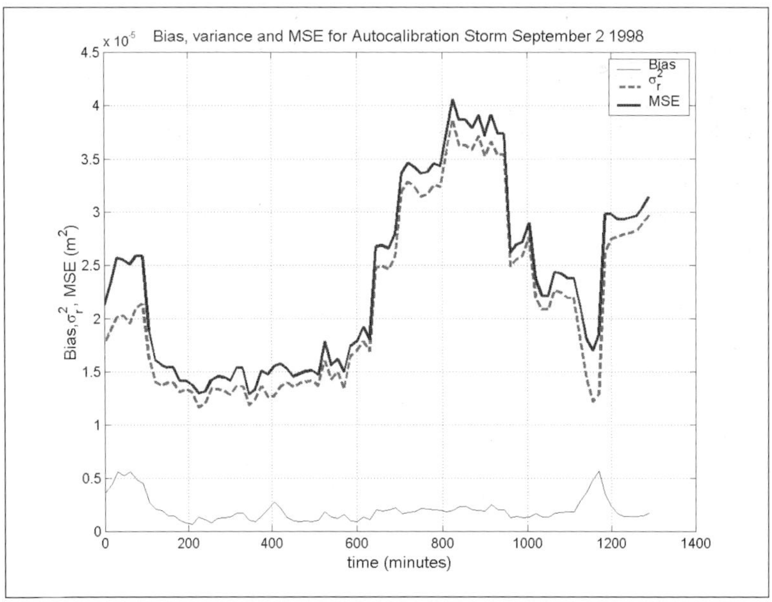

Figure 124: Bias, variance and mse for gauging point 1, auto calibration storm September 2 1998. As can be seen bias is at a low level throughout the whole time domain.

5.9 Artificial examples using the Nelder-Mead Algorithm

In order to demonstrate the Nelder-Mead algorithm (or Simplex- algorithm) it is applied to the artificial example for the 'DIAMOND given in Section 5.8. The discussion is kept limited because it is hard to make a clear general distinction between the benefits and drawbacks of this algorithm as compare with e.g. the Levenberg-Marquart algorithm.
For the purpose of demonstration the runs 4,5 and 6 as discussed in Section 5.7 are presented. using an artificial measured signal at which white noise was added with $\sigma_h=5*10^{-3}$m.

	start value C	start value k_n	End value C	End value k_n	Iterations
Run 4	0.5	5.0	1.88	1.11	47
Run 5	5.0	15.0	2.00	9.22	95
Run 6	0.2	15.0	1.99	9.21	100

Table 64: Results of runs 4,5 and 6.

As can be seen the Nelder-Mead algorithm needs less iterations to come to a halt when compared to the Levenberg-Marquart algorithm, but it isvulnerable for slow convergence in locations with small gradients. Furthermore, it has the disadvantage that in order to obtain statistical information on the parameter values found, additional calculations are necessary. Nevertheless, the algorithm is useful when used in conjunction with other algorithms, especially in domains in which the object function shows relatively large gradients.
Figure 125 shows the paths followed by the algorithm (run 6).

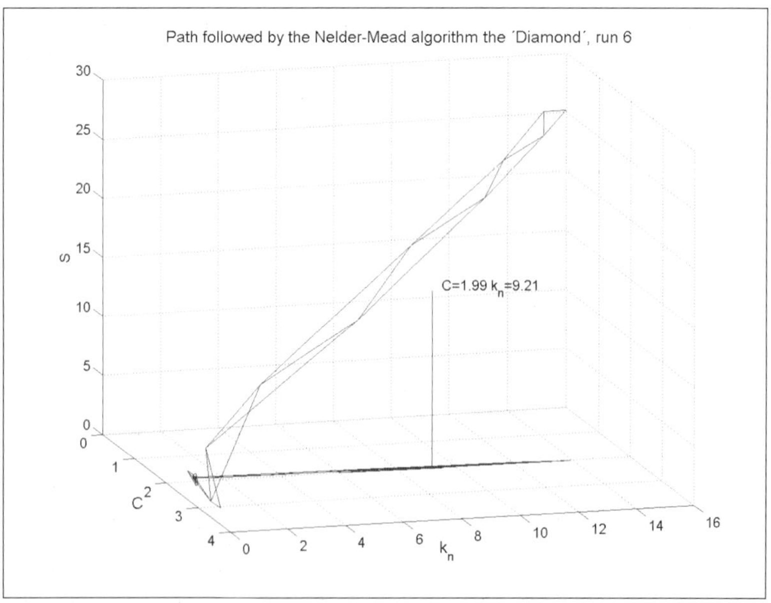

Figure 125: Path followed by the Nelder-Mead algorithm in run 6 for the DIAMOND. As can be seen the goal function has a narrow 'ally' at C=2.00 over the whole domain for k_n, which makes it difficult to focus in on the correct value for k_n. This is a direct consequence of the model being relatively insensitive for variation in this parameter.

5.10 Artificial examples using a genetic algorithm

5.10.1 General

As outlined genetic algorithms are an interesting alternative when dealing with a goal function having several local minima. A first example (application on an analytical function having 11^5 local minima of which one is the absolute minimum) was discussed already in paragraph 5.6.4.4. As a second example the results obtained using the artificial example of Section 5.7 (the DIAMOND) in which two minima in the goal function were identified are discussed. The genetic algorithm is compared with the Levenberg-Marquart method and the Nelder-Mead algorithm with respect to:

- Information obtained.
- The potential to identify the minimum in the goal function.

For the example using the 'DIAMOND' in which two local minima in the goal function evolved results of a genetic algorithm in this particular case are shown.

5.10.2 The Diamond

The results obtained for the artificial case for the 'DIAMOND' using the GA as described in the Section 5.6.4.4. Each generation consisted out of 10 individuals.
The 'real' values are C=2.0 and k_n=10.0. In order to create some distortion in the objective function, a 'measuring error' with σ=0,005 m and mean value 0 was introduced. The optimisation results are shown in Table 65. In total 181 generations were used, as can be seen after 11 generations the best parameter combination was found.

Generation number	Objective function value S	CC	k_n
1 (initial values)	-639.082	1.6002	3.9196
4	-2113.66	2.0000	9.0031
12	-2113.41	1.9860	9.0507
166	-2117.591	1.9975	9.7737
'real' values		2.0	10.0

Table 65: optimisation results for the 'DIAMOND'

Figure 126 and Figure 127 show the results obtained compared with those of the Levenberg-Marquart algorithm, as can be seen the GA results in the global location of the local minima. Using the algorithm described in Section 5.10.4, two local minima were identified.
The relative insensitivity for the roughness value results in a 'string' of points along the value of $C \approx 2$ (see Figure 128). The measuring error can not distract the optimisation routine from finding values, close to the 'real' value of C. The values for k_n show a far larger variation, and higher densities of points are found at k_n values less than 10. This may be an indication for the existence of local minima, introduced by, e.g., a measuring error.

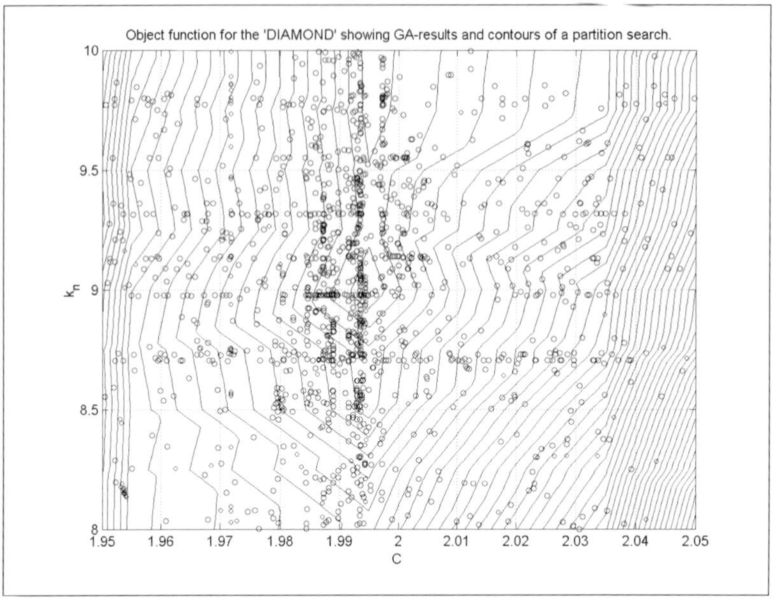

Figure 126: results GA , as can be seen both local minima are identified.

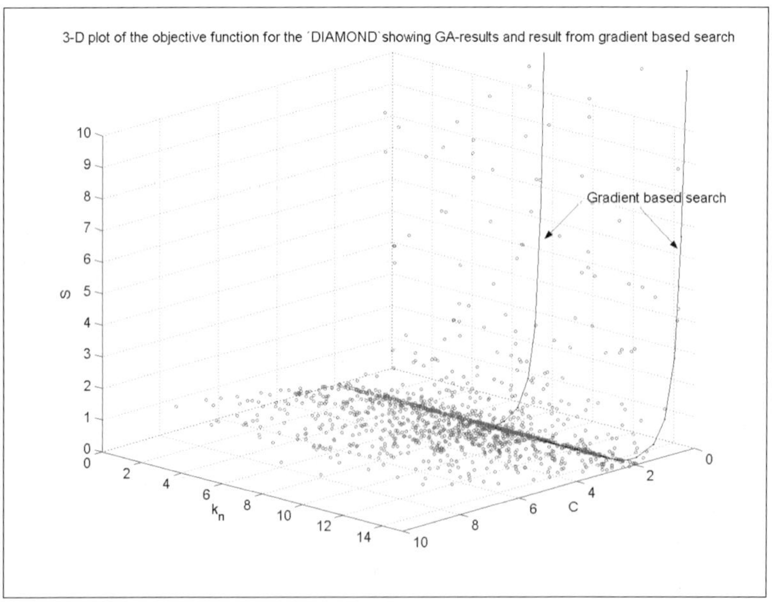

Figure 127: Results GA computation and gradient based search.

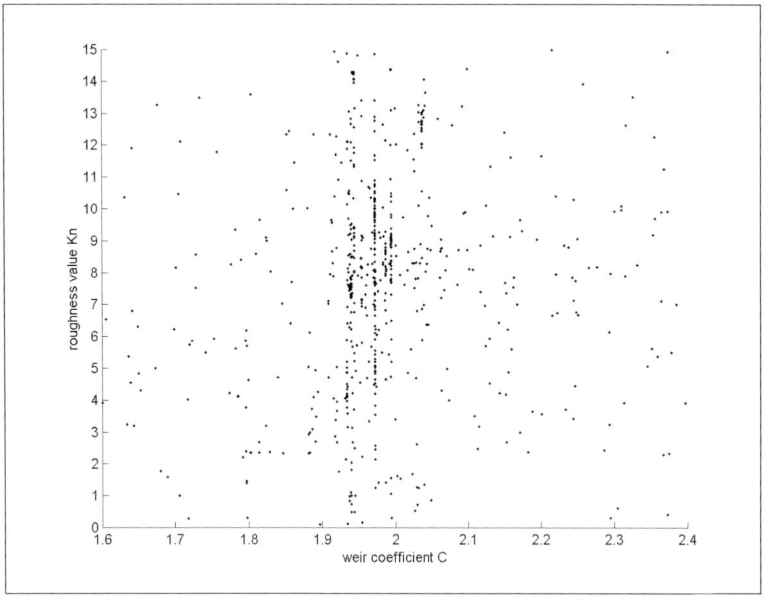

Figure 128: Results of the GA. As can be seen a pattern of lines parallel on the k_n axis evolve.

5.10.3 De Hoven

The genetic algorithm has been applied to an artificial model calibration procedure, called an 'auto-calibration'. The model to be calibrated is a small sewer system for urban drainage (De Hoven, Deventer, the Netherlands), which is processed by the hydrodynamic software HydroWorks®. The calibration parameters are the weir coefficients of the three overflow structures in the system. The auto-calibration procedure comprises the following steps when applying a GA:

1. Model simulation using defined weir coefficients;
2. Generating time series of water levels in 'measuring' locations close to the weirs. These simulated water levels are considered 'measured' values;
3. Initialisation of calibration procedure by generating n input files with different sets of randomly generated weir coefficients;
4. Running n model simulations and generate n time series of water levels;
5. Determination of n goal function values based on the difference between simulated and 'measured' values;
6. Running GA procedure and generation of n new sets of weir coefficients;
7. Build n input files with new weir coefficients;
8. Continue from 4;

If the auto-calibration succeeds, the known weir coefficients of step 1 should be found or values close to them. In this particular case a population of 10 was applied, while 3 parameters were varied being the 3 weir coefficients. The weir coefficients have a value between 0 and 10.
The coefficients linked to the 5 lowest function values S are shown in Table 66.

Iteration number	Objective function value S	Cc1	Cc2	Cc3
1(initial values)	38693.25	1.4232	0.2371	7.4004
104	-3685.43	0.6985	0.5892	0.5133
173	-3684.95	0.6997	0.6132	0.4919
172	-3683.99	0.7011	0.6208	0.5033
174	-3678.63	0.6917	0.6095	0.4874
238	-3670.07	0.6992	0.6509	0.4959
'real' values		0.7000	0.6000	0.5000

Table 66: Results of auto-calibration of 3 weir coefficients

The shape of objective function S cannot, within reason, be determined, as it would take extremely long computation times. For example, using a spatial partition of 0.1 in the interval [0,10] for three coefficients, 1 million simulation runs should be made.

5.10.4 Identifying local minima

The result of a GA computation is a set of vectors $(x_1,\ldots,x_m,S)$ in which $x_1,\ldots x_m$ are the parameter values and S is the corresponding function value.
It is possible to calculate the chance that a certain vector is a local minimum. Suppose there are N vectors, a vector is stated to be a (local) minimum when its function value is less than its k neighbours, further supposing every vector has a unique function value, a ranking is created such that:
$R_s=v_1,v_2,\ldots v_N$ in which v_1 has the smallest function value and v_N has the largest function value.
The probability that a vector v_i is a local minimum, is calculated from (see Whitley et al (1998)):

$$P_i = \frac{\binom{N-i}{k}}{\binom{N-1}{k}}, \text{ and } [1 \le i \le (N-k)]. \qquad \text{(eq. 5. 66)}$$

One could use this information as a criterion which vectors are candidates for further investigation, in practice though it is difficult to define a limit for P_i above which such an investigation is to be performed. The following procedure for identification of local minima is applied:
A vector v_i in search space is considered a local minimum when the following conditions are met:

- In a hyper sphere with radius r in the ^{10}log-transformed search space at least 2D([49]) points are present (this is excluding the point v_i itself);
- For every individual parameter axis at least one vector v_k must be present for which $x_k-x_i>0$ and one point m for which $x_m-x_i<0$;
- The function value belonging to x_i is the smallest of all points in the hyper sphere.

The ^{10}log-transformation must be imposed to obtain an effective scaling for parameters that differ in order of magnitude.

([49]) D is the dimensionality of the search space.

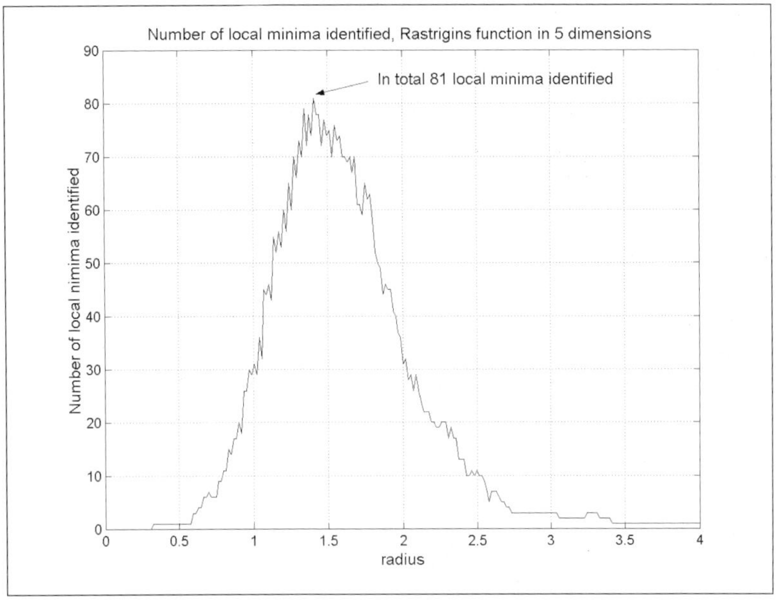

Figure 129: The number of identified local minima in Rastrigin's function in 5 dimensions, eventually 81 local minima were identified.

This criterion is applied on each individual result from the GA for a range of radii. For small radii no local minima are found, with increasing radius a variation in the number of minima is seen, while above a certain value only the global minimum is left fulfilling all the criteria set. It is well known that the results obtained by this procedure improve with an increasing number of simulation runs; if the density of results in the direct vicinity of a local minimum is too small, the minimum will not be identified. For two-dimensional cases the location of local minima can be judged by visual inspection (see e.g. Figure 126 which shows the results for the DIAMOND using a GA, Section 5.10.2). For the sake of demonstration this algorithm has been applied on the artificial case for Rastrigin's function in 5-D parameter space (see Section 5.6.4.4). The results are shown in Figure 129 in terms of identified local minima as a function search radius. The percentage may seem low (81 located out of a number of 11^5), but these positively identified minima are all located in the direct vicinity of the global minimum. The one closest to the global minimum was (0,0,0,1,0). Note that the maximum number of local minima identified is at a radius $r \approx \sqrt{2}$, which is the minimum distance between two local minima.

5.11 Working sequences

5.11.1 General

Based on the several topics discussed and illustrated in this chapter working sequences (or procedure) are defined for:

- Judging measuring setups on their ability to produce information fit for calibration.
- The calibration process itself.

5.11.2 Judging measuring set-ups

5.11.2.1 Working sequence

In Chapter 4 the problem of defining the sampling interval was discussed, but the problem of spatial distribution was not addressed. As was pointed out (see Section 4.2.3.3), methods normally applied to define a criterion for the spatial distribution of monitoring locations fail when applied on an urban drainage system. Limiting the discussion to the monitoring objective 'calibration', the following working sequence is proposed:

1. Identify the locations in the system available for monitoring (selection on practical criteria like accessibility). This results in N_{max} locations.
2. Define the measuring accuracy and the sampling interval in time using the methods described in Section 4.2.4.
3. Define the maximum number of gauging locations (normally limited by available budgets and manpower). This results in M_{max} locations, generally $N_{max}>>M_{max}$.
4. Identify the model parameters that are candidates for parameter optimisation, P_{max}.
5. Run a hydrodynamic model $P_{max}+1$ times in order to be able to calculate the Jacobean (see Section 5.6.3.3).
6. Construct the Jacobean for N_{max} monitoring locations and P_{max} model parameters. (This Jacobean is referred to as $\underline{\underline{J}}_{total}$).
7. Make a singular value decomposition for $\underline{\underline{J}}_{total}$ and judge the rank of $\underline{\underline{J}}_{total}$ (see Section 5.3).
8. If the rank of $\underline{\underline{J}}_{total}<P_{max}$ or $\underline{\underline{J}}_{total}$ is nearly rank deficit for some parameters then identify these parameters and remove them from the parameter list, this results in P_{max}-$P_{rankdef}$ parameters.
9. Since M_{max} is normally significantly smaller than N_{max}, choices have to be made on which monitoring locations are to be selected; this can be done in two manners.
   a. Study the contribution for each monitoring location for each parameter on the Jacobean and select the M_{max} locations having the largest contribution. Construct the Jacobean based on the selected monitoring locations and make a singular value decomposition in order to verify if this Jacobean is (nearly) rank deficit. If it is not, the parameterisation and the measuring setup match. If the Jacobean is (nearly) rank deficit, a further reduction in the number of parameters is needed. Alternatively step 2 can be reconsidered; an increase in measuring accuracy could help decrease the sampling interval making more information in the time domain available.
   b. Judge every possible combination of M_{max} out of N_{max} monitoring locations by the results of the singular value decomposition of the Jacobean associated with

it. Pick the one combination with the largest minimum singular value. If this combination is still judged to be nearly rank deficit, either M_{max} should be increased (resulting in an increase of investments en operational costs) or a further reduction of the number of parameters must be induced. Alternatively step 2 can be reconsidered; an increase in measuring accuracy could help decrease the sampling interval making more information in the time domain available.

10 Perform auto calibration runs (see Sections 5.8, 5.9 and 5.10) with different initial parameter vectors using the parameterisation and the measuring setup obtained in the preceding 8 steps. This will reveal information on the shape of the object function. If problems related to stability or non-uniqueness occur, a further scrutiny of the steps 5-8 is advised.

11 Proceed with detailed design of the monitoring network.

It is stressed that following this procedure is no guarantee that when the actual measuring data become available these hold enough information to identify all parameters that follow from the initial estimate obtained in the working sequence 1-9. This estimate must be seen as a maximum, since the sensitivity analysis (which is basically what the working sequence is) is performed on unbiased data, for only a limited number of storms and for a limited number of parameter values. It is stressed that due to non-linearity the analysis of the Jacobean for a given set of parameter values cannot be extrapolated to another set of parameters. This implies that if any prior information on the parameters is available this must be used, thus enhancing the predictive value of the results from the working sequence proposed. Of course, the vulnerability of the monitoring setup for systematic errors can be judged by introducing artificial errors in the auto-calibration process. The construction of the Jacobean matrixes in step 9^b does not make additional model runs necessary, these can be constructed directly from $\underline{\underline{J}}_{total}$ because the latter matrix is given the following form:

$$\underline{\underline{J}}_{total} = \begin{bmatrix} \underline{\underline{J}}_1 \\ \underline{\underline{J}}_2 \\ \cdot \\ \cdot \\ \underline{\underline{J}}_{N\max} \end{bmatrix} \qquad \text{(eq. 5. 67)}$$

In which $\underline{\underline{J}}_i$ is the 'sub' Jacobean matrix defined by the results obtained at monitoring location i. and has the following form:

$$\underline{\underline{J}}_i = \begin{bmatrix} \frac{\partial x_{t_0}}{\partial p_1} & \cdot\ \cdot\ \cdot & \frac{\partial x_{t_0}}{\partial p_n} \\ \frac{\partial x_{t_1}}{\partial p_1} & \cdot\ \cdot\ \cdot & \frac{\partial x_{t_1}}{\partial p_n} \\ \cdot & \cdot\ \cdot\ \cdot & \cdot \\ \cdot & \cdot\ \cdot\ \cdot & \cdot \\ \frac{\partial x_{t_{\max 0}}}{\partial p_1} & \cdot\ \cdot\ \cdot & \frac{\partial x_{t_{\max}}}{\partial p_n} \end{bmatrix} \qquad \text{(eq. 5. 68)}$$

In which x is the model result as a function of time. Since x becomes available as a discrete series the rows in $\underline{\underline{J}}_i$ are valid for successive time steps, these are to be taken equal to the sampling interval in time as applied in the monitoring setup.

5.11.2.2 Example

Suppose the 7 monitoring locations are selected as shown in Figure 115 while setting out to limit the number of locations actually to be used to 3, so N_{max}=7 and M_{max}=3. This leads to a number of $\begin{pmatrix} N_{max} \\ M_{max} \end{pmatrix} = \frac{7!}{4!3!} = 35$ combinations for the Jacobean associated with these 35 possible measuring setups([50]). The initial parameterisation is chosen as:
for two types of surfaces (type 1 and 2): the (constant infiltration rates I1 and I2, the routing constants F1 and F2 and the depression storage B1 and B2). So the total adds up to 6 parameters. As storm event standard storm no. 6 and the historical storm of September 2 1998 were applied. In Figure 130 the singular values are shown for the total Jacobean and for the Jacobean defined per measuring location for storm no. 06. The parameter number referred to is the parameter that is predominant in the SVD.
It is clear that the parameters 2,4 and 6 have a relatively low identifiability while parameters 1,3 and 5 are relatively well identifiable. Furthermore, it is seen measuring location 4 is important, since it has large singular values (when compared to the other locations) for most parameters.

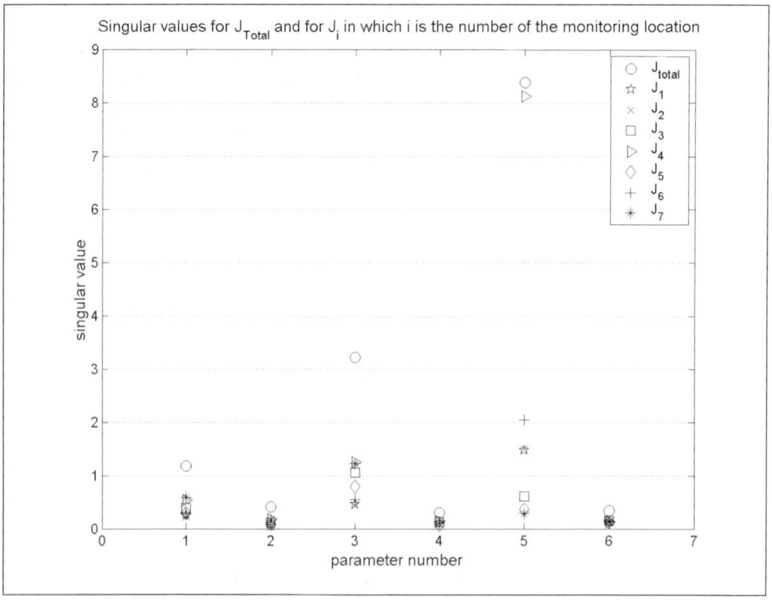

Figure 130: Apparently the parameters 2,4 and 6 are relatively less identifiable. Furthermore, it is seen that measuring location number 4 (close to the pumping station) reveals an important part of the information.

([50]) It must be noted that for large numbers of possible measuring locations (N_{max}) proceeding the calculation quickly becomes unpractical due to the extremely large number of possible combinations. In such a case, it is wise to eliminate locations showing small singular values for most parameters. In this manner N_{max} is reduced. Application of genetic algorithms for identifying favourable combinations is regarded a possibility also.

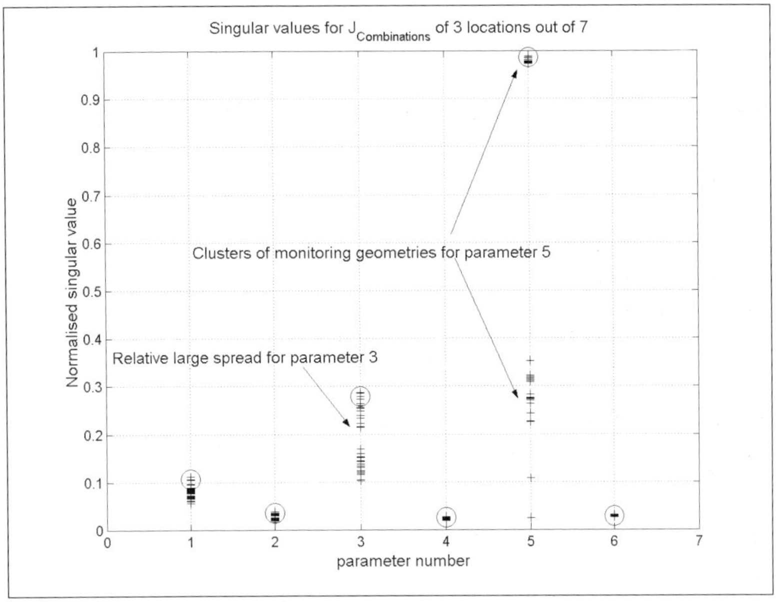

Figure 131: Spread of the normalised singular values for the 6 parameters for all possible measuring configurations using 3 locations out of 7. As can be seen parameters 2,4 and 6 are relatively insensitive for the measuring setup. Parameters 3 and 5 show a larger sensitivity for the setup, while for parameter 5 two clusters evolve; one with a small normalised singular value and one with a high normalised singular value.

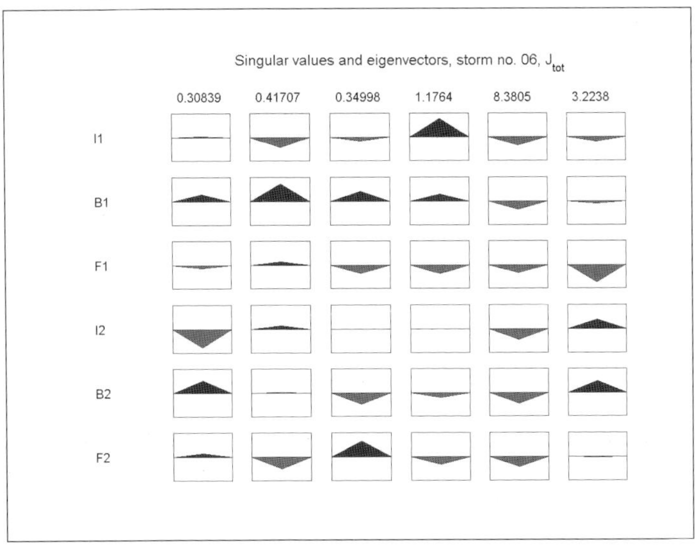

Figure 132: Eigen vectors and singular values for storm no 06. As can be seen non of the parameters can be identified separately. The eigenvector with the largest singular values has components that are about equal for all parameters (parameter 5 though, is dominant).

In Figure 131 the normalised singular values for the 35 combination of measuring locations are shown. From this Figure , it is clear that the choice on the combination of measuring locations has a significant influence on the identifiability of the parameters. As was to be expected this influence is small for parameters 2,4 and 6, whereas especially parameters 3 and 5 are sensitive for the combinations of measuring locations. The best combination (marked with 'O', and judged by having the largest value for the sum of the singular values (tr(S))), was location 2,4 and 7, second best was 2,4 and 5. It must be stated though, that especially location 4 is dominant in the set of best combinations. In Figure 132 the eigen vectors and singular values are shown, as can be seen the eigen vector with the largest singular value has significant components for all parameters. This implies that none of the parameters can be identified separately for storm no. 06 using any measuring setup

results storm September 2 1998

Since it is known that the load on a system has a large influence of the processes taking place and on their relative importance (see also the sections in chapter 4 on characteristic time) it is easily understood that there is a fair possibility that different individual storms will result in different characteristics of the Jacobeans. As a result, the 'optimal' measuring setup and the set of identifiable parameters is basically to be defined per individual storm.

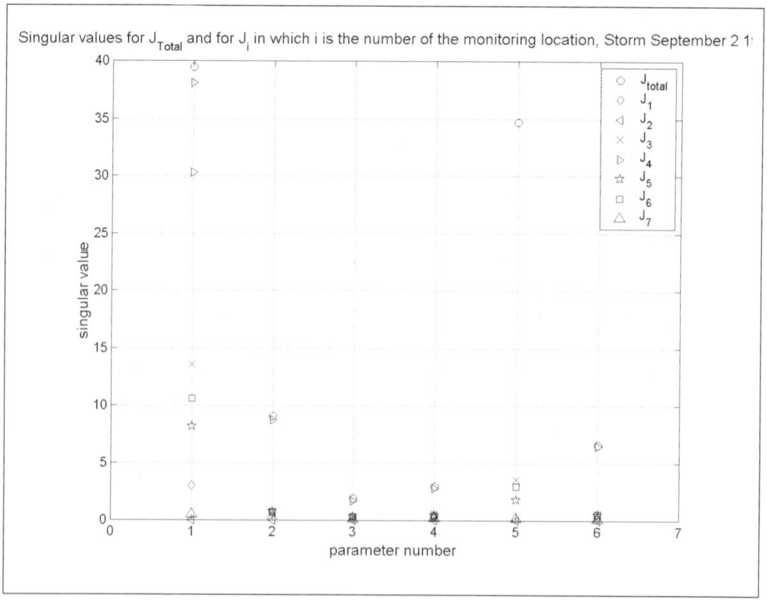

Figure 133: For the storm on September 2 1998, the parameters 1 and 5 have the largest identifiability, while for all parameters the singular values are significantly larger than for storm event 06. This is explained from the fact that storm no. 06 is short and shows high rain intensities. This implies that the processes are strongly dominated by (quick) filling and emptying of the system. Since the historical event had a duration of about 24 h showing low rain intensities, this implies that the run-off process had a far greater influence on the process in the drainage system. In this case also location number 4 is by far the most important.

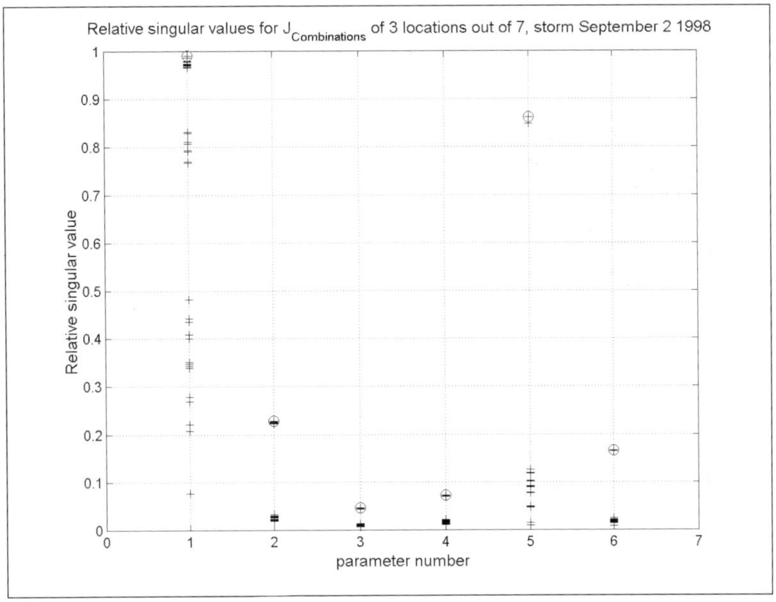

Figure 134: Normalised singular values for the 35 combinations for the storm event on September 2 1998.

In order to illustrate this point the results of the working sequence for the catchment 'De Hoven' is shown also for the historical storm of September 2 1998. In Figure 133 the singular values are shown. When compared to the results for storm no. 06, the singular values for all parameters are significantly larger implying an overall increase of the identifiability. Furthermore, it becomes clear that in this case location number 4 is the most important location. The best combination was 2 4 5, in this case also, the best combinations were dominated by the presence of location 4. When studying the eigen vectors (eigen values for $\underline{\underline{J}}^T\underline{\underline{J}}$), see Figure 135, it is seen that only the parameter 1 (infiltration for area type 1) and the routing constants F1 and F2 can be identified more or less individually, all other eigen vectors have almost equal components for two or more variables.

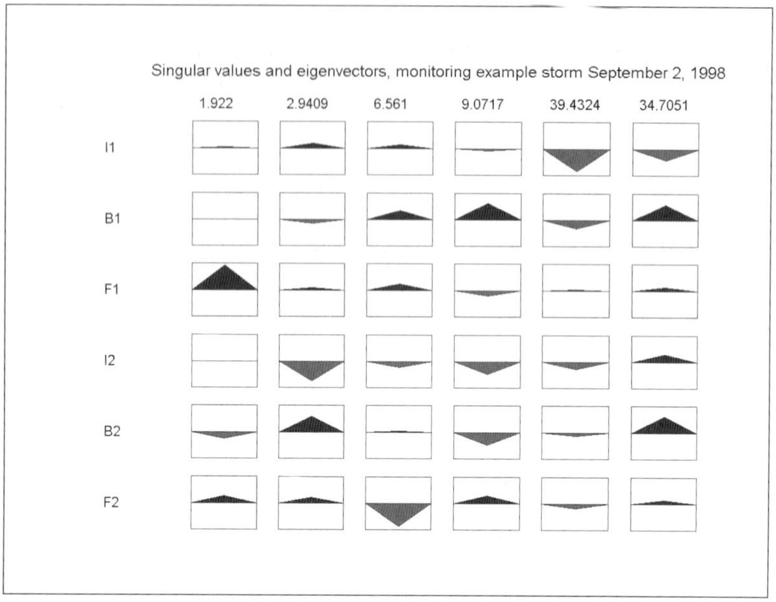

Figure 135: Singulars values and eigen vectors for J_{total} storm on September 2, 1998.

5.11.3 Calibration process

The calibration process is best done in parallel with the monitoring project, as the intermediate results from calibration during the monitoring can give valuable insight into possible adjustments of the measuring setup and its operation. Following the working sequence described in the previous section, now using real data, results in an enhancement of the over-all result obtained. An absolute prerequisite for a successful calibration is that the geometrical/structural database of the urban drainage network is correct (see Chapter 3 on this particular subject). The calibration process itself is done by application of the following working sequence:

1 Eliminate systematic errors from the measuring data as well as possible, using the information available in the logbook (see Section 4.1.2).
2 Perform a number (e.g. 3) of parameter optimisation runs.
3 If the resulting parameter vectors in step 3 are (almost) equal proceed with step 5, if not proceed with step 4.
4 Check for the existence of local minima using e.g. a genetic algorithm (see Section 5.10.4). Identify the location of the global minimum and return to step 2 using initial vectors circumventing local minima. If a situation occurs in which an 'infinite loop' evolves in steps 2-4, the problem is ill defined and either the parameterisation is to be adjusted or step 1 was not done accurately enough and should be repeated.
5 Analyse the residuals with respect to([51]):
  - Cross correlation between the gauging points.

([51]) It must be mentioned that no hard criteria can be given for acceptance levels with respect to bias, correlation structures or deviations from the Gaussian distribution. In practice (see e.g. Chapter 6) the practitioner will have to accept some degree of subjectivity in this.

- Auto correlation in time for every individual gauging point.
- Bias in a shifting time window (the width of the window is to be chosen proportional to the characteristic time scale, see Sections 4.2.2 and 5.4.
- The probability distribution, it should be (nearly) Gaussian in order to be in line with the underlying assumptions when applying the MLE method (see Section 5.2.2.).

If the residuals show bias, significant unexplained cross-correlation, correlation in the time domain or deviate substantially from being Gaussian, some systematic error is present. Such a systematic error can be found in either:

- The measuring data.
- The geometrical/structural database for the catchment.
- An erroneous process description.
- A software induced error.

When using commercial software it is in most cases not possible to correct for the latter two causes. The degree of success in removing the first two sources mentioned depends on the accuracy of the measuring program (in terms of logbook keeping and verification of synchronity and zero drift of sensors) and the data acquisition that has been done prior to building the database.

6 Judge the quality of the parameters obtained by analysing the covariance matrix, the eigen vectors and eigen values. (see Section 5.3).

5.12 Conclusions chapter 5

The following main results and conclusions from the study in this chapter are:

- A systematic way of approaching the calibration problem in urban drainage is to be preferred to the commonly applied trial and error method. It is proposed to follow the working sequences as described in Section 5.11 in this respect.
- The instrument of auto calibration is a tool that gives valuable information when designing a monitoring network with the objective to calibrate a model. Combined with a sensitivity analysis the parameterisation in the calibration and the monitoring network can be tuned. In cases of existing monitoring data, the information content with respect to the calibration is to be judged.
- Two simple parameter optimisation methods based on gradient search are demonstrated and a method to operate them in conjunction with commercial software for hydrodynamic calculations is presented.
- Genetic algorithms show a relatively slow convergence rate, but are useful to reveal the possible existence of local minima.
- The accuracy with which parameters passing between the optimising algorithm and the calculation model can be effected has an influence on the quality of the calibration results.
- The choice of measuring locations in a urban drainage system has potentially a major influence on the obtainable results in model calibration. Using a hydrodynamic model to judge the information content of a measuring setup (using singular value decomposition) can reveal the locations holding most information. Since this information content depends also on the storm the calculation was based on, several storms should be used in the design process.
- The methods proposed in this chapter are to be regarded as a starting point for further research. It has been demonstrated that existing techniques and simple means can lead

to an increase in insight. When calibration becomes more commonly included in day-to-day practice the following topics call for further attention:

- The accuracy of parameter passing between hydrodynamic software and the optimisation routine, to this end the only thing to be done is increasing the number of decimals in the user accessible databases (input as well as output).
- The incorporation of faster and more advanced algorithms for parameter optimisation. In this respect the application of Kalman filtering and the application of adjoint models are regarded as promising for parameter optimisation, sensitivity analysis and monitoring network design.

CHAPTER 6: CASE STUDY

6.1 Introduction

The area under study is situated in the Netherlands directly on the banks of the river 'de IJssel' and is known geographically as 'De Hoven' in the municipality Deventer. The drainage system is of the combined type and was built in the early seventies using concrete as main constructing material. The system comprises one pumping station transporting the sewage to the treatment plant at Terwolde. There are three external weirs present incidentally spilling into small ditches during storms. The DWF consists solely out of municipal wastewater since industrial discharges are absent. The catchment area is 12.69 ha. and includes roofs, and semi-impervious road areas. The number of inhabitants is circa 2200. The total storage volume of the system is 865 m^3 including small diameter pipes in municipal and gullypot manifolds. This Figure is corrected for lost storage, i.e. the volume caught in local depressions in the system due to subsidence or poor construction practice. The capacity of the main pumping station is 119 m^3/h. An important issue is the fact that it is known that the system has a noticeable leakage with respect to groundwater. The ground water table responds quickly to the water table in the nearby river IJssel. During the summer period the groundwater level is below the lowest level in the system, while during autumn and winter the groundwater level is above the crest level of the majority of the pipes in the system, resulting in a noticeable increase in DWF.

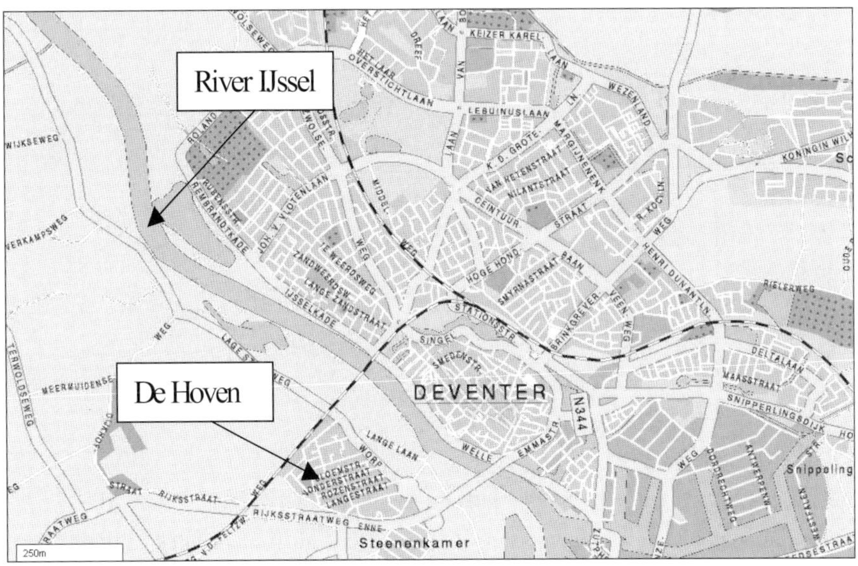

Figure 136 : Geographical situation.

A detailed field survey was done in order to acquire geometrical data and the structure of the system. The area contributing to run-off was quantified by means of a digital method based on GIS techniques (see e.g. van Capelleveen (1999)), and the results were checked by a field survey. In this context it must be mentioned that it was necessary for the project to follow the

Dutch standard method prescribing a differentiation in type of run-off areas. This method may lead to a total of 12 different types of run-off area each characterised by a set of model-parameters with respect for the run-off model. Following strictly this procedure and at the same time stressing the use of calibrated models will lead to unpractical situations.

6.2 Measuring setup

The original goal of the measuring project was to obtain data on the functioning of the three external weirs in the system. Therefore, the water levels are measured at these locations only. In the pumping cellar a level gauge was present for the control of the pump. This gauge however, proved to show a strong non-linear systematic error that could not be quantified, therefore the data from this gauging station are left out. This has an important negative effect on the information content of the measuring setup as a whole. As was made clear in Section 5.11, especially the location in the direct vicinity of the pumping station is of importance in this respect. The location of the measuring gauges is shown in Figure 137.
The recordings from this gauging point at the pumping station could only be used to obtain information with respect to the time at which a pumping action was taken (the pumping station is of the type switch on- switch off).

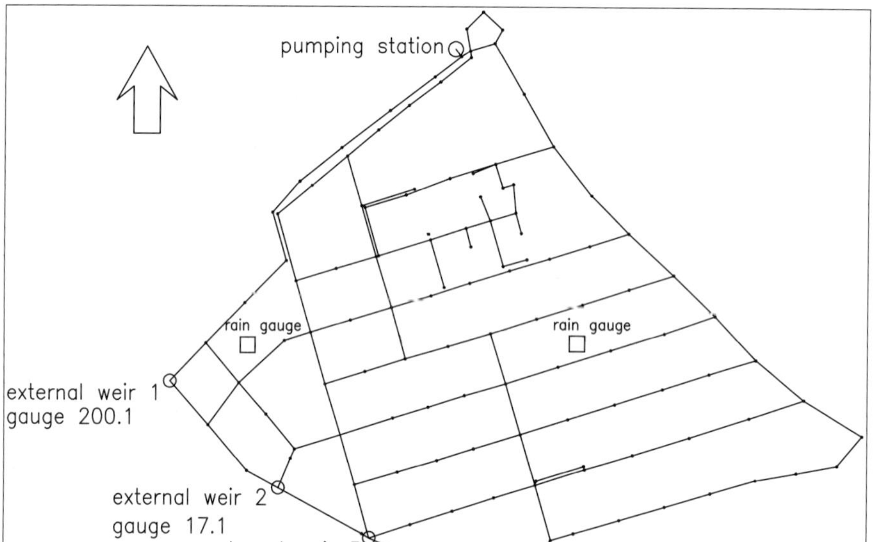

Figure 137: Locations of the measuring gauges.

The rain gauges applied are tipping buckets (type Casella with a catchment area of 400 cm^2, and a rain intensity resolution of 0.2 mm/tip); the measuring data were corrected for rain intensity. It is known from literature (see Overgaard et al (1998)) that the type of rain gauges applied shows a systematic error in their readings when not corrected afterwards. The sampling frequency of the water level gauges was made dependent on the water level in the system; during dry weather periods a frequency of 4/h was used, as soon as the water level

reached 5.25 m NAP([52]) the sampling frequency was increased to 1/min. These frequencies proved to be in line with the requirements discussed in Section 4.2.4. and no adjustments in this respect were made during the project.
The water level gauges were installed in such a manner that the measuring range was between lowest invert level and street level. In this manner the effect of minor rain events could be monitored also. The water level sensors used are Campbell Scientific, type PDCR 1830, having an estimated absolute accuracy of circa 6 mm (95% confidence interval) at the measuring range of 2 m. During the project a detailed logbook was kept, and updated every 2 weeks after the routine field survey. During these surveys the clocks of the measuring equipment were synchronised with the system time of the office computer, discrepancies being recorded. During the project the adjustments for synchronicity stayed at an acceptable level (maximum of 30 seconds shift in 14 days), except for the rain gauge no. 2, that needed a more regular adjustment. The data obtained were analysed directly with respect to:

- Completeness.
- Zero drift.
- Logical errors (e.g. water level changes parallel on a storm event at 2 locations and 1 location that did not respond.)

The results of these routine analyses were recorded in a database and visualised in graphs, this source of information proved to be an important one after finishing the monitoring project

6.3 Model database and error removal

Systematic errors were removed from the obtained measuring data. It turned out that a systematic error in the reference level for gauge 200.1 was made of 0.05 m. This was corrected in the data analysis. A number of errors in the geometrical database used, like wrong invert levels, were traced in the process of calibration. At all gauges during the nightly hours occasionally the measured level dropped at the lowest invert level present in the databases. This offered the possibility to verify the sensors on zero-drift and helped to trace and remove some errors in invert levels in the database
One of the most important systematic errors found during the calibration process was the incorrect representation of house manifolds, gully pots and small-bore pipes connecting them to the main sewers in the geometrical/structural database. For practical reasons these objects are normally not incorporated in the database, since they would more than double the number of objects in the database. This would result in an enormous extra effort to obtain a correct database. As was discussed in Chapter 3, it is no trivial task to obtain a correct database incorporating the main objects. Adding all small objects while maintaining a high level of accuracy is judged to be hardly feasible in a technical and economical sense.
Normally, the volume held by these objects is therefore either compensated for in the geometry of the manholes or is neglected altogether and looked upon as a, non-quantified, 'safety factor' in the calculations. When setting out to calibrate, systematic errors showed up when the water level in the system rose above circa 5.00 m^+ NAP. Several explanations for this are possible. It could be caused by the increased pressure at the suction side of the pumping station (resulting in an increase of the pumping capacity). An other possible cause is an incorrect geometrical representation of small objects that were omitted from the database. Since the exact locations of the small-bore pipes are unknown, an estimate based on the available knowledge was implemented:

- There are 834 house connections.
- These are connected to the main drainage system by pipes leaving the house at 0.70 m below ground level.

([52])NAP is the national Dutch reference level.

- The diameter of these pipes is 120 mm.
- The average length is estimated to be 10 m
- A gully pot is present at both sides of the street every 25 m' and is connected directly to the main sewer.

Based on these assumptions, the known ground levels and the geometrical database an estimated water level-volume relation was constructed (see Figure 138). After implementation of the additional storage, an important portion of the systematic errors disappeared.
The practical implementation of the additional storage was difficult to get correctly in the database. The additional storage was circa 67 m^3 ([53]) concentrated between 4.55 m^+ NAP and 5.6 m^+ NAP, more or less uniformly spread over the system.

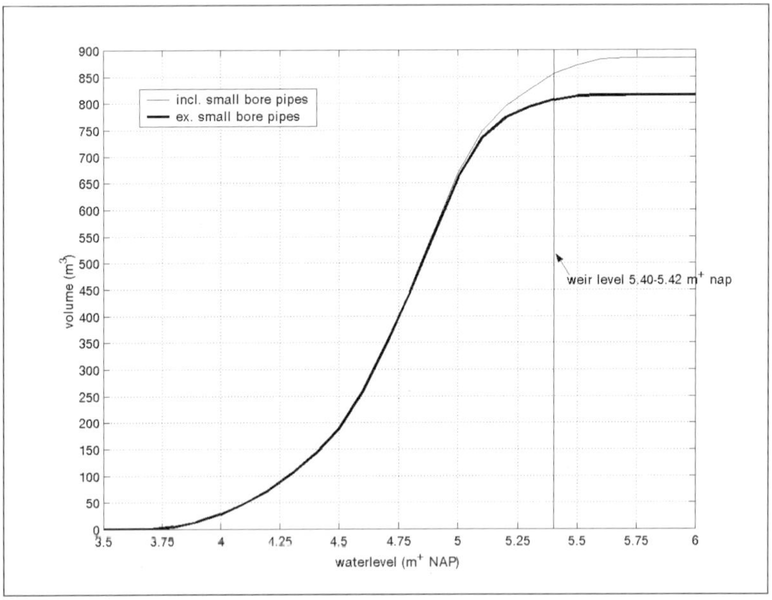

Figure 138 : The water level-volume relation for 'De Hoven'. As can be seen the additional storage is a significant factor at water levels above circa 5.00 m^+ NAP.

When setting out to model this additional storage by adapting the manhole geometry, this proved to be impossible due to the limitations in the number of digits that could be introduced in the database. Eventually a workable solution was found; after some experimentation it proved that the volume held by the Preissman slot was taken into account in the model. For the sake of demonstration, the influence of this software-related parameter is shown for a width of 7 mm and 70 mm (see Figure 139)([54]). The only difference between both model runs is the width of the Preismann slot. As can be seen the results for a slot of 70 mm width deviate significantly from the results for 7 mm width as well as from the measured results. As can be seen from the falling limb of the curve, the emptying rate is influenced significantly by the

([53]) The extra storage of 67 m^3 is significant when compared to the volume held by the main sewers and manholes (815m^3). At weir crest level these values are 49.9 and 806 m^3 respectively, from which it is concluded that the storage volume above weir crest is significantly influenced by the volume in the small-bore pipes and objects.
([54]) Normally the slot width is chosen to be very small relative to the dimensions of the conduit, a slot width in the order of magnitude of 10^{-4} is mostly applied.

width of the Preissmann slot. During the emptying stage, the flow velocity reached a maximum value of about 5 cm/s. Therefore, the effect cannot be explained by the (known) error made in the momentum equation. The only explanation left is that the volume in the Preissmann slot is taken into full account in the software.

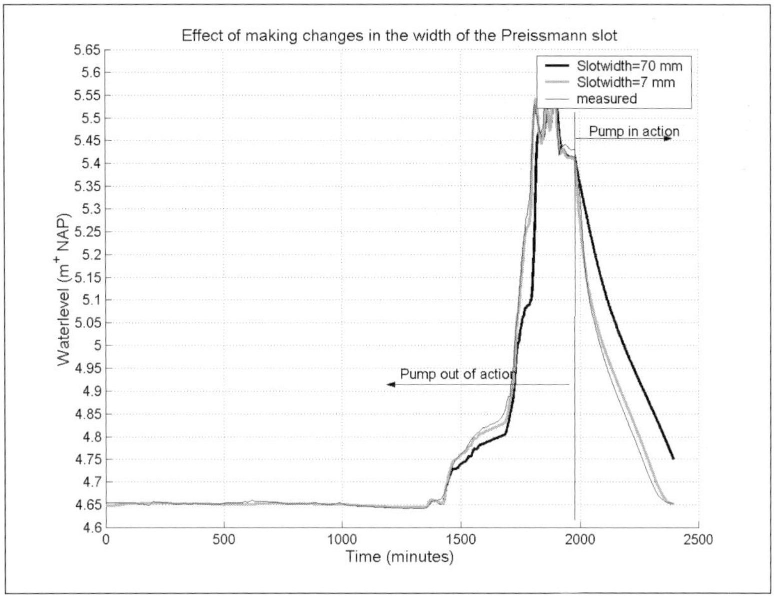

Figure 139 : The effect of the width of the Preissmann slot on calculation results in Hydroworks.

This particular characteristic of the software applied offers the possibility for a more accurate modelling of the estimated volume-water level relation. Eventually a width of 7 mm for the Preissmann slot results in an acceptable reproduction of the rate at which the system was emptied in the range above 4.55 m^+ NAP. It is well understood that a Preissmann slot having a width of 7 mm does not comply with the theoretically correct value as was explained in Chapter 2, this results in a, theoretically, incorrect modelling of the dynamics. The effects of this however, become significant only if high streaming velocities occur and only when the conduits are surcharged. Since the streaming velocities in the cases considered here stay well below 0.5 m/s, the correction obtained by the increased slot width on the mass balance equation is considered more important than the correct modelling of the momentum equation. The overall water level-volume relation obtained using the algorithm described in annexe I.3 is shown in Figure 138. The slot width of 7 mm resulted in an extra volume of 67 m^3 located between the crest level of the conduits and the crest level of the weirs present in the system. In order to verify the distribution of this extra volume the modelled and measured emptying rate of the system were compared.

As can be seen in Figure 140 only in the domain between 5.15 m and 5.22 m a significant deviation occurs between observed and modelled emptying rates when applying a slot width of 7 mm. For some reason, concentrated in this range, some volume is missing (the volume in reality is less than modelled in the database).

The amount of volume missing is calculated to be circa 10 m^3, or about 1.15 % of the total volume; it proved impossible to correct this in the small range of levels in the geometrical database.

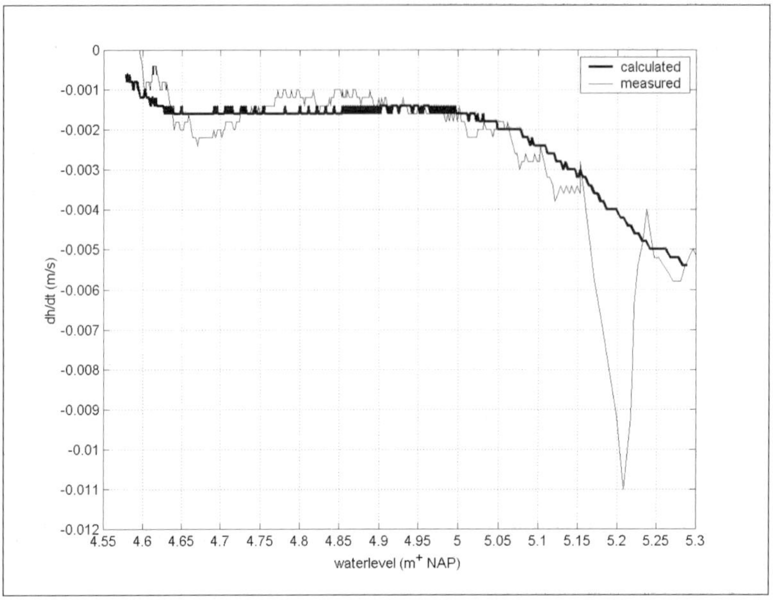

Figure 140: Emptying rate (dh/dt) as a function of time, measured as well as modelled with a Preissmann slot with a width of 7 mm (Storm event on August 25 1998).

Since no physical evidence for this missing volume was found, the geometrical model is accepted as it is. It is well realised that the geometrical discrepancy will result in a systematic error, bias, in the final calibration results.

6.4 Model and parameterisation

'De Hoven' is a small system having 123 manholes, 141 conduits, 3 weirs and 1 pumping station, while 2 types of contributing surfaces are distinguished. This leads, in theory, to a number of 274 parameters to be introduced in the model. As was explained in Chapter 5, it is not feasible to introduce all these parameters in the optimising process for practical reasons. Main consideration in this respect is the fact that the number of measuring data should be several orders of magnitude larger than the number of parameters to be optimised. Therefore, choices have to be made in order to limit the number of model parameter. One way to do this is clustering, meaning that e.g. Nikuradses roughness is supposed to be equal for groups of conduits (selected on the basis of e.g. material, age). All measures of the object comprised in the system and the structure of the system are assumed to be exactly known. Possible errors in the structure of the system should become clear in the calibration process.
The software package applied was HydroWorks® version 3.3. With respect to the hydrological model this software tool offers various possibilities of which a simple variant was chosen:

- A constant infiltration rate model, incorporation evaporation losses also.
- A constant value for depression storage, including initial losses.

- A linear reservoir model: S=FQ, in which F(s) is the (constant) reservoir parameter, S (m^3) is the water volume stored in the (sub)catchment and Q (m^3/s) is the run-off discharge into the drainage system.

Main reason for applying this simple hydrological model is to limit the number of parameters. Furthermore, it is anticipated that this simple model offers enough flexibility. The parameterisation for the model as a whole (hydrological and hydraulic) is as follows:

- Since the hydraulic roughness depends probably on the materials used and the age of the system it is postulated that the roughness in the model can be represented using one single value for all pipes. Because local hydraulic losses in nodes have exactly the same influence on the calculation results, this parameter is not taken explicitly into the model. In small, flat systems variation in discharge and water level are relatively insensitive to the hydraulic roughness. Therefore, the hydraulic roughness was set to a fixed value of 3 mm during the calibration process. This is also justified by the fact that the velocities in the system stay well under 1.0 m/s during all phases. Under these conditions, the influence of the hydraulic resistance in a sewer system is minimal.
- The three external weirs present differ substantially from one another with respect to construction details and flow-geometry. Therefore, the overflow coefficients for the three weirs are to be calibrated individually. Apart from design data, no prior knowledge on their values was available.
- The pumping capacity is known from separate measurements and is therefore not used as a parameter to be calibrated.
- The linear reservoir constant, initial storage losses and the infiltration capacity, on which no prior knowledge is available, are optimised using the measuring data obtained.
- The mean value for the DWF is determined using measuring data.
- DWF patterns are constructed based on information obtained on the functioning of the pumping station during dry weather periods.

For the inventory of the geometrical data of the system, the Dutch standard procedure was followed (see Chapter 1) indicating that 12 sub-types of run-off area are distinguished. In the area under study only 3 of them were present. This implies that a total of 18 Run-off parameters should be optimised (3 routing coefficients, 3 initial storage loss values and 3*4 parameters for the Horton infiltration model (see Horton, 1940)). Since neither specific run-off measurements, nor any separate indications on for instance infiltration capacities were available, the presumption was that based on the presented measuring setup, it is impossible to get correct estimates for all these parameters. During several trial runs using artificial optimisation runs, it turned out that it was impossible to obtain correct results using all three surface types. This is due to the information contained by the measurements. Moreover, the information content differs from storm event to storm event. This forces to investigate which combination of parameters leads to unique and identifiable parameter values for each individual storm. The working sequence outlined in Section 5.11 was followed for each individual storm. Eventually it was possible to distinguish two types of contributing area; a 'slow' type and a 'quick' type. The 'quick' type was identified with roof areas and impervious pavement, a total of about 5.65 ha., and 7.04 ha. 'slow' area. The transformation from rain intensity into a run-off hydrograph was modelled using a linear reservoir model and an initial loss term, the slow area was modelled using a linear reservoir model, initial loss (including storage) and an infiltration factor. The infiltration was assumed to be constant with time. Evaporation was neglected in the model; the mean evaporation rate in the period the measurements were done had an average value of 0,8 mm/day. Any contribution from this process will be present in the other run-off parameters. This approach is consistent with the

Dutch standard. In line with the statements made on this subject in Section 5.1.2, an approach is followed in which only the process-related parameters are subject to adjustment in the parameter optimisation process. As was explained in Section 5.11, the identifiability of the parameters has to be judged for every individual storm, since the characteristics of the storm (like duration and variation in rain intensities) do have a significant influence on the identifiability.

6.5 Selecting storm events for calibration

Analysing the processes taking place during a storm event in a drainage system, it is possible to quantify the number of measuring data available for the optimisation of groups of parameters given a certain set of measuring data. In Figure 141 a typical measuring result at a gauging station is represented schematically.

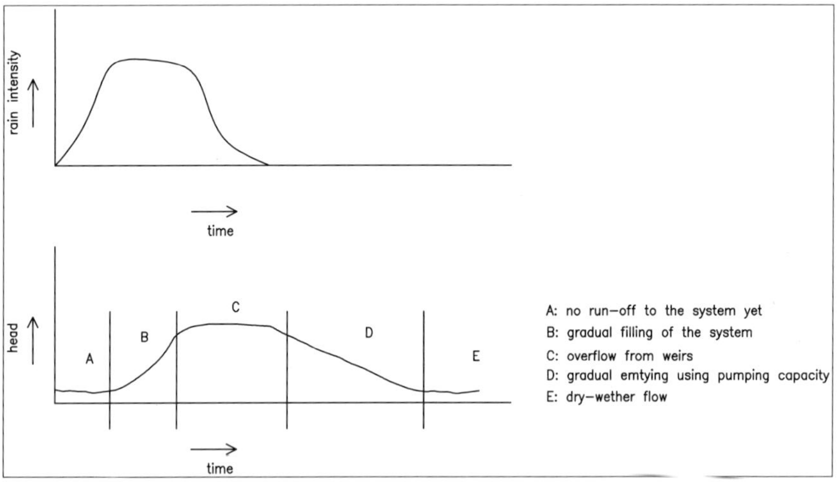

Figure 141: Different phases in water level development during a storm.

In the phase A (storm event starting at t=0) no effect is seen in the drainage system, this is because the surfaces are not generating any run-off yet. In this phase information is obtained with respect to infiltration rates and the initial losses and depression storage of the catchment. As soon as the catchment generates run-off, this results in an increase of water level with time at the measuring locations. Phase B is entered in which the water levels are influenced by hydraulic losses and the geometry of the system (water level-volume relation). At a certain water level, one or more weirs will come in action (phase C). In phase C the measured water levels are mainly influenced by the overflow coefficients and the geometrical data of the individual weirs. So, during phase C information is obtained with respect to the weir coefficients. At a certain moment the pumping capacity becomes larger than the total inflow, phase D is entered. Assuming the pumping capacity to be exactly known (and as such not being a parameter to be optimised) the measuring data mainly depend on the linear reservoir constant, the DWF and the pumping capacity. At a given moment, the phase in which the heads are depending on dry-weather flow (phase E) is entered. This analysis should be done when judging the measuring data obtained in conjunction with the parameters taken into account in the optimisation process. Every recorded storm event contains valuable information with respect to at least a subset of the parameters to be optimised. (E.g. it is meaningless to try to obtain a value for the overflow coefficient if in the measuring data no overflow situation of a substantial duration is present.). The characteristic time scale varies with the phases mentioned, during phase A and E a T_{char} is in the order of magnitude of hours, while in phase B and C T_{char} is in the order of magnitude of minutes. See also Section 4.2.2 on this particular subject.

The measuring data obtained during the study period were selected for use in the calibration process using the following constraints:

- Absence of suspicion of systematic measuring errors.
- Completeness of the data set.

During the measuring period (August 3 1998 to November 1 1998), in only 2 storms an overflow situation occurred. Furthermore there were 3 storm events which did not lead to overflow situations and held reliable information for calibration runs. These events were used to quantify the DWF and the parameters in the hydrological model.

storm	Start	End	Total rain gauge 1	Total rain gauge 2	Remarks
1	25/8 00.00	26/8 23:55	16.0 mm	16.4 mm	pumping station disabled from 25/8 10.00 h until 26/8 16:55 h, external overflow occurred
2	2/9 21:20	3/9 21:00	18.8 mm	20.0 mm	-
3	7/10 17:55	8/10 18:40	14.6 mm	14.2 mm	-
4	9/10 18:00	13/10 00:20	33.0 mm	27.8 mm	-
5	24/10 17:35	25/10 12:00	21.8 mm	21.2 mm	external overflow occurred

Table 67: Selected storms for calibration.

The parameters in the calibration process are:

- 2 linear reservoir constants
- 2 values for constant infiltration rate
- 2 values for depression storage (including initial losses)
- averaged value for the dry weather flow
- 3 overflow coefficients

So, initially, the parameterisation consists out of 10 parameters.

6.6 Determination of the DWF pattern

In the period between August 6 until August 13 no precipitation occurred, in order to get information on the DWF pattern the number of pumping actions per hour were counted. Since the volume per pumping action was known to within an accuracy of 1%, the hourly average DWF arriving at the pumping station could be estimated. Since a clear periodic pattern was observed, a Fourier analysis was applied from which an averaged 24-hour pumping pattern was derived. It is stressed that the pumping pattern is not equal to the DWF discharge pattern. A distortion (i.e. a time shift and a change in amplitude) between DWF discharge pattern and the pumping action pattern exists due to travelling time and damping in the system. Using the amplitudes in the discharge pattern as unknown parameters an inverted model was made, the results are shown in Figure 142. A clear time shift between DWF-discharge and pumping action is seen as well as the fact that the maximum amplitude of the pumping pattern is less than the maximum amplitude in the discharge pattern. It is stressed that the obtained DWF pattern is not equal to the actual pattern but merely represents an acceptable model used in the further calibration process. Especially when calibrating on storms having a relative long (i.e. > 20 h) duration and low rain intensities, the correct modelling of the DWF is of importance when calibrating a model. The daily average value for the DWF production from this analysis was 32,9 m^3/h with a maximum value of 38,0 m^3/h and a minimum value of 30,7 m^3/h. Since the pumping capacity was known, the dry-weather flow could be estimated using data from rain event nr.1 from the moment on the rain stopped.

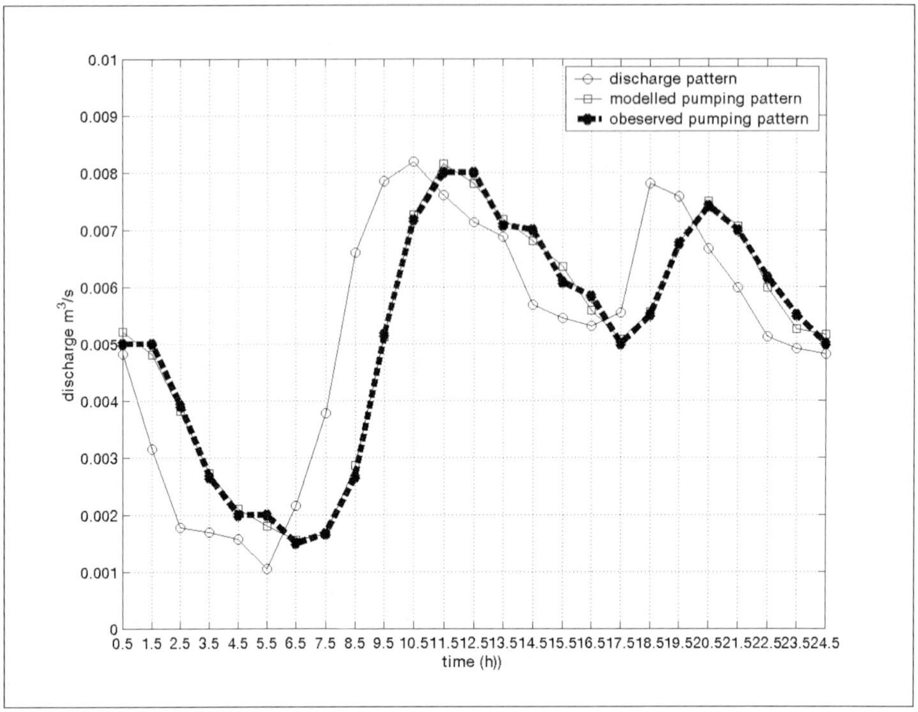

Figure 142: DWF pattern for 'De Hoven'.

This led to an estimate for the DWF of 26.16 m^3/s This is almost exactly the same value as expected from standard data (12 l/h.inh, with 2200 inhabitants adds up to 26.4 m3/h). As will be seen from the calibration results, this amount increases over the measuring period. Probably this is due to groundwater leaking into the system since a relation is seen between the water level in the river IJssel and the DWF.

6.7 Calibration results

6.7.1 General

In this section the methods applied and results obtained from the calibration of the 5 selected storm events are presented. It is impossible to show and discuss all intermediate results. Therefore, the results for one storm are discussed in some detail.
The results of the other 4 storms are only discussed briefly, since they do not contribute any further to the illustration of the methods applied and the overall results obtained. Therefore, storm no. 1 is discussed in detail while the results for the remaining 4 storms are only presented briefly, detailed information on these storms are found in annexe VII.

6.7.2 Methods applied

During the study a total of about 100 calibration runs were made, in this process several systematic errors were removed and the parameterisation was adjusted for every individual storm.
The genetic algorithm as described in Section 5.10 was applied to identify possible local minima and to obtain a first insight into the sensitivity of the results for every individual parameter. Once the global location of the global minimum was identified, a combination of the Nelder-Mead and Levenberg-Marquart algorithm was applied to define the exact location and further properties of this minimum.

The following notation with respect to the parameters is applied:

- N1: mean value for the dry weather flow (m^3/h)
- Bx: depression storage (initial losses included) (mm)
- Ix: constant infiltration rate (mm/h)
- Fx: linear reservoir constant (s)
- CCy: overflow coefficient in the relation:

$$Q = L.*CCy*h_w^{\frac{3}{2}} \qquad \text{(eq. 6. 1)}$$

In Which:

L	weir length	(m)
CCy	overflow coefficient	($m^{0.5}/s$)
h_w	water depth above weir level	(m)
Q	discharge	(m^3/s)

The value of 1.5 for the power is fixed in the software used. As was discussed in Section 4.4.3.2 in practical cases a relation like:

$$Q = L*CCy*h_w^{\beta} \qquad \text{(eq. 6. 2)}$$

in which β may have a user defined value, is more appropriate.

- For 'x', '2' is used to identify the 'slow' surface, '7' for the 'quick' surface and '*' for a 'wildcard'. A 'wildcard' implies that no differences between both surface types are made.
- The value for y (in CCy) refers to the identity of the individual weirs.

Furthermore, the singular values in the tables and graphs presented are defined as $sigval_i = \frac{1}{\sqrt{\xi_i}}$, in which ξ_i is the eigen value for parameter i. in the parameter covariance matrix (see Section 5.3).

6.7.3 storm event August 25 1998

6.7.3.1 General

In Figure 143, the recorded cumulative rain depth for both rain gauges is shown. This storm event was by far the most interesting since the pumping station was out of order for most of the time (it started after 1971 minutes) due to maintenance activities. This implies that the pumping capacity was ruled out as a factor for a long period. Since it is impossible to show and discuss all results obtained from the various calibration runs made, only the most important results are reported.

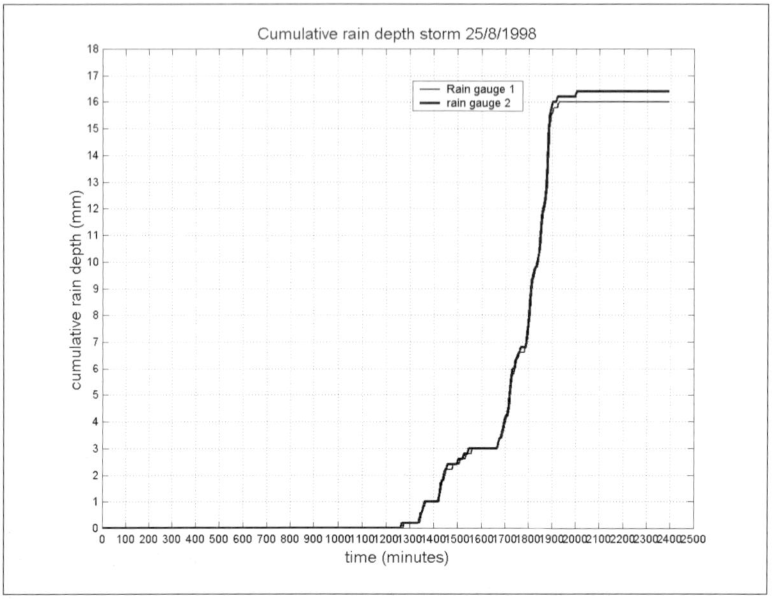

Figure 143: Cumulative rain depth recorded on August 25 1998 (raw data).

6.7.3.2 Results Genetic algorithm

The genetic algorithm as described in Section 5.6.4.4 was applied using 15 individuals and a total of 127 generations. This resulted in a quick identification (see Figure 144) of the global position of the global minimum in the goal function. Eventually 9 local minima were identified (see Table 69), some of them located closely together. It is noticed that the 'best' result (see Table 68) was not identified as a local minimum, probably because of too little function evaluations in its direct vicinity. This illustrates the fact that a genetic algorithm, combined with an algorithm for identifying local minima, is not to be applied as sole parameter optimisation technique. Application of the genetic algorithm was useful to avoid picking local minima as a starting vector for the more classical methods like the Levenberg-Marquart or Nelder-Mead algorithm. As is seen in Figure 144, the algorithm relatively quickly identifies parameter combinations resulting in small goal function values. Since the main goal for applying a genetic algorithm was to obtain information on the shape of the goal function, a total of 128 generations was allowed to evolve.

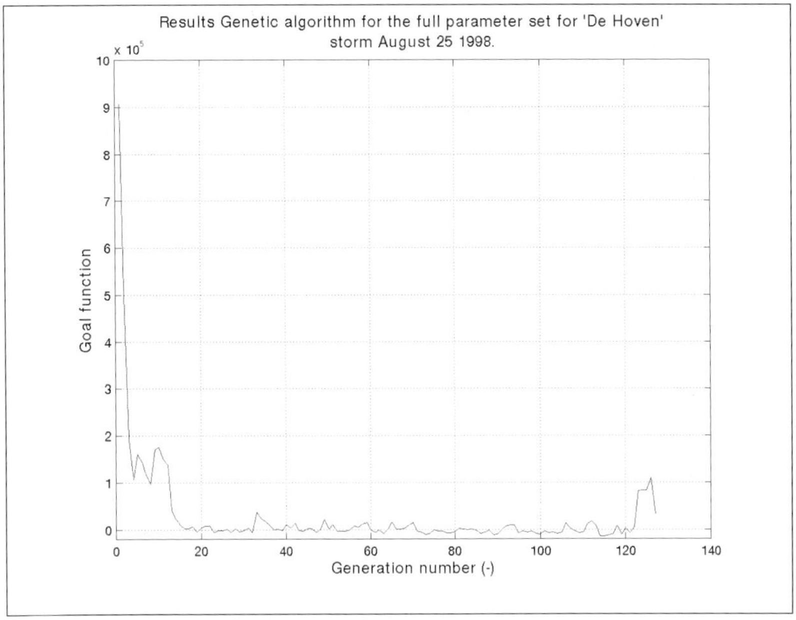

Figure 144: Results obtained using the GA on a full parameterisation for storm on August 25 1998. As can be seen the algorithm quickly reached a low level for the goal function (after circa 18 generations), no significant improvement is obtained in the following generations.

s	n1	b2	i2	f2	b7	i7	f7	c108	c17	c200
-14367.958	25.89832	0.3278	0.2046	356.2925	0.4619	0.3036	225.0506	0.3836	2.2295	1.703
-11853.118	25.95617	0.6091	0.1532	346.5549	0.1968	0.4719	200.211	0.797	1.3893	1.7516
-11358.438	26.37899	0.504	0.2492	317.2928	0.3181	0.2415	515.0255	0.4496	1.4279	1.7216
-10365.518	25.95895	0.6091	0.1564	241.764	0.3323	0.3767	193.8094	0.7976	2.0488	1.5927
-10046.359	25.96001	0.6153	0.1639	495.8122	0.2559	0.3929	54.1977	0.8957	1.9254	2.413

Table 68 : The 5 best results obtained from the GA on a full parameterisation.

s	n1	b2	i2	f2	b7	i7	f7	c108	c17	c200
-6593.2378	27.218215	1.2163	0.162	222.1185	0.2742	0.2539	500.7308	0.5242	3.0881	1.3209
-3904.5979	25.948756	0.6452	0.1022	655.1898	0.4584	0.3916	356.19	0.1937	2.7718	2.0238
-3732.718	27.867136	0.8788	0.1303	124.447	0.4326	0.4205	399.0006	0.6875	0.8662	2.9033
-9626.0381	27.188286	0.3208	0.2707	151.4815	0.4111	0.2826	346.7547	0.8207	0.4994	1.4432
-1645.678	27.147838	0.4334	0.2762	168.9754	0.3203	0.2565	563.883	2.0538	0.8359	1.2649
669.16302	27.148957	0.5588	0.2763	71.11821	0.2679	0.2563	330.1227	2.3891	0.8393	2.2884
-10020.038	26.923473	0.5787	0.1687	814.5138	0.2916	0.371	208.0716	0.4393	0.9769	3.0524
-11853.118	25.956178	0.6091	0.1532	346.5549	0.1968	0.4719	200.211	0.797	1.3893	1.7516
-10532.718	25.964799	0.6049	0.1751	104.6892	0.3574	0.3533	164.363	0.7983	1.2151	1.7751

Table 69: The 9 identified local minima in the goal function.

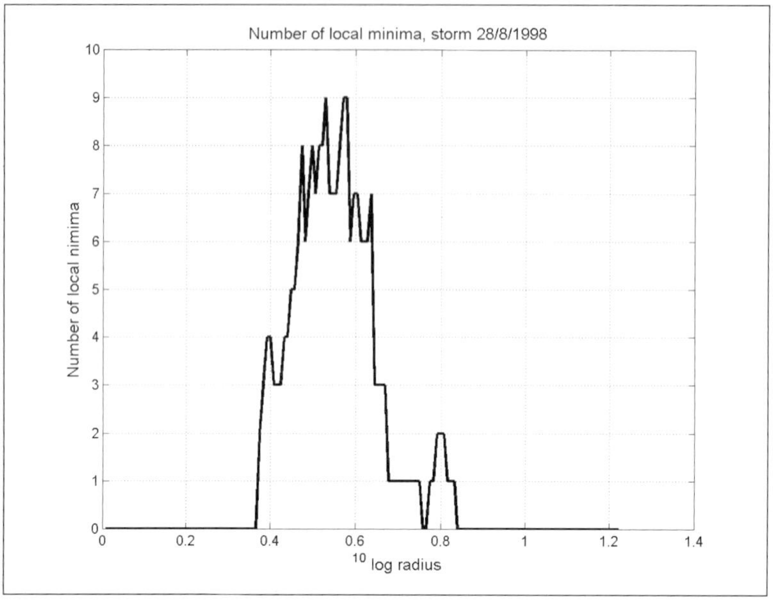

Figure 145: Number of identified local minima, as can be seen the minimum value was not identified as a local minimum, at radius > 0.83 no local minima are identified. From this Figure it can also be concluded that the local minima are at a distance of about 0.6 - 0.8 (in log transformed space).

Figure 145 the number of possible local minima as a function of radius is shown, as can be seen at a radius of 0.83 (in log10 transformed parameter space) no local minima are identified. This implies that, apart from the fact that the 'best' individual was not seen as a minimum, all minima found lie within a hyper sphere with a radius of $10^{0.83}$. The variation in magnitude of the parameters involved was 10^{-1}-10^{3}, this implies that the minima are located at a relatively large distance from one another. Another interesting observation is the difference in 'flatness' of the goal function along the different parameters axes. For instance, the goal function proves to be relatively sensitive for the DWF and the infiltration rates (see Figure 146 and Figure 147 respectively), while being far less dependent on the value for the linear reservoir constant (see Figure 148). This observation is a first indication for differences in identifiability for the parameters mentioned. It is to be expected that the linear reservoir constants can be quantified within a wide 95% confidence range only.

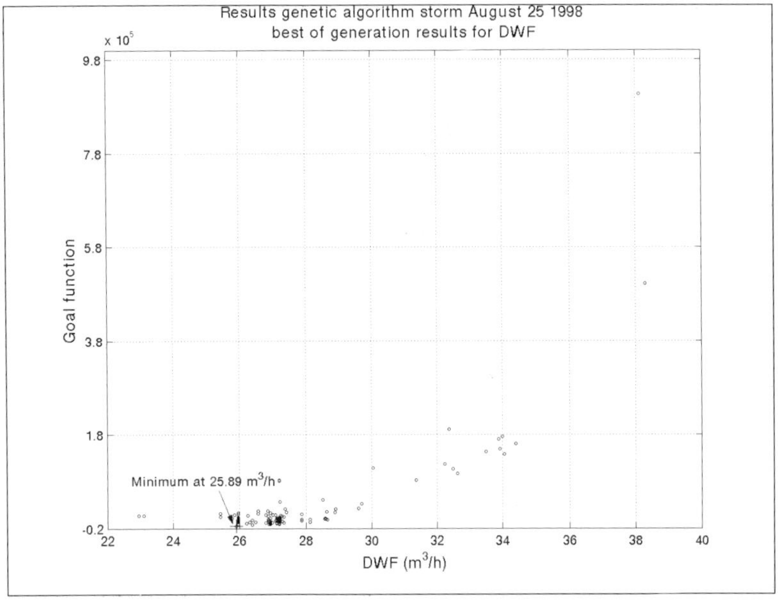

Figure 146: Results for best of generation for the DWF, as can be seen the goal function is relatively sensitive for this parameter. This is easily understood, since during a significant period the DWF is the only hydraulic load of importance working on the system.

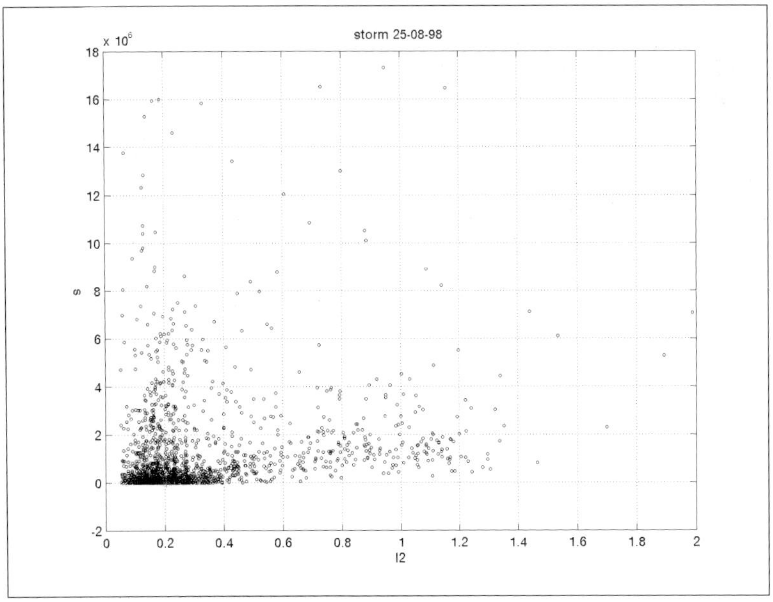

Figure 147: Results obtained from the genetic algorithm for the infiltration capacity I2.

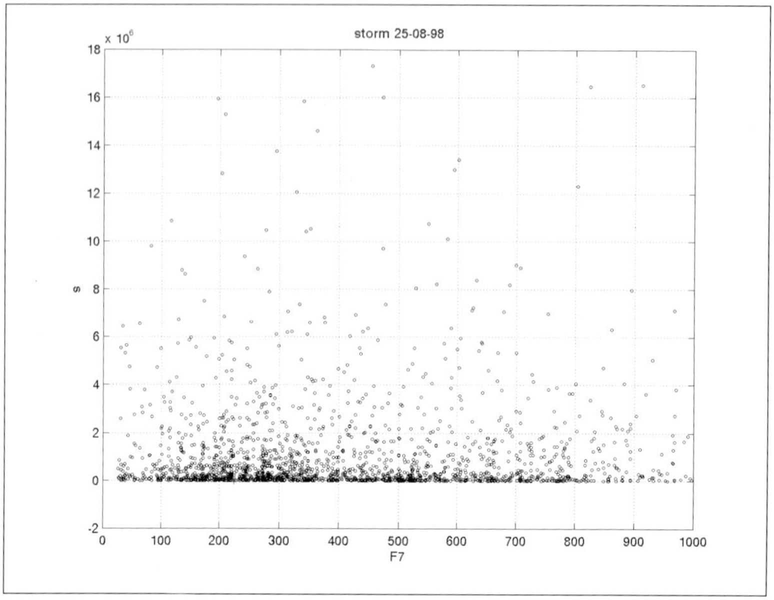

Figure 148: Results obtained from the genetic algorithm for the linear reservoir constant F7.

6.7.3.3 Results full parameter set

Taking the results obtained by the genetic algorithm as a starting point, 10 vectors (coinciding with the local minima and the best parameter combination) were further evaluated using the Levenberg-Marquart and Nelder-Mead algorithms. Eventually the starting vector obtained at the 'global minimum' identified by the genetic algorithm resulted in the best fit between model and measured results. Eventually the three 'local' minima in the direct vicinity of the best parameter set proved to result in the global minimum also, it is concluded that some unnoticed 'corridors' in the goal function were present.

6.7.3.3.1 Correlation between parameters

The correlation matrix obtained using the full parameterisation is shown in Figure 149. As can be seen correlation of any importance is only present between the parameters for the run-off model for the two types of surface on the one hand and the correlation between the parameter values for the weir constants at the other. These correlation are all negative, as can be seen the negative correlation exists between I2 and I7 ($\rho_{I2,I7}$=-0.8544), B2 and B7 ($\rho_{B2,B7}$=-0.55288), F2 and F7 ($\rho_{F2,F7}$=-0.63257) and between CC108, CC17 and CC200. This implies that when either one of the parameters is increased, the corresponding other parameters must be decreased in order to obtain a similar fit between model and measuring results.

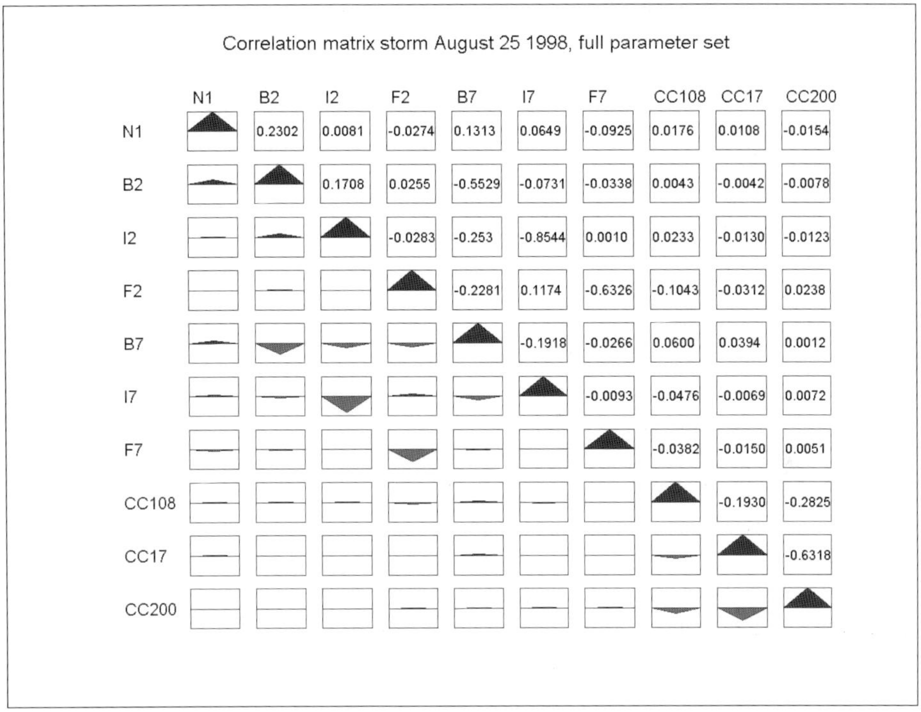

Figure 149: Correlation matrix storm August 25 1998, full parameter set.

6.7.3.3.2 Singular values and eigen vectors

As was argued in Chapter 5, the results from the singular value decomposition of the Jacobean at the minimum of the goal function and the eigen vectors of the parameter covariance matrix will give information on the identifiability of the parameters.
The results are shown in Figure 150 and comprise some interesting points. As can be seen the DWF (parameter N1) is the most important parameter since it dominates strongly the eigen vector with the largest singular value ($sigval_{N1}$=2137.057, the third column in Figure 150). This implies that the DWF is the parameter that is the most identifiable, an observation coinciding with the analysis of the results obtained using the genetic algorithm. This eigen vector has also small components along the parameter axes associated with the infiltration rates. These components have a positive sign whereas the component along the DWF axis has a negative sign. This implies that the residues, and thus the goal function, is sensitive for a situation in which the DWF is in- or decreased and the values for the infiltration rates are de- or increased. This is easily understood since both forms of variation result in a change in the volume of water entering the system. In this respect it is interesting to discuss the role of eigen vector 4. This vector has components along the axes associated with DWF, I2 and I7 also. These components have identical signs, implying that the goal function is sensitive for a simultaneous in- or decrease of the parameters mentioned (though somewhat less pronounced that reflected in vector 3 because of the smaller singular value). This is explained by the fact that the DWF is working on the system during the whole period taken into consideration whereas the infiltration is only of importance during (and shortly after) the storm period. E.g., an increase in DWF combined with an increase in the infiltration rates results in a situation in

which during the storm event the effects may cancel one another. In the period after the storm however, the increased DWF cannot be compensated by the increased infiltration, resulting in a deviation. Effectively vector 4 is additional to vector 3. The latter vector represents the 'overall' effect of parameter variation in DWF and infiltration while the former vector identifies a secondary effect that is due to the sequences of processes that influence the goal function. The interdependence between the overflow coefficients is beautifully illustrated in the eigen vectors 2,5 and 9. Vector 9 has the largest singular value ($sigval_{CC108,CC17,CC200}$=103.573), the three overflow coefficients are all equally dominant (i.e. the components of these parameters in the eigen vector are almost equal) and have an identical sign. This implies that the goal function is sensitive for a situation in which all overflow coefficients are either in- or decreased. In this respect the eigen vector 5 reveals that (though less pronounced because of the smaller singular value) the goal function is sensitive for a situation in which CC108 is increased and the values for CC17 and CC200 are decreased. A similar reasoning holds for the eigen vector no. 2, which shows the interdependence for CC17 and CC200. This can be explained as follows: In- or decreasing all overflow coefficients will result in an overall increase of the residues (and thus in a larger goal function value); this is the main effect, therefore is has a relatively large singular value. Vector 5, having a singular value that is about half that of eigen vector 9, shows that the overall residues (and thus the goal function) are sensitive for increasing CC108 and decreasing CC17 and CC200 simultaneously. Such a parameter variation results in a different distribution of the residue values over the measuring locations.

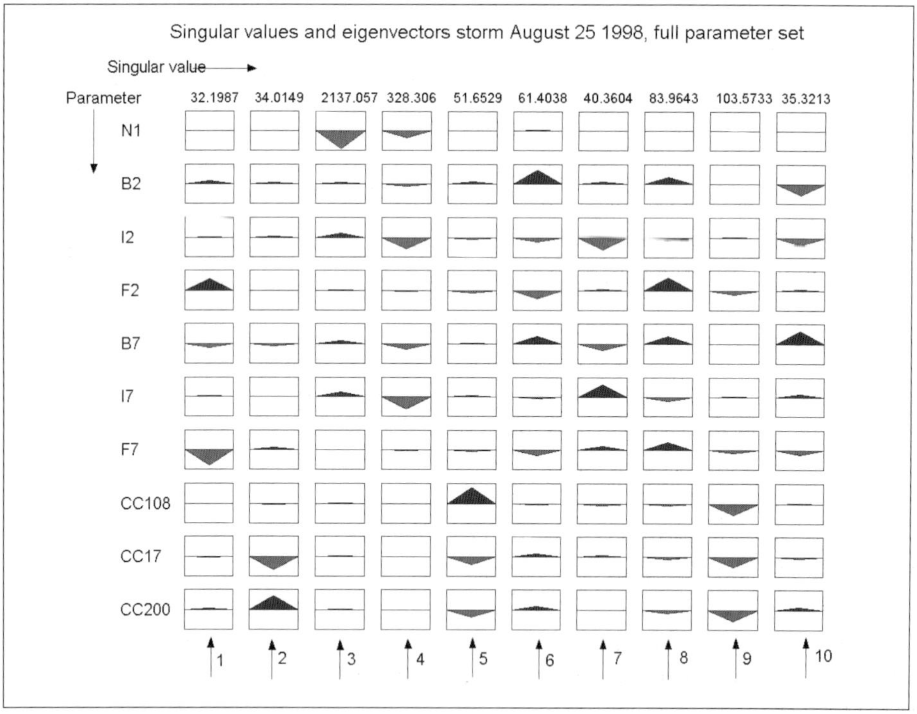

Figure 150: Singular values and eigen vectors, storm August 25 1998.

Less simple, but equally logical are eigen vectors 6 and 8 with singular values of 61.4038 and 83.9643 respectively. These eigen vectors essentially show the sensitivity for the combination of B2, B7, F2 and F7. In this case also there is a main effect and a secondary effect. The main effect is given by eigen vector 8, the goal function is sensitive for a simultaneous in- or decrease in all initial and depression losses and all linear reservoir constants. This is easily explained by the fact that such a parameter variation results in a combination of a change of total inflow and a shift in time for (storm)water entering the drainage system. The secondary effect (vector 6) is that the goal function is also sensitive (though less than for the primary effect) for a simultaneous in- or decrease in losses (B2 and B7) and a de- or increase in the linear reservoir constants (F2 and F7). Both parameter types do have an effect on the moment at which run-off starts; in other words, they influence the moment at which phase B (as defined in Section 6.5) starts. Phase B is the phase in which the water levels in the system are influenced by the run-off. Logically, the infiltration rates do have a role in this respect also; this is reflected in the (small) contributions these parameters have in eigen vectors 6 and 8.
It is concluded that the model parameters for the hydrological model can not be identified for both types of surfaces individually. An increase in parameter identification can only be obtained by either increasing the information obtained from the monitoring network (i.e. more measuring locations and/or a different combination of locations) or by adding prior information on these parameters obtained from separate measurements. A same reasoning is valid for the overflow coefficients.

6.7.3.4 Results reduced parameter set

From the conclusions obtained in the analysis of the eigen vectors for the full parameterisation, it worthwhile to try to reduce the number of parameters while maintaining the achieved level of MSE.
Since the weir coefficients show a strong mutual correlation (as was confirmed by the analysis of the eigen vectors), these are clustered (i.e. only one weir coefficient is taken into account, without differentiation between the individual weirs.). No distinction is made with respect to the type of surface.
So, the alternative parameterisation is: DWF, surface storage, infiltration rate, linear reservoir constant and weir coefficient, implying a reduction from 10 to 5 parameters.

Full parameter set		Reduced parameter set	
parameter	Stdfac	parameter	Stdfac
N1	1.0029	N1	1.0026
B2	1.0597	B*	1.0180
I2	1.0410	I*	1.0056
F2	1.0489	F*	1.0291
B7	1.0645	CC*	1.0180
I7	1.0412		
F7	1.0666		
CC108	1.0378		
CC17	1.0546		
CC200	1.0524		

Table 70: Standard deviation factors for the parameter values for the full and the reduced parameter set. The standard deviation factors defined the 95% accuracy interval as was defined in chapter 5.

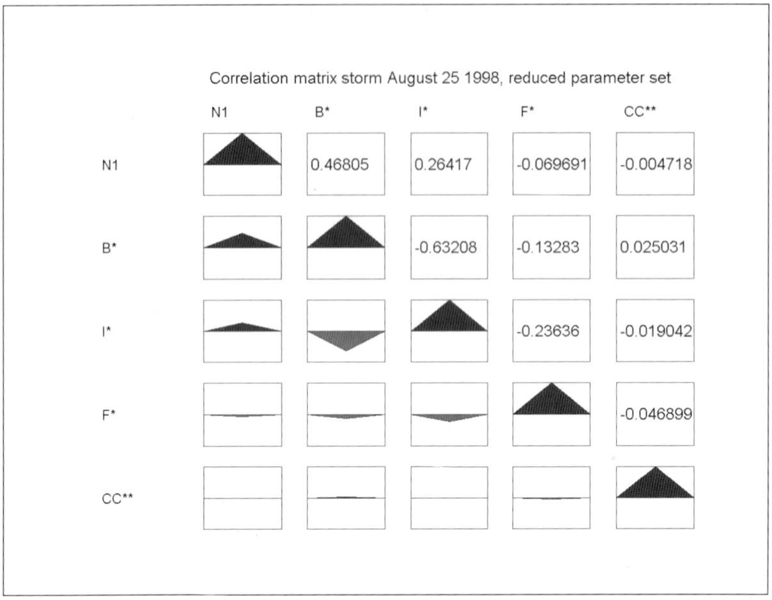

Figure 151: Correlation matrix reduced parameter set, storm August 25 1998.

In the full parameter set the correlation between the parameters was easily explained. When studying the correlation matrix for the reduced parameter set (see Figure 151) it is seen that some correlation is introduced between infiltration and surface storage and between DWF and surface storage and infiltration rate. On the other hand the linear reservoir constant F* and the weir coefficient CC* show no significant correlation with any of the other parameters.

Full parameter set		Reduced parameter set	
Singular value	Dominant parameter(s)	Singular value	Dominant parameter(s)
2137.0575	N1	2603.4696	N1, I*
328.306	I2,I7	525.9795	N1,I*
103.573	CC (all)	130.061	CC*
83.9643	F2,F7	127.1748	B*
61.4038	B2	79.7958	F*
51.6529	CC108		
40.3604	I2, I7		
35.3200	B2,B7		
34.0149	CC17, CC200		
32.1987	F2, F7		

Table 71: Singular values and dominant parameters in the eigen vectors for the full parameter set and the reduced parameter set. As can be seen the parameter reduction results in a (slight) increase in the singular values. This implies an increased identifiability.

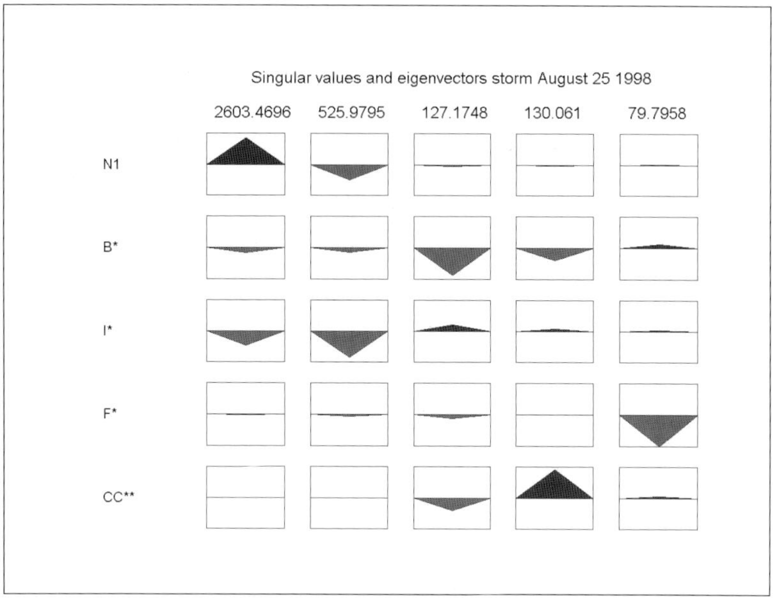

Figure 152: Singular values and eigen vectors, storm August 25 1998, reduced parameter set.

When comparing the results obtained after reducing the parameter set, it is concluded that a slight increase in identifiability of the parameters is obtained. As can be seen in Table 71 and Figure 152, especially the identifiability for the linear reservoir constant and the weir coefficient is increased. Which parameter set is to be preferred, can only be decided based on the analysis of the residuals.

6.7.3.5 Analysis of the residuals

The residuals are analysed with respect to:

- Probability distribution (compared to the normal distribution).
- the standard deviation.
- the root mean squared error.
- Bias.

In Figure 153 and Figure 154 the probability distributions obtained using the full and the reduced parameter set respectively are shown. As can be seen, the interval for the residues is somewhat smaller for the full parameter set than that for the reduced parameter set; (-0.146,0.112) versus (-0.150,0.120). Both distributions have the same shape and deviate from Gaussian, especially in the range (-0.01, -0.035). This deviation is caused by a systematic error in the period from 1971 minutes onwards to the end of the time interval when the system is being emptied. This is due to small geometrical deviations in the database as compared to the geometry of the real system. As was pointed out in Section 6.3, a deviation in the water level-volume relation is present in the water level range 5.15 m^+ NAP and 5.22 m^+ NAP. This leads to a deviation in water levels in model and measurements of circa 3 cm. This illustrates the high demands that are to be put on the database and the quality of the measuring results.

Since neither the software used nor the available information on pumping action, wastewater discharges etc. offer the possibility to match the volume held by the system in the small level range in which the deviation was found, no further improvement could be obtained.

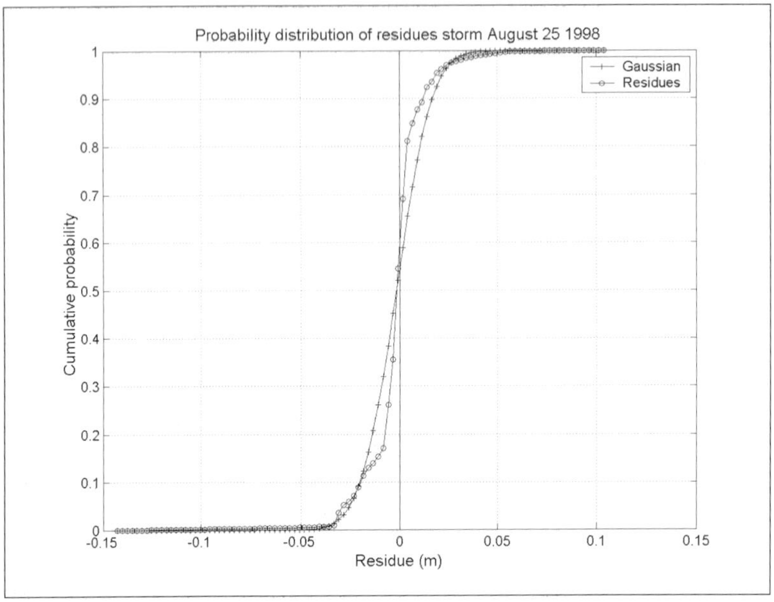

Figure 153: Probability distribution for the residues obtained using the full parameter set.

In Table 72 the statistical properties of the residues for the full and the reduced parameter set are summarised. As can be seen no significant differences occur. In Figure 156 the distribution in time for the variance, mean squared error and bias in a shifting time window of 200 minutes for gauge 108 is shown. As can be seen the variance and the MSE increase as soon as the storm starts (t=1253 minutes), Bias is especially present in the DWF mode (period before t=1253) and during the emptying stage after t=2001 minutes.)

	gauge 108 full parameter set	gauge 108 reduced parameter set	gauge 17 full parameter set	gauge 17 reduced parameter set	gauge 200 full parameter set	gauge 200 reduced parameter set	total of residues full parameter set	total of residues reduced parameter set
MSE	$2.196*10^{-4}$	$2.624*10^{-4}$	$2.424*10^{-4}$	$2.528*10^{-4}$	$1.708*10^{-4}$	$1.664*10^{-4}$	$2.112*10^{-4}$	$2.253*10^{-4}$
VAR	$2.195*10^{-4}$	$2.628*10^{-4}$	$2.303*10^{-4}$	$2.352*10^{-4}$	$1.700*10^{-4}$	$1.667*10^{-4}$	$2.102*10^{-4}$	$2.258*10^{-4}$
RMSE	$1.482*10^{-2}$	$1.623*10^{-2}$	$1.557*10^{-2}$	$1.59*10^{-2}$	$1.307*10^{-2}$	$1.29*10^{-2}$	$1.453*10^{-2}$	$1.513*10^{-2}$
STD	$1.482*10^{-2}$	$1.622*10^{-2}$	$1.518*10^{-2}$	$1.53*10^{-2}$	$1.304*10^{-2}$	$1.29*10^{-2}$	$1.449*10^{-2}$	$1.501*10^{-2}$

Table 72: Statistical key- values for the residues obtained after applying the full and the reduced parameter set. As can be seen there are no significant differences between the full and the reduced parameter set.

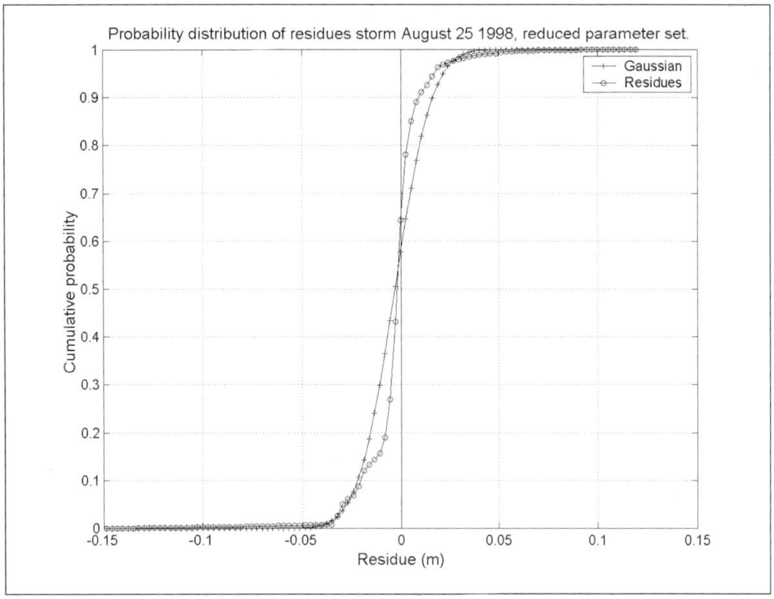

Figure 154: Probability distribution for the residues obtained using the reduced parameter set.

Bias is present at all three gauges and over substantial time intervals. For both parameter sets the following conclusions with respect to bias are drawn:

- Relative bias (see Figure 157 and Figure 158) does not comply with the demands that are to be put (i.e. bias<0.2 σ_r^2).
- Relative bias is especially present in the DWF period (prior to t= 1262 minutes) and after t=1971 when the pumping station has come into action.
- In an absolute sense the variation (as well as MSE and Bias, see Figure 155) are at their maximum during the overflow and during the emptying of the system.

As was argued before, bias especially occurs in the period after t=1971 minutes for reasons mentioned. On the other hand, substantial bias is present in the DWF period prior to the storm event. This is possibly caused by:

- The presence of sludge layers not accounted for in the geometrical database
- Relatively high relative measuring accuracy (circa 6 mm versus water depths of 3 to 5 cm).
- A possible temporally variation in pumping capacity due to variation in pressure head on the suction side of the pump.
- The 'shallow water depth' problem encountered in modelling (see Chapter 2).
- The effect of incidental local DWF discharges into the system.

When comparing bias and the distribution of it in time and over the three locations no significant differences between the full and the reduced parameter set are found.

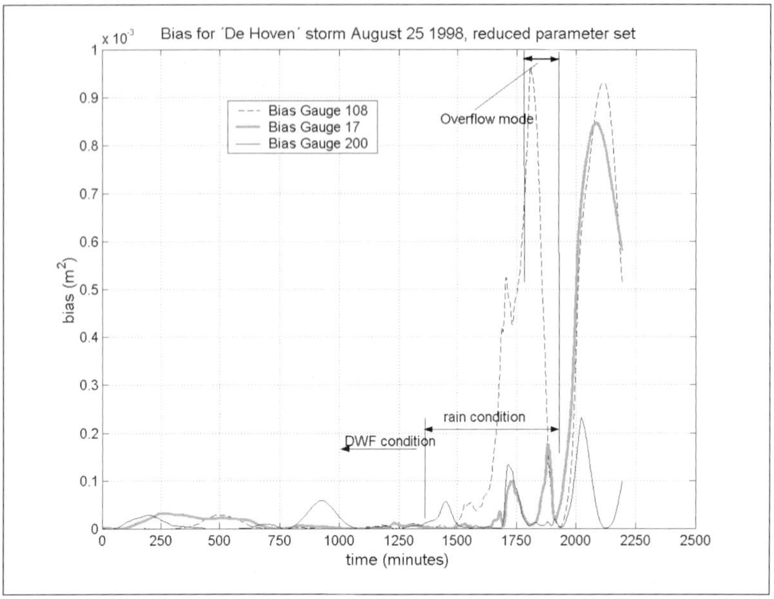

Figure 155: Bias for the three gauging locations in a shifting time window of 200 minutes.

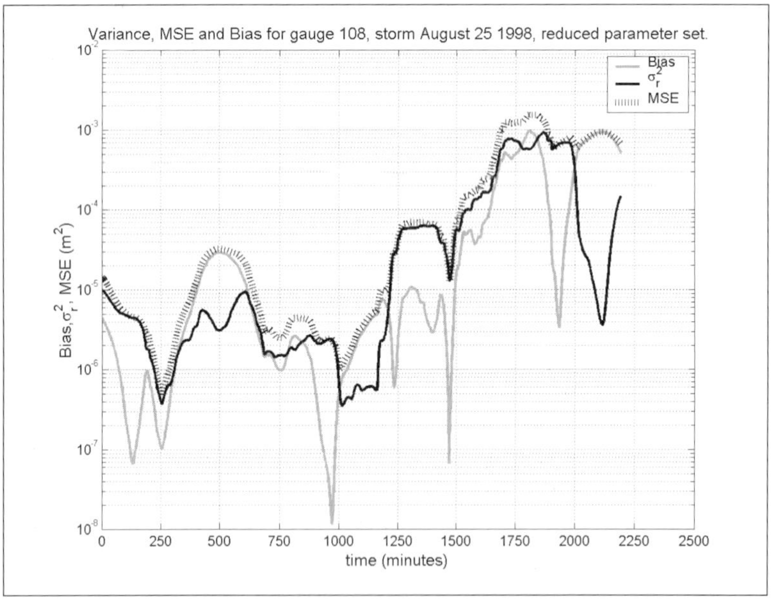

Figure 156: Variance, MSE and BIAS for gauge 108 in a shifting window of 200 minutes. As can be seen substantial bias is present over some periods.

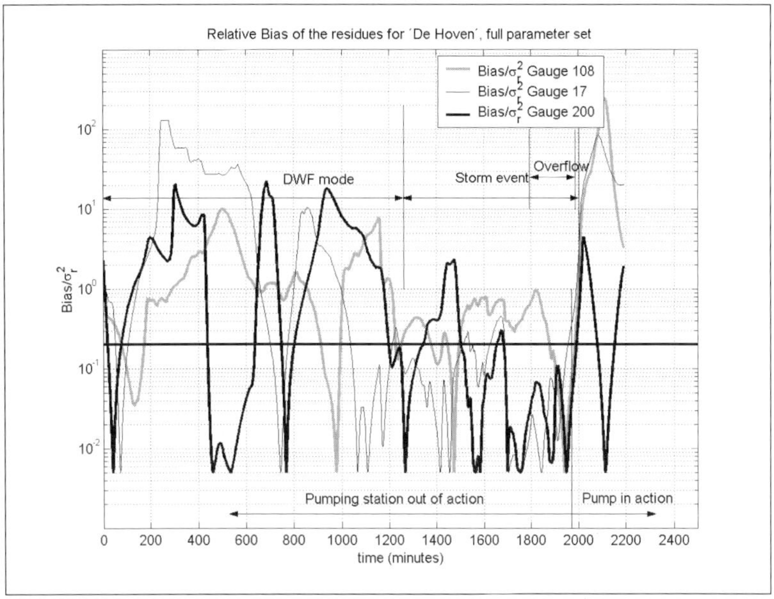

Figure 157: Relative bias for the full parameter set, it must be concluded from this graph that especially in the period between start (t=0) and t=1050 minutes and the period from t=1971 minutes substantial bias is present.

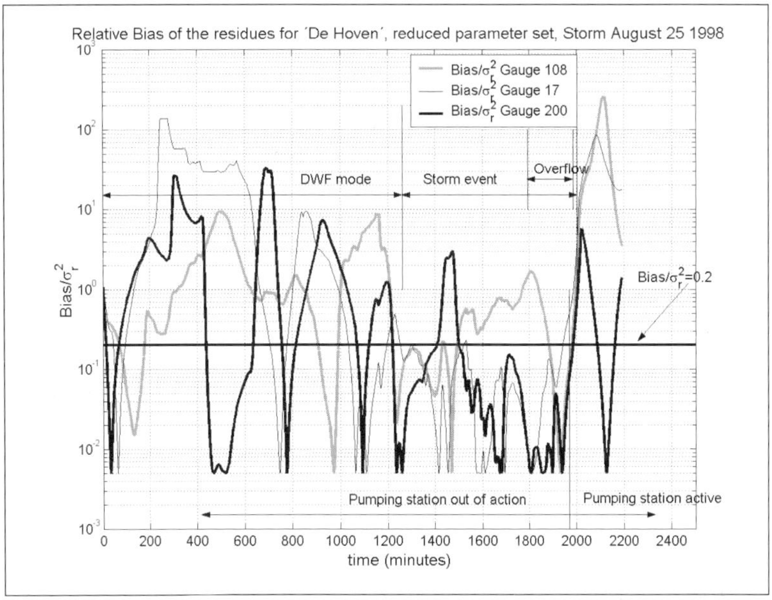

Figure 158: Relative bias for the reduced parameter set. When compared to the Bias resulting from the full parameter set (see Figure 157) it is seen that the reduction in parameters did not increase bias.

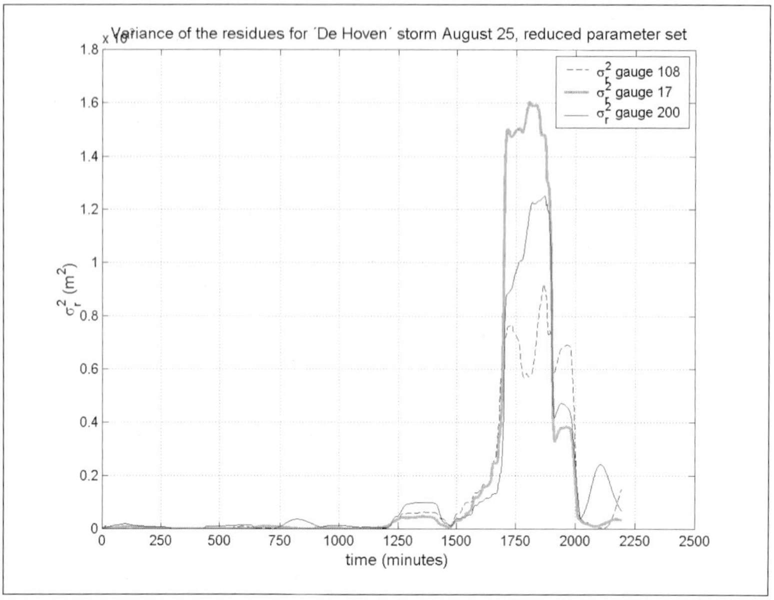

Figure 159: Variance in a shifting window of 200 minutes, reduced parameter set.

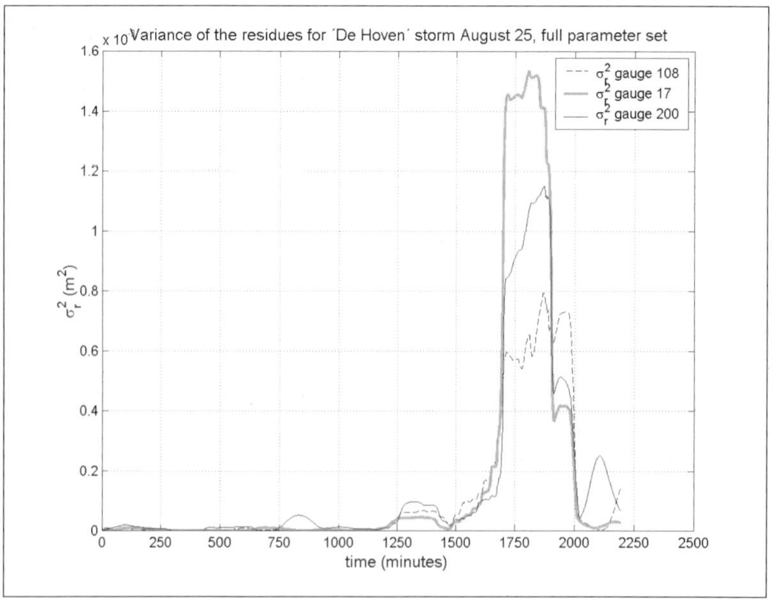

Figure 160: Variance in a shifting window of 200 minutes, full parameter set.

Since no significant differences exist between the results obtained using either the full or the reduced parameter set, application of the latter is to be preferred, since using the reduced set results in smaller 95% confidence intervals for the parameters.

The differences are, however small. Furthermore, it must be noted that the results obtained for this particular storm are significantly better than those for the other storms, as will be described in the next sections.

In Table 73 the results for the two sets of parameters are shown, as can be seen the standard deviation factors for the reduced parameter set are somewhat smaller that those for the full parameter set.

Full parameter set			Reduced parameter set		
Parameter	parameter vector	$Stdfac_{corr}$	Parameter	parameter vector	$Stdfac_{corr}$
N1 (m^3/h)	25.95	1.0029	N1 (m^3/h)	26.16	1.0026
B2 (mm)	0.364	1.0597	B* (mm)	0.490	1.0180
I2 (mm/h)	0.186	1.0410	I* (mm/h)	0.229	1.0056
F2	347.86	1.0489	F* (s)	276.58	1.0291
B7 (mm)	0.504	1.0645	CC* ($m^{0.5}/s$)	1.0342	1.0180
I7 (mm/h)	0.311	1.0412			
F7	205.31	1.0666			
CC 108 ($m^{0.5}/s$)	0.383	1.0374			
CC 17 ($m^{0.5}/s$)	2.156	1.0546			
CC 200 ($m^{0.5}/s$)	1.751	1.0524			

Table 73: Results calibration storm 1.

The resulting 95% confidence intervals are shown in Table 74 for both parameter sets.

Parameter	95% confidence interval, full parameter set	Parameter	95% confidence interval, reduced parameter set
N1 (m^3/h)	25.80-26.10	N* (m^3/h)	26.02-26.30
B2 (mm)	0.324-0.409	B* (mm)	0.473-0.508
I2 (mm/h)	0.172-0.202	I* (mm/h)	0.226-0.232
F2	316.18-382.71	F*	261.16-292.91
B7 (mm)	0.444-0.570	CC * ($m^{0.5}/s$)	0.998-1.072
I7 (mm/h)	0.287-0.337		
F7	180.47-233.57		
CC 108 ($m^{0.5}/s$)	0.356-0.412		
CC 17 ($m^{0.5}/s$)	1.939-2.398		
CC 200 ($m^{0.5}/s$)	1.581-1.939		

Table 74: 95% confidence interval parameters for storm 1.

In order to illustrate the results obtained, the measured and calibrated water levels are shown as a function of time in Figure 161, Figure 163 and Figure 165. In Figure 162, Figure 164 and Figure 166 the residues are shown. In the initial vector referred to in these graphs, is the result obtained by the 'best' parameter vector from the genetic algorithm.

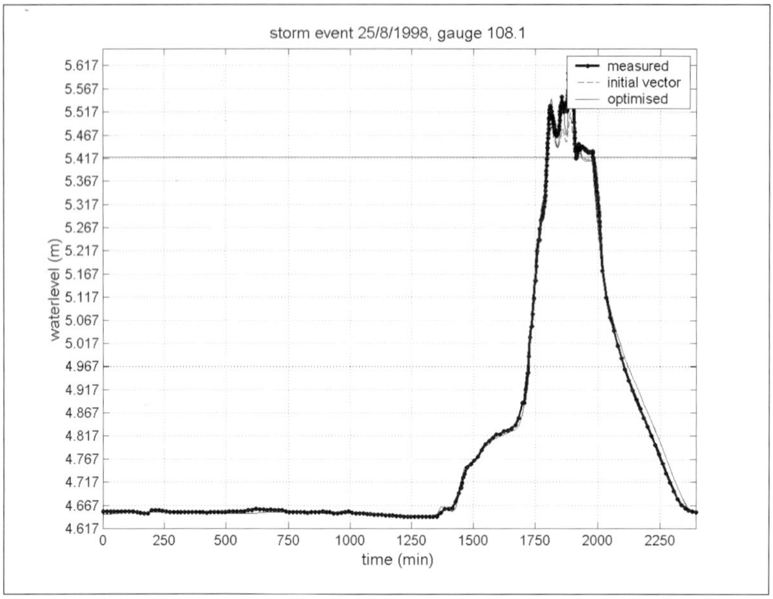

Figure 161: Calibration result for the storm on August 25, gauge 108.

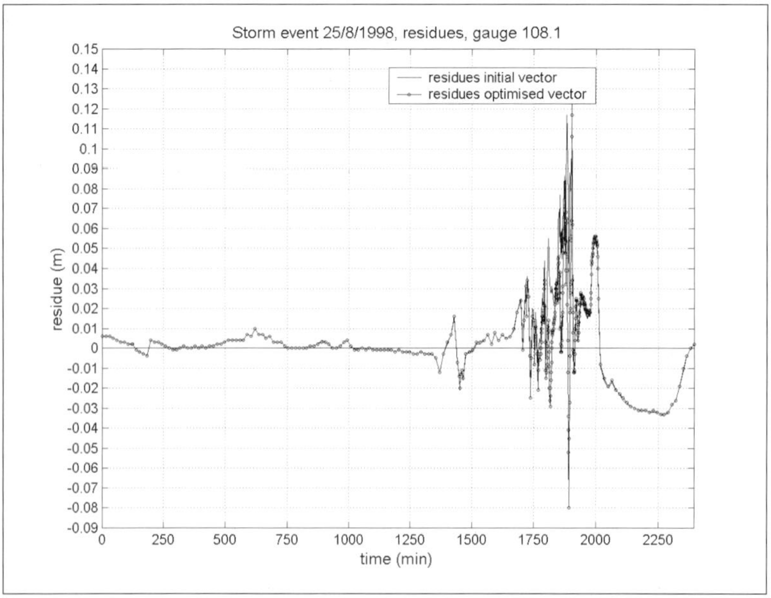

Figure 162: Residues for gauge 108, storm on August 25 1998.

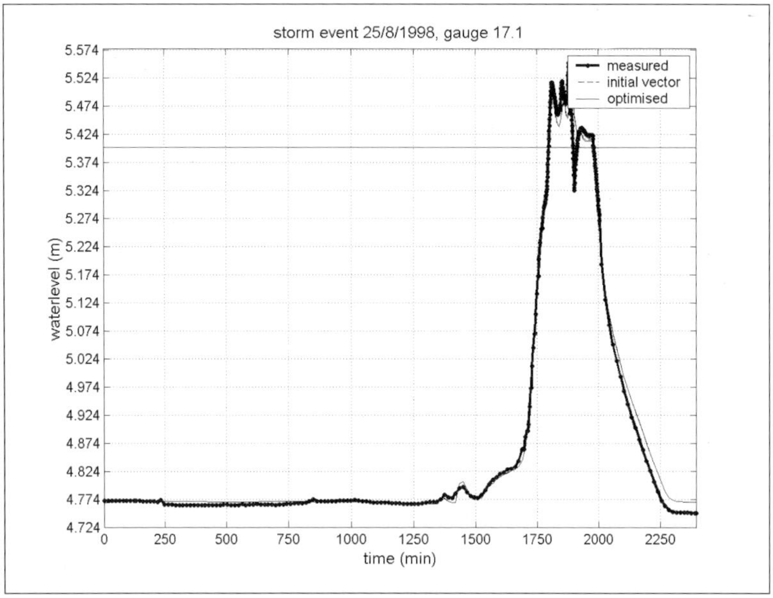

Figure 163: Calibration result for the storm on August 25, gauge 17.

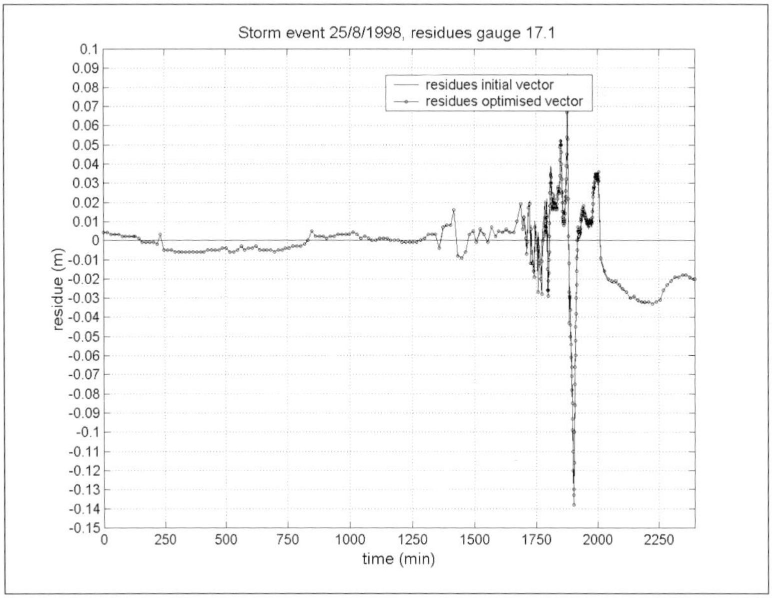

Figure 164: Residues for gauge 17, storm on August 25 1998.

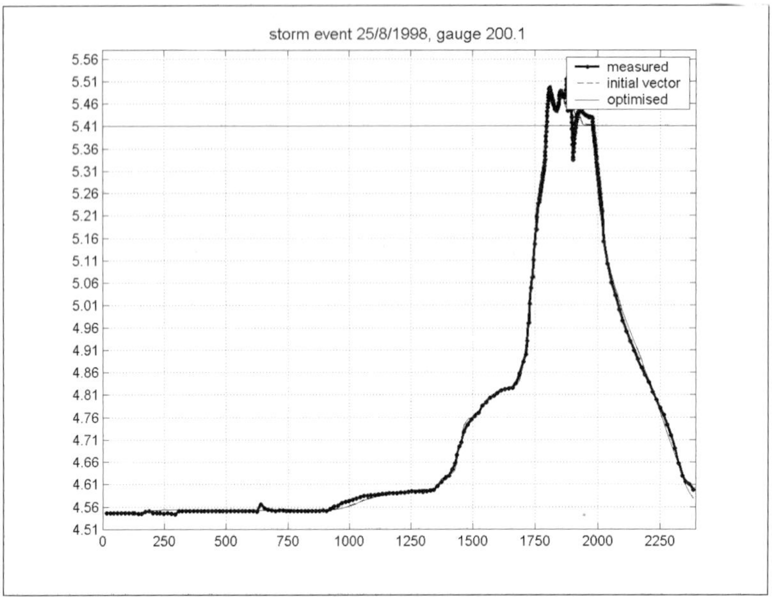

Figure 165: Calibration result for the storm on August 25, gauge 200.

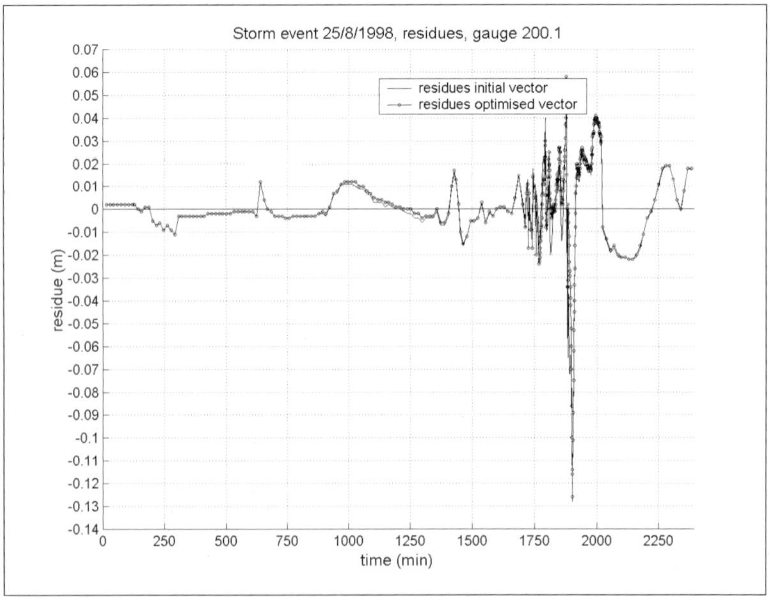

Figure 166: Residues for gauge 200, storm on August 25 1998.

When studying the residues for every gauging location, it is seen that at t=1905 minutes a sudden drop in water level (dropping even below weir crest) shows up in the measuring data. There is no proven explanation for this, suspect number 1 is the pumping station. The water level drop is easily explained if the pumping station had come into action at that moment. No proof of such an event was found in the recordings.

6.7.4 Storm event September 2 1998

The rain intensities measured at the two rain gauges are depicted in Figure 167. The storm is characterised by the fact that it contains several periods of precipitation and dry periods in between them. The maximum intensity was 15.4 l/s.ha., while the total amount of rain was 19.4 mm (averaged value for the two rain gauges). This storm event did not lead to an external overflow but resulted in a maximum rise in water level at the measuring locations of about 0.30 m. Following an identical procedures as was explained in the discussion on the storm on August 25 1998, the parameterisation [N1,B2,I2,B7,I7,F*] was found to be the best, judging by the mutual correlation, eigen vectors and confidence intervals obtained.

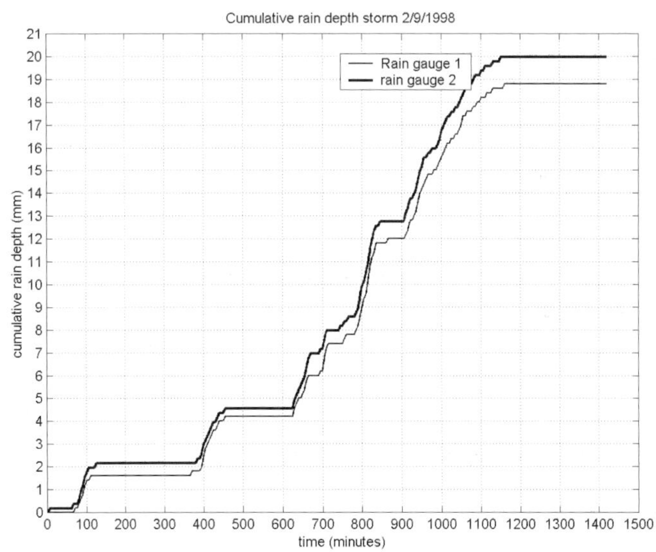

Figure 167: Cumulative rain depth storm event 2: September 2 1998.

Parameter	vector	$Stdfac_{corr}$	Singular value	Dominating parameters
N1 (m^3/h)	48.79	1.06589	51.3331	N1
B2 (mm)	1.96	1.12292	22.5596	N1, I2 ,I7
I2 (mm/h)	0.42	1.07955	77.8298	I2, F*
B7 (mm)	0.30	1.14788	19.8278	I7, B7
I7 (mm/h)	0.26	1.14749	12.3219	I7, B7, I2, B2
F* (s)	226.44	1.02015	860.9652	F*

Table 75: Calibration results for storm 2.

Parameter	95% interval
N1 (m^3/h)	42.94-55.43
B2 (mm)	1.55-2.47
I2 (mm/h)	0.36-0.49
B7 (mm)	0.23-0.40
I7 (mm/h)	0.20-0.34
F* (s)	217.57-235.66

Table 76: 95% confidence interval for storm 2.

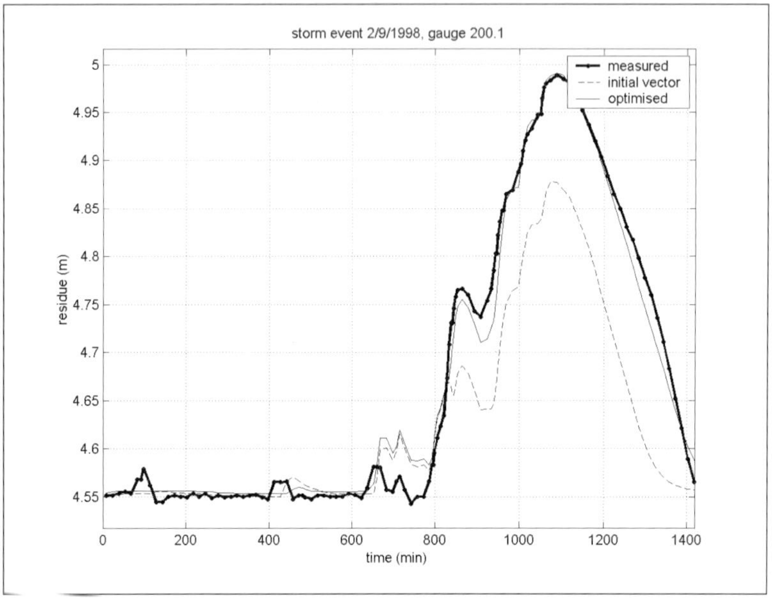

Figure 168: Results at gauge 200.

6.7.5 Storm event October 7 1998

In terms of information held, the storm of 7th October does not differ essentially from the storm on 2nd September. The main differences however, are the conditions with respect to ground water levels and the fact that in the period before the event smaller rain events took place that may lead to a difference in initial conditions with respect to infiltration capacity and depression storage. The recorded cumulative rain depth is shown in Figure 169.

Parameter	vector	$Stdfac_{corr}$	Singular value	dominating parameters
N1 (m^3/h)	65.25	1.04267	89.9172	N1
B2 (mm)	3.34	1.06313	357.9525	B2
I2 (mm/h)	0.13	1.17071	18.2625	I2,I7,B2,B7
B7 (mm)	0.28	1.17275	13.3677	B7
I7 (mm/h)	0.10	1.19591	11.1154	I7,I2
F* (s)	742.64	1.10881	23.9708	F*

Table 77: Calibration results for storm 3.

Parameter	95% confidence interval
N1 (m^3/h)	60.02-70.94
B2 (mm)	2.96-3.77
I2 (mm/h)	0.095-0.18
B7 (mm)	0.20- 0.39
I7 (mm/h)	0.07-0.14
F* (s)	604.04-913.04

Table 78: 95% confidence intervals for storm 3.

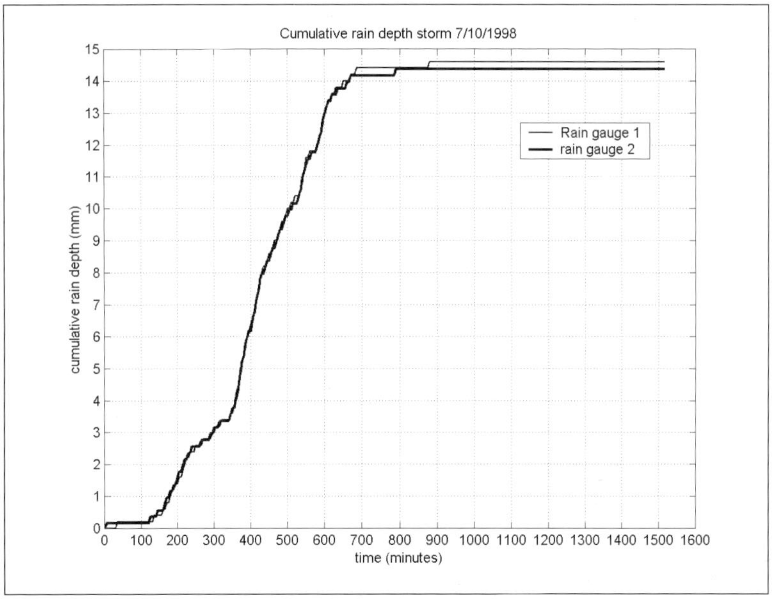

Figure 169: Recorded cumulative rain depth, storm 3: October 7 1998.

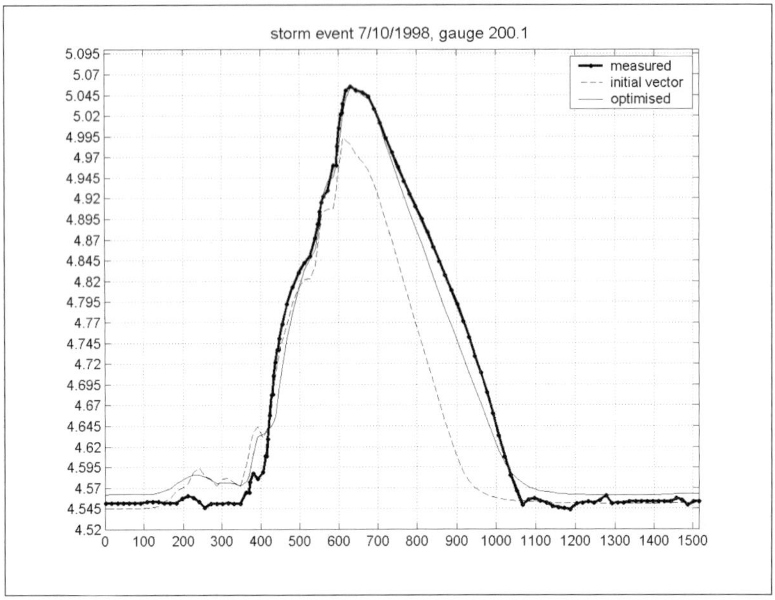

Figure 170: Example of the results obtained for storm 3: October 7 1998.

In this case a reduction in the parameter set did not result in an increased identifiability. The confidence intervals, however are relatively large, see Table 78.

6.7.6 Storm event October 9-13 1998

This storm event is in fact not one single event but merely a concatenation of relatively short periods of low rain intensities (see Figure 171). The differences between the recorded rain depth are significant in this case (a total of 33.0 mm versus 27.8 mm). As was explained in Section 4.4.4, the rain gauges applied are vulnerable for errors at low rain intensities. So the observed difference may stem for instrumental error, there was no proof of this in the logbook however. In spite of the doubts with respect to the rain gauges, this period was incorporated in the study to see if the simple model assumptions applied with respect to surface routing, infiltration and surface storage can lead to usable results.

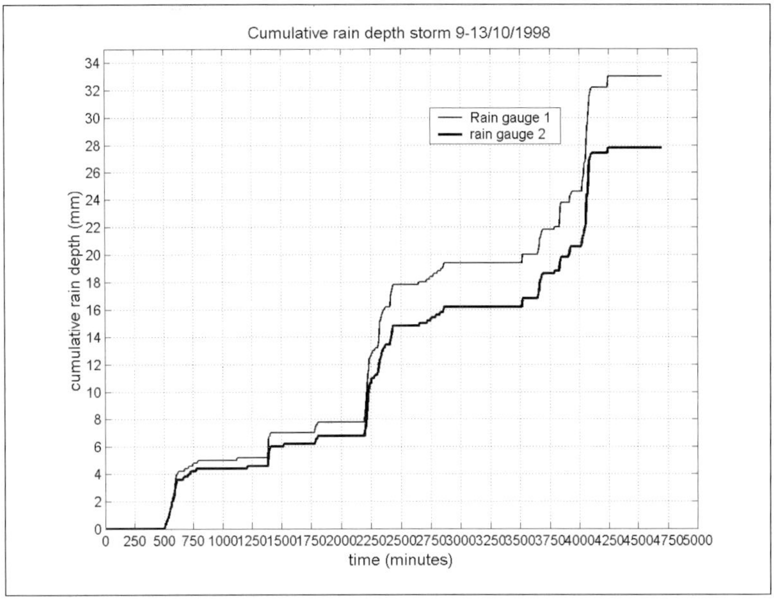

Figure 171: Recorded cumulative rain depth in storm 4: October 9-13 1998. As can be seen there is a noticeable difference between both rain gauges.

Parameter	parameter vector	$Stdfac_{corr}$	Singular value	Dominant parameters
N1	63.83	1.021980	17.1041	B7,I7,F*
B2	2.30	1.034688	18.0572	B7,I7,F*
I2	0.30	1.079244	28.5394	B7,I7,F*
B7	0.13	1.115253	87.4581	N1,I2
I7	0.09	1.124541	1343.2180	N1,I2,I7,F*
F*	63.57	1.098322	356.1142	N1,I2

Table 79: Calibration results for storm 4.

Parameter	95% confidence interval
N1 (m^3/h)	61.11-66.67
B2 (mm)	2.15-2.46
I2 (mm/h)	0.25-0.35
B7 (mm)	0.10-0.16
I7 (mm/h)	0.07-0.11
F* (s)	52.70-76.68

Table 80 : 95% confidence intervals for storm 4.

Prior to the discussion on the parameter values in Section 6.8, it must be noticed that especially the parameter value for the linear reservoir constant is doubtful. The low value of 63.56 s, in comparison with the values of 200-700 s obtained for the other storms, combined with a relative high value for the DWF points at a 'shortage' of run-off. The optimisation algorithm tries to compensate for this by adding DWF and 'speeding up' the run-off by decreasing the linear reservoir constant.

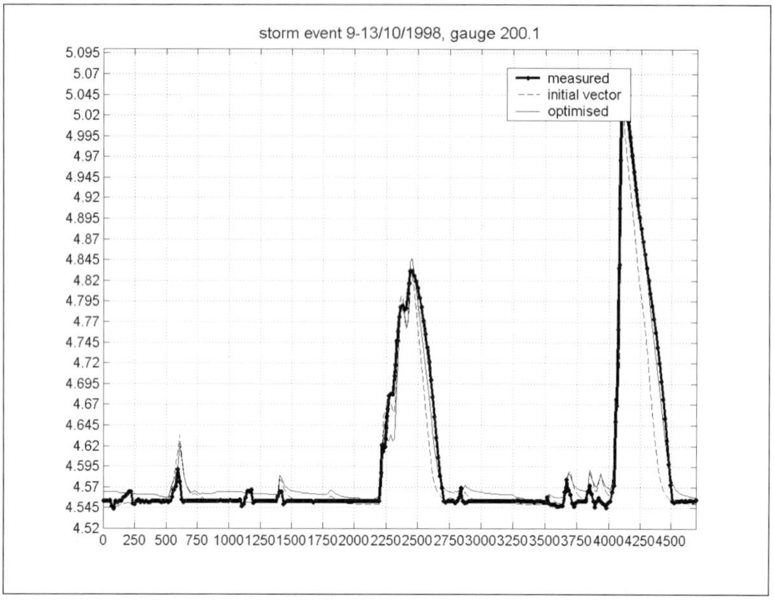

Figure 172: Example of the results obtained for storm 4: October 9-13 1998.

This 'shortage' of run-off probably stems from a systematic error in rain gauge 2. The assumption for a too large value for the DWF is further strengthened by the observation that during the DWF periods the model systematically produces too large water depths (see Figure 172).

6.7.7 Storm event October 24 1998

The cumulative rain depth on 24 October is shown in Figure 173. This event led to an external overflow, and was therefore fit to be used to calibrate the overflow coefficients of the weirs. Furthermore, the run-off parameters were incorporated in the calibration process. In this storm a similar situation with respect to the parameterisation evolved as was found for the storm on August 25. Therefore, the results obtained on the full parameter set are shown also.

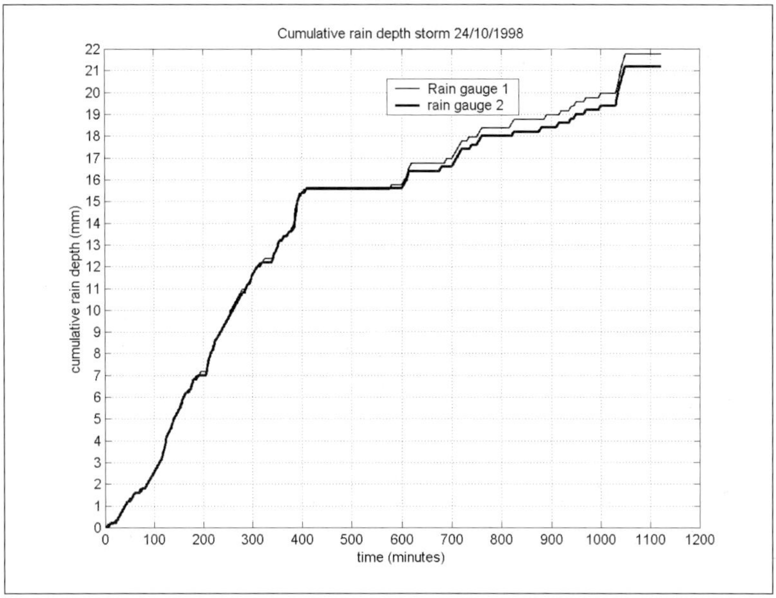

Figure 173: Recorded cumulative rain depth for storm 5, October 24 1998.

Full parameter set			Reduced parameter set		
Parameter	vector	$Stdfac_{corr}$	Parameter	vector	$Stdfac_{corr}$
N1 (m^3/h)	49.83	1.02355	N1 (m^3/h)	51.96	1.01301
B2 (mm)	0.5447	1.05278	B* (mm)	0.4471	1.03473
I2 (mm/h)	0.2284	1.03565	I* (mm/h)	0.1833	1.01849
F2 (s)	878.45	1.02094	F* (s)	669.21	1.01793
B7 (mm)	0.1943	1.08127	CC* ($m^{0.5}/s$)	1.2750	1.05745
I7 (mm/h)	0.0720	1.08055			
F7 (s)	196.52	1.07689			
CC 108 ($m^{0.5}/s$)	1.1395	1.24098			
CC 17 ($m^{0.5}/s$)	1.1721	1.22840			
CC 200 ($m^{0.5}/s$)	1.0605	1.18151			

Table 81 : Calibration results for storm 5, full and reduced parameter set.

Full parameter set		Reduced parameter set	
Parameter	95% interval	parameter	95% interval
N1 (m^3/h)	47.56-52.20	N1 (m^3/h)	50.62-53-32
B2 (mm)	0.4914-0.6037	B* (mm)	0.4176-0.4787
I2 (mm/h)	0.2129-0.2450	I* (mm/h)	0.1767-0.1901
F2 (s)	842.79-915.63	F* (s)	645.84-693.43
B7 (mm)	0.1662-0.2272	CC* ($m^{0.5}/s$)	1.1402-1.4257
I7 (mm/h)	0.0617-0.0841		
F7 (s)	169.21-227.91		
CC 108 ($m^{0.5}/s$)	0.7399-1.7548		
CC 17 ($m^{0.5}/s$)	0.7767-1.7687		
CC 200 ($m^{0.5}/s$)	0.7597-1.4804		

Table 82 : 95% confidence interval for storm 5.

Analogous to what was seen in storm 1, the 95% confidence intervals are significantly reduced after reducing the parameter set, in this case also the effect of the parameter reduction on the residues was small. In the obtained results, bias was predominantly present, for the full parameter set as well as for the reduced parameter set. Furthermore, for the full parameter set the overflow coefficients and the parameters associated with area type 7 were less identifiable than the other parameters. In this case also the DWF proved to be the dominant parameter also.

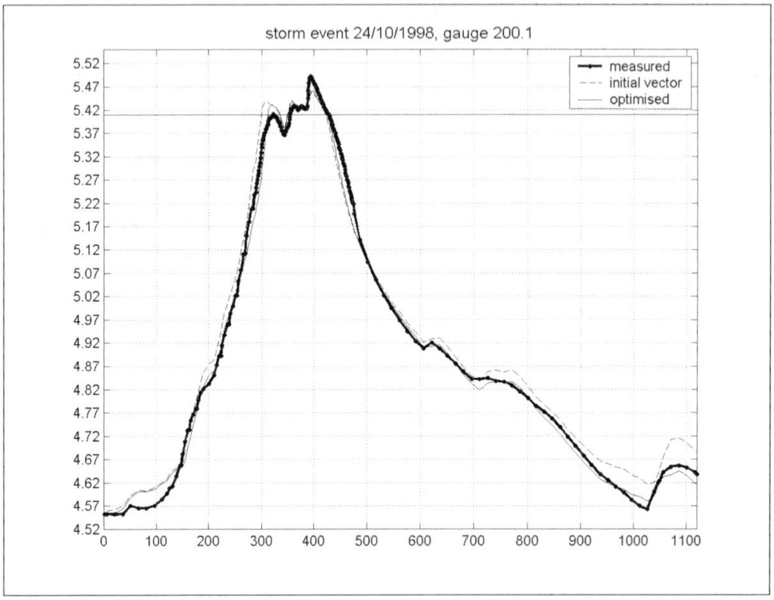

Figure 174: Example of the results obtained for storm 5, October 24 1998.

6.8 Portability

Formally, when calibrating models, a distinction is made between calibration and verification. Verification implies that the parameter values found in the calibration process are tested on storm events not taken into account in the calibration process. Lei et al (1999) argue this classical approach is to be abandoned since the portability of the parameter values obtained in the calibration process is poor.

In the case presented, the optimal parameterisation and the parameter values obtained differed per storm event (see Table 83). It is therefor clear that there is a problem with respect to portability of the parameter values obtained; on this issue the argument by Lei et al is agreed upon. With respect to abandoning the classical calibration/verification approach however some reservations are to be made. The calibration process reveals clues were to look for causes of the poor portability. Tackling them and trying to improve on either models applied, databases used or measuring setups imposed with the aim to eventually obtain a proper calibration/verification procedure should be a goal in scientific development.

The causes for a poor portability are;

- The processes modelled are not correct. Especially the processes incorporated in the hydrological model are strongly simplified.

- Small database errors, especially with respect to the anticipated distribution of the run-off areas over the manholes and anticipated the occurrences of local, incidental wastewater discharges.
- The presence of sludge layers in the sewers, changing in dimensions and location with time, not accounted for in the model.
- Restrictions due to the software applied, for instance the modelling of the volume distribution over the depth range with respect to small-bore pipes poses a problem. (see discussion on the results of storm 1). Another point in this respect is the use of the Kindsvater-Carter equation for overflows imposing a fixed value of 1.5 as value for the power relation.
- The mean value for the DWF varies due to groundwater draining in and out the system.

Parameter	storm 1	storm 2	storm 3	Storm 4	storm 5
N1 (m^3/h)	25.80-26.10	42.94-55.43	60.02-70.94	61.11-66.67	47.56-52.20
B2 (mm)	0.324-0.409	1.55-2.47	2.96-3.77	2.15-2.46	0.4914-0.6037
I2 (mm/h)	0.172-0.202	0.36-0.49	0.095-0.18	0.25-0.35	0.2129-0.2450
F2 (s)	316.18-382.71				842.79-915.63
B7 (mm)	0.444-0.570	0.23-0.40	0.20- 0.39	0.10-0.16	0.1662-0.2272
I7 (mm/h)	0.287-0.337	0.20-0.34	0.07-0.14	0.07-0.11	0.0617-0.0841
F7 (s)	180.47-233.57				169.21-227.91
CC 108 ($m^{0.5}$/s)	0.356-0.412				0.7399-1.7548
CC 17 ($m^{0.5}$/s)	1.939-2.398				0.7767-1.7687
CC 200 ($m^{0.5}$/s)	1.581-1.939				0.7597-1.4804
F* (s)	-	217.57-235.66	604.04-913.04	52.70-76.68	
Reduced parameter set for storms 1 and 5					
N* (m^3/h)	26.02-26.30				50.62-53-32
B* (mm)	0.473-0.508				0.4176-0.4787
I* (mm/h)	0.226-0.232				0.1767-0.1901
F* (s)	261.16-292.91				645.84-693.43
CC* ($m^{0.5}$/s)	0.998-1.072				1.1402-1.4257

Table 83 : 95% intervals for the parameter values obtained for storms 1-5.

As can be seen the values obtained for the overflow coefficients show a wide 95% range while the mean values obtained in storms 1 and 5 differ substantially.
During part of the measuring period the groundwater table and the water level in the river IJssel was monitored. In Figure 175 these levels are shown together with the values found for the DWF in the drainage system.
In the period between August 4 and September 9 the groundwater table was at an almost constant level. Still a serious difference between the mean DWF in the dry weather period August 6-13 and the storm on September 2 on one hand and the storm on August 25 on the other hand was observed. This may be explained from the fact that during the storm on August 25 1998 the water pressure in the system was relatively high (up to 2.5 m water column([55]) during a relatively long period, whereas during the storm on September 2 1998 this pressure was during a short period only 0.5 m water column. A possible explanation for the low DWF is exfiltration of water out of the drainage system into the groundwater. After September 13 1998 an increase in the groundwater table as well as in the water level of the IJssel is seen. The large DWF during the storm on October 7 and in the period between 9 and

([55]) Measured relative to the groundwater table.

13 October 1998 ([56]) may be explained by infiltration of groundwater into the drainage system.

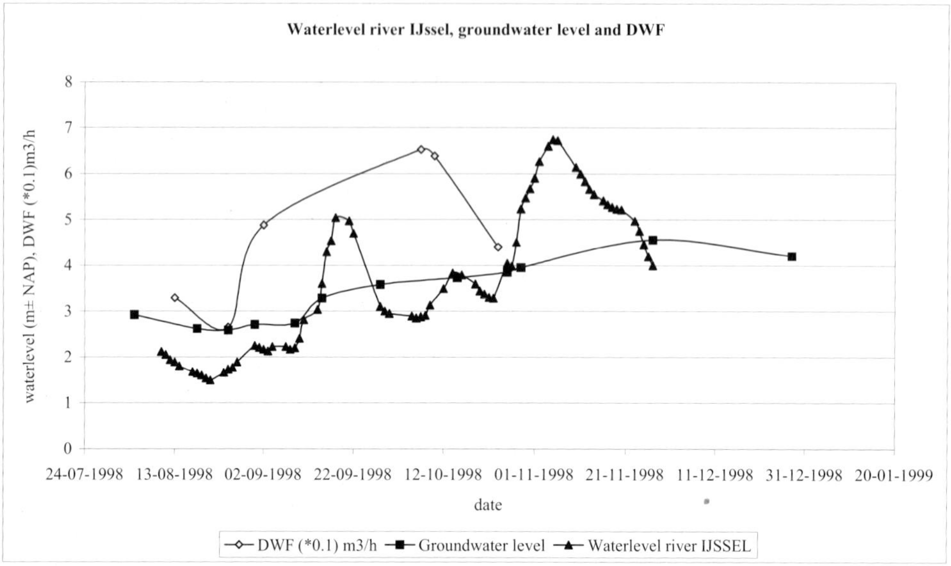

Figure 175 : Ground water levels, water levels in the river IJssel and the results for the DWF in the calibration runs.

The suspicion of leakage was backed up by some physical evidence when inspecting the pumping cellar, a noticeable amount of groundwater was leaking into the system at that point. In any case, the DWF is an important factor in the hydraulic behaviour of the system and is suspected to be influenced to a large extent by the groundwater level. The number of observations on the DWF is far too small to draw consistent conclusions on the relationship between ground water level and extra DWF caused by infiltration.

6.9 Conclusions

The main conclusions drawn from applying a mathematical optimisation technique for calibration of an urban drainage networks are:

- In the cases presented it was not possible to meet the constraints put on uniqueness and identifiability for the parameter values obtained. In every storm discussed, bias in the residues is present, although every practically possible effort was done to remove systematic errors from database and measuring data. It was possible however, to identify the probable causes of bias in the results. The hydrodynamic software applied however, did not offer the possibility to eliminate the underlying systematic error.
- Relative bias is larger during dry weather periods when compared to periods in which a noticeable rain intensity occurs. This implies that the model is more capable of a correct description of hydrodynamics under storm loads than under dry weather loads.

([56]) It must be noted that the calibration result for the storm period October 9-13 1998 is not trusted completely in terms of the absense of systematic errors in the rain recordings, see Section 6.7.6 also.

- When studying the residues it turns out that in every storm the residues are correlated in time and in space (i.e. the residues as a function in time are more or less the same for every individual gauging station). This implies that either there is a systematic error in the measuring data or a systematic error exists in the model. Since a fairly simple model describing initial losses, routing and infiltration was applied, it is suggested to seek the cause here. This implies that, using the applied model, the portability of model parameters is limited. In this sense a more complex process description of the run-off may possibly enhance the results. This however, will result in many more model parameters to be calibrated, resulting on its turn in more severe demands on the amount and quality of the measuring data.
- The parameter values obtained, especially for the run-off model, vary from storm to storm.
- The working sequence proposed in Chapter 5, using simple methods and widely accessible tools, for calibrating a hydrodynamic model proved to be applicable for a practical case. Using this working sequence results, apart from a set of model parameters also in a clear understanding of the shortcomings of the software, the model and the geometrical database. This knowledge is of importance when applying the model for judging the effects of measures to be taken in the system.
- In environmental regulations the main focus is on discharges from sewer overflow on surface waters. It must be stated, however that without an in situ calibration of the discharge-water level relation or direct discharge measurement, no accurate values for discharges or spilled volumes can be obtained by applying a calibrated model as was discussed in this chapter.

In the author's opinion several issues should be investigated further in order to stimulate the practical application of model calibration in urban drainage:
- The correct modelling of shallow water depths is of ominous importance when calibrating a model. An important source of bias is due to shortcomings in the modelling of flow under these specific conditions.
- A simple run-off model was applied, an investigation into the best run-off model has to be performed. Under 'best' should be understood a model that optimises between number of parameters, constraints on the data set and the portability of parameter values.
- At this moment the parameterisation used in the calibration process is largely based on expert judgement and posterior criteria put on calibration results. In the opinion of the author, however, a more rational and repeatable technique for parameterisation should be developed. As reported by Speed & Ahlfield (1996) it is possible to use singular value decomposition techniques in order to quantify the maximum number of parameters that can be identified given the measuring data and the model. These methods however, are only valid at the optimum of the goal function and under constraints with respect to linearity.

CHAPTER 7: SUMMARY

7.1 English summary

7.1.1 General

The main goal of this thesis is to report on working methods related to the use of (computer) models in the field urban drainage engineering, which is neither purely scientific nor purely a craft. Over the last two decades, especially the wide availability of cheap computer power and software has had an impact on the possibilities for practising engineers to study the hydrodynamic behaviour of drainage systems. The possibilities open by modelling seem unlimited and led to the idea that 'the whole world' can be put into a computer. In the 1980's and 1990's this resulted in a wide application of so-called 'pollutional load' models making prediction on the pollutional load discharged by sewer overflows into surface waters possible. In this respect, calibration of the models applied was only seldom effected. Several reasons account for this, apart from a lack of measuring data it was realised that the underlying processes were only poorly understood while the modelling accuracy of the quantity processes (water flows) was not known. Therefore, in this thesis the ability of the modern generation hydrodynamic models to describe reality is studied. This reality in its turn can only be known by means of measuring data. Where model results suffer from inaccuracies in data bases and assumptions with respect to the modelled processes and process parameters, the accuracy of measuring data is limited because of the inherent measuring inaccuracies and the limitations there are in terms of resolution in time and space and the duration of a measuring project. Combining the different characteristics of model and measurement, with the aim to obtain an increased accuracy and reliability of the model, is the process of calibration. This is the main item in this thesis. An important prerequisite for the methods applied in this thesis is that they are either readily available or are relatively easy to manufacture by any specialist in the field of urban drainage. The thesis discusses the relevant processes when modelling the hydrodynamics in a drainage system, the translations of these into mathematical formulae and the implementation of these in a (computer) model (Chapter 2). In Chapter 3, the effects of errors and inaccuracies in databases on the model results are discussed. Chapter 4 discusses measuring projects in which especially the aspects measuring accuracy, measuring frequency and measuring methods are related to each other. In Chapter 5, the calibration process, matching model and measuring results is discussed. In this chapter, the wide applicability of the methods described is of ominous importance. The calibration process, which comprises more than just 'fitting' a model, is illustrated by means of some examples. Chapter 6, finally, reports on a case study in which the methods described in chapter 5 are applied under practical conditions. In the remainder of this summary a short description of the individual chapters along with the main conclusions is given.

7.1.2 Chapter 1: Introduction

Chapter 1 describes briefly the history of urban drainage and the role of hydraulic models in engineering practice. In a separate paragraph, the 'National Dutch guidelines for hydrodynamic modelling' are discussed. The main ideas and working methods stipulated in these guidelines are followed in the thesis. In the thesis however, it is tried to incorporate measured reality into modelled reality.

7.1.3 Chapter 2: Hydrodynamic models

In chapter 2 the theoretical foundations being applied in modern hydrodynamic models are briefly discussed. Discrimination is made between hydraulic and hydrological sub-models, describing in-sewer processes and run-off processes respectively. An important issue in this chapter is the stability of the Preissmann scheme; a numerical scheme that enjoys great popularity. A well-known problem with this scheme is that it tends to show instability like behaviour at shallow water depths. In literature, the cause of this is thought to be the non-linear geometries applied in urban drainage. In this chapter it is shown that this explanation is at least incomplete, the occurrence of $2\Delta x$ waves is an inherent property of the Preissmann scheme. Furthermore, it is shown that increasing the hydraulic roughness, suggested as a remedy to avoid $2\Delta x$ waves, may in some cases even aggravate the problem. This is just one example of the limitations a computer model has in describing reality; avoiding zero divisions, limiting CPU time and the inherent limitations in accuracy a computer has are being circumvented by so called 'programmers solutions'. In a separate paragraph the influence of these programmers solutions on the overall result are discussed. To this end, two software tools based on identical principles but developed in two strictly separate environments are applied on one well-defined database. The differences found are relatively small when compared to the differences found between models and measured reality. The conclusion in this respect is that when applying the commonly accepted principles the resulting software is at least internally consistent.

7.1.4 Chapter 3: Data and data errors

In Chapter 3 the focus is on the structure and geometry of an urban drainage network that has to be incorporated into a database in order to make is accessible for hydrodynamic calculations. Generally it is stated that the exact description of all dimensions (profile dimensions, street levels, conduits lengths etc.) and the structure of the drainage system are known with a limited accuracy only. The limitation in this accuracy is because of several causes; loss of information, poor archivation of revised plans, measuring inaccuracies, negligence of data bases management over the years. In this chapter methods developed in day-to-day engineering practice to check (large) databases on errors and the efficient elimination of them are described. Still, even after thoroughly checking a database a certain percentage of the data will prove to show errors. By means of Monte Carlo simulation on two catchments the extent to which a databases should be error free is set at 1 %. This implies that when 1% of the data suffer from errors of such a nature they are not traceable without detailed fields survey the influence of these errors is in the same order of magnitude compared to other sources of model uncertainty like parameter uncertainly and the measuring inaccuracies that are present in all dimensions introduced in the database. Another important source of model inaccuracy is the catchment area. Although modern digital based methods using standardised digital maps are accurate in itself, deviations occur with reality. This is because of:

- Not all pavement on private terrain is incorporated.
- Recent changes made in the urban infrastructure may are not always incorporated on the available maps.

7.1.5 Chapter 4: Measuring programs

Chapter 4 deals with monitoring of drainage systems. The design of measuring setups in relation to the main measuring objective is the main item in this chapter. Measuring accuracy, measuring frequency and the choice of the measuring locations, all influence the information content a monitoring network is capable to produce. Depending on the original purposes with which a monitoring network is designed, different demands are put on the variables mentioned. It is shown that apart from the measuring objective, the specific characteristics of the system in which the monitoring network is to operate have a significant influence also on the design. This leads to a situation in which a monitoring network is designed by application of a non-calibrated model, the measuring data obtained from the networks are eventually used to enhance the model. In the ideal case this would lead after some iteration to a 'perfect' measuring network and a 'perfect' model, in practice though this is hard to achieve.
In the last few paragraphs in chapter 4 the available measuring methods and achievable measuring accuracies are briefly discussed. Generally, it is concluded that flow measurement is more expensive and less accurate than water level measurement. Therefore, when calibrating a model water level measurement is to be preferred, this is more so when many (temporary) monitoring locations are to be arranged.

7.1.6 Chapter 5: Calibration

Chapter 5 deals with calibration of hydrodynamic models in urban drainage. First, the terms validation, calibration and verification and their mutual relation are explained since these terms are in practice often mixed up. Validation comprises the tests needed to see if the processes incorporated in a model are in line with the observable reality. Calibration is the 'tuning' of the model in such a manner that for a specific case the model and reality are identical to within a quantified difference. Verification is testing if the calibrated model is able to produce results for a measured case to within a certain accuracy. Parameter optimisation is just one, albeit important, part of the calibration process. At least as essential is the analysis of the residuals (differences between model and measurements), after 'optimal' parameter values are obtained. The residues contain important information with respect to imperfections either in the measurements (systematic measuring errors) or in the model (e.g. incorrect or incomplete process, software bugs) or in the geometrical data available. Apart from a number of 'classical' (gradient based) parameter optimisation methods and more recently developed genetic algorithms, the focus is on the analysis of the residues.
It is difficult predict which optimisation method is 'best' in a given practical case; based on the cases discussed it is concluded that combining 'classical' optimisation techniques with a genetic algorithm is successful. A genetic algorithm is able to quickly give a estimated location of the optimum, while gradient 'classical' methods give detailed information on the identifiability and mutual correlation of the model parameter obtained. The last two paragraphs working sequences are described by which to calibrate a hydrodynamic model in urban drainage. These methods can relatively easily be applied by anyone having access to hydrodynamic software and some third or fourth generation programming language.
In addition, a first suggestion for a working method to use a hydrodynamic model for designing a measuring network when the measuring objective is 'to obtain measurements for calibration purposes' is given. The method is based on a generalised sensitivity analysis of the model combined with the information content of the individual measuring locations with respect to the model parameters that are to be estimated.

7.1.7 Chapter 6: Case study

In chapter 6 a case study in which the process from data acquisition on to the calibration of a hydrodynamic model is described. The case deals with a relatively small catchment in which a combined sewer system is present. The measuring project was performed in the autumn of 1998. Eventually, using the methods described in the previous chapters, it proved to be possible to obtain a 'fit' between model and measurements to within a small (order of magnitude of cm) range of differences in modelled and measured water level. The model parameters obtained differ in their absolute value from individual storm to individual storm. For the DWF, this is explained by the occurrence of in- and exfiltration of water. For the other parameters (especially the parameters associated with run-off processes) the actual processes are probably more complicated than is assumed in the hydrological models applied. Because of the limited information content of the monitoring data available, application of more refined hydrological models (implying still more model parameters to be incorporated in the calibration) was not opportune. As 'by-product' of the calibration process, some geometrical errors in the database were recognised and eliminated, these errors were of the type not be found by normal control procedures as described in chapter 3. Furthermore the effect the storage volume contained in 'small structures' (like gully pots) was found to be crucial when striving for a calibrated model, eventually the storage volume of the drainage systems was quantified to within 1.2 % accuracy. In addition the location at which the discrepancy between 'real' volume and modelled volume was present was identified to be present in a water level range of about 15 cm. Due to limitations in the format of the database applied, it proved not possible to correct this. In order to achieve an enhancement of the results of model calibration in urban drainage in the future, the following recommendations are given:

- Design the monitoring system using methods like described in chapter 5 of this thesis.
- It is essential to keep databases up to date, small errors in geometry or structure of the system can have a large influence on the results obtained by model. This will also effect the quality of decision for (re) construction based on these model results.
- Relatively minor improvements in the hydrodynamic models (software) applied will enhance the possibilities for calibration :
    - elimination of the shallow water depth problem (see chapter 2) by applying numerical techniques that are able to cope with special problems involved here.
    - the possibility to introduce data in the database with higher numerical accuracy (i.e. more decimals).
    - to treat the value for the exponent in the overflow formula as a user defined variable.

7.2 SAMENVATTING

7.2.1 Algemeen

Het voornaamste doel van dit proefschrift is het toegankelijk maken van werkmethoden gericht op het gebruik van hydrodynamische (computer)modellen binnen de rioleringstechniek. Een vakgebied dat noch fundamenteel wetenschappelijk noch puur ambachtelijk is en waarbinnen in de laatste twee decennia grote veranderingen zijn opgetreden in de ingenieurspraktijk. Met name de beschikbaarheid van goedkope rekenkracht en relatief goedkope software voor de modellering van de dynamische processen in rioolstelsels zijn in dit verband noemenswaardig. De mogelijkheden van de modellering lijken ongelimiteerd en hebben velen de overtuiging gegeven dat 'de hele wereld' kan worden uitgerekend. Dit idee leidde in de jaren tachtig en negentig van de twintigste eeuw tot een praktijk waarin op grote schaal zogeheten vuiluitworpmodellen werden toegepast zonder dat een deugdelijke toetsing aan de praktijk plaatsvond. In de overtuiging dat een bruikbare beschrijving van stoftransportprocessen (een essentiële voorwaarde voor vuiluitworpmodellen) eerst dan mogelijk is indien het kwantitatieve aspect (ofwel de dynamiek van de waterstroming) binnen gekwantificeerde betrouwbaarheidsgrenzen kan worden gemodelleerd, is in dit proefschrift getracht na te gaan in hoeverre de moderne generatie hydrodynamische modellen in de praktijk in staat is de 'werkelijkheid' van de hydrodynamische verschijnselen in een rioolstelsel te beschrijven. Deze werkelijkheid op haar beurt is alleen kenbaar door waarnemingen en metingen aan de processen die zich afspelen in een rioolstelsel. Waar modelberekeningen lijden aan gebreken ten aanzien van onzekerheden in de gehanteerde gegevensbestanden en de gedane aannamen ten aanzien van rekenparameters, wordt de betrouwbaarheid van meetresultaten beperkt door meetnauwkeurigheden en beperkingen die er in de praktijk zijn in termen van ruimtelijke resolutie en meetduur. Het samenbrengen van de kwaliteiten van modellen en meetresultaten met als doel een betrouwbaarder beschrijving van de werkelijkheid te bereiken is het resultaat van de calibratie, of ijking, van een model. Het calibreren van modellen is het hoofdthema van dit proefschrift. Hierbij is als uitgangspunt de alledaagse ingenieurspraktijk genomen; alle beschreven methoden en gehanteerde gereedschappen zijn ofwel direct voorhanden of met enige inspanning zelf te produceren. Het proefschrift gaat in op de processen die van belang zijn bij het beschrijven van de hydrodynamica in een rioolstelsel, de vertaling hiervan naar wiskundige betrekkingen en de implementatie in numerieke rekenmodellen (hoofstuk 2). In hoofdstuk 3 wordt beschreven op welke wijze de geometrie en de structuur van een rioolstelsel in een rekenmodel wordt geïntegreerd en wat de effecten van onzekerheden hierin zijn op de uiteindelijk bereikte rekenresultaten. Hoofdstuk 4 gaat in op meetprojecten, waarbij meetmethoden, meetnauwkeurigheden en meetfrequentie met elkaar in verband worden gebracht. In hoofdstuk 5 wordt ingegaan op het calibratieproces, het samenbrengen van model en gemeten 'werkelijkheid'. Hierbij is als uitgangspunt de brede toepasbaarheid van de gepresenteerde werkmethoden genomen. Het calibratieproces (hetgeen aanzienlijk meer omvat dan alleen het 'fitten' van parameterwaarden) wordt aan de hand van diverse voorbeelden geïllustreerd. Hoofdstuk 6, tenslotte, doet verslag van een praktijkvoorbeeld waarin aan de hand van metingen een hydrodynamisch model is gecalibreerd. Hierna wordt een korte samenvatting van de diverse hoofdstukken en de belangrijkste conclusies gegeven.

7.2.2 Hoofdstuk 1: Inleiding

Hoofdstuk 1 beschrijft in het kort de historie van de riolering en spitst zich toe op de rol die rekenmodellen in de praktijk zijn gaan spelen. In een aparte paragraaf wordt ingegaan op de momenteel in Nederland algemeen geaccepteerde wijze van hydrodynamische berekeningen De uitgangspunten en algemene methoden zoals deze zijn neergelegd in deze werkwijze zijn gehanteerd in de overige hoofstukken van het proefschrift, met dit verschil dat de gemeten 'werkelijkheid' als leidend wordt gezien boven de gebruikte rekenmethodiek.

7.2.3 Hoofdstuk 2: Hydrodynamische modellen

Hoofdstuk 2 beschrijft de theoretische uitgangpunten die worden gehanteerd in moderne hydrodynamische rekenmodellen. Hierbij is een onderscheid gemaakt tussen zogenaamde inloopmodellen en de modellen die het dynamisch gedrag van waterstromen in een rioolstelsel beschrijven. Een belangrijk deel van dit hoofdstuk handelt over de stabiliteit van numerieke rekenschema's die worden gebruikt om de (grote) stelsels van partiële differentiaalvergelijkingen op te lossen die resulteren uit de theorie. Belangrijkste resultaat van dit hoofdstuk is de conclusie dat het op grote schaal gebruikte Preissmann schema belangrijke inherente beperkingen kent bij de modellering van kleine waterdiepten. De in de vakliteratuur gedane suggestie dat dit zijn oorsprong vindt in de niet-lineariteit van de geometrie van gesloten leidingen blijkt tenminste onvolledig. Dit geldt ook voor de vaak geopperde suggestie dat een (kunstmatige) vergroting van de hydraulische weerstand dit probleem oplost; het ontstaan van zogenaamde 2Δx oscillaties is inherent aan het numerieke schema. Dit is slechts één voorbeeld van de beperkingen die ontstaan in de beschrijving van de werkelijkheid met een (computer)model; het voorkomen van nuldelingen, het beperken van excessieve rekentijden en de inherente beperking aan de nauwkeurigheid die het gebruik van computers met zich meebrengt worden omzeild doormiddel van 'programmeursoplossingen'. In een afzonderlijke paragraaf wordt ingegaan op de effecten van dergelijke 'programmeurs oplossingen' door een eenduidig gedefinieerd rioolstelsel met twee, in strikt van elkaar gescheiden omgevingen geprogrammeerde, rekenpakketten door te rekenen. De gevonden afwijkingen blijken aantoonbaar doch relatief gering (vergeleken met de afwijkingen die worden gevonden tussen model en gemeten 'werkelijkheid'). De conclusie is dan ook dat bij een consciëntieuze toepassing van algemeen aanvaarde uitgangspunten, de resulterende software intern consistente resultaten oplevert.

7.2.4 Hoofdstuk 3: Gegevens en effecten van fouten in gegevens

In hoofdstuk 3 wordt ingegaan op de geometrie en de structuur van rioolstelsels die moet worden ingevoerd in een rekenmodel om voor een gegeven situatie modellering te kunnen bedrijven. In het algemeen kan worden gesteld dat de exacte beschrijving van de maatvoering (afmetingen van leidingprofielen, hoogteligging en lengte van leidingen, afmetingen van rioolputten, maaiveldhoogten etc.) en de structuur van rioolstelsels bij beheerders slechts met een beperkte betrouwbaarheid bekend is. Deze onbekendheid heeft diverse oorzaken: het verloren gaan van gegevens, het niet archiveren van revisiegegevens na afronding van uitgevoerde werkzaamheden, meetonnauwkeurigheden en de verwaarlozing van de riolering als te beheren infrastructuur gedurende vele jaren. In dit hoofdstuk wordt een beschrijving gegeven van werkmethoden die in de ingenieurspraktijk zijn ontwikkeld om databestanden op fouten te controleren en vervolgens deze fouten te elimineren. Desondanks zullen er vrijwel altijd (kleine) afwijkingen blijven bestaan tussen de werkelijke geometrie en de representatie in een gegevensbestand. Door middel van een Monte Carlo simulatie met twee rioolstelsel is nagegaan in hoeverre een databestand foutloos moet zijn. Hoofdconclusie is dat een foutenpercentage van circa 1% in termen van afwijkingen in de gegevens toelaatbaar is, mits deze fouten van een aard zijn dat ze niet zonder gedetailleerd veldwerk zijn op te sporen. Een tweede belangrijke onzekere factor is de hoeveelheid afvoerend oppervlak. Hoewel moderne digitale methoden, gebaseerd op bijvoorbeeld de grootschalige basiskaart Nederland, op zich zelf uiterst nauwkeurig zijn, ontstaan afwijkingen met de praktijk. Dit heeft de volgende oorzaken:

- Het niet in alle detail bekend zijn van door particulieren aangelegde verhardingen.
- Wijzigingen in de stedelijke infrastructuur die niet zijn verwerkt in het beschikbare kaartmateriaal.

7.2.5 Hoofdstuk 4: Meetprogramma's

In hoofdstuk 4 wordt ingegaan op 'meten aan riolering'. Hierbij staat het ontwerp van een meetprogramma centraal. Meetnauwkeurigheid, meetfrequentie en meetlocatie hebben in onderlinge samenhang invloed op de hoeveelheid informatie, en de kwaliteit hiervan, die met een meetnet kan worden verkregen. Afhankelijk van het oorspronkelijke doel van het meetprogramma worden verschillende eisen gesteld aan de genoemde drie grootheden. In het hoofdstuk wordt aan de hand van een aantal voorbeelden gedemonstreerd dat, behalve het meetdoel, ook de specifieke eigenschappen van een rioolstelsel van belang zijn bij het ontwerpen van een meetnet. Dit leidt tot een situatie waarin met een ongecalibreerd model van een rioolstelsel een meetnet wordt ontworpen, de gegevens die hieruit worden verkregen worden op hun beurt gebruikt om het model te verfijnen. Idealiter zou daarom lopende het meetprogramma steeds tussentijds het gebruikte model moeten worden gecalibreerd aan de hand waarvan het geïnstalleerde meetnet eventueel kan worden verfijnd. In de laatste paragrafen van hoofdstuk 4 wordt ingegaan op de beschikbare meetmethoden en die hiermee haalbare meetnauwkeurigheden. In het algemeen kan worden gesteld dat een debietmeting in een rioolstelsel aanzienlijk kostbaarder en onnauwkeuriger is vergeleken met een waterstandmeting. Zeker als veel (tijdelijke) meetpunten noodzakelijk zijn, wordt de voorkeur gegeven aan het meten van waterpeilen t.b.v. de calibratie van een rekenmodel boven het inrichten van meetpunten voor debietmeting.

7.3.6 Hoofdstuk 5: Calibratie

Hoofdstuk 5 behandelt de calibratie van hydrodynamische rekenmodellen. Allereerst worden de begrippen validatie, calibratie en verificatie in hun onderlinge samenhang gedefinieerd, in de praktijk worden deze drie aan elkaar gerelateerde begrippen dikwijls door elkaar gebruikt. De validatie behelst de toetsing of de processen die in het model zijn opgenomen overeenstemmen met de waarneembare werkelijkheid. Als voorbeeld; een model dat het gedrag van een vallend voorwerp beschrijft waarbij niet wordt voldaan aan de waarneembare werkelijk (het voorwerp stort ter aarde) is niet valide. De calibratie omvat het 'inregelen van het model' zodanig dat voor een specifiek geval het model en de waarneembare werkelijkheid zo goed mogelijk met elkaar overeenstemmen. Verificatie, tenslotte, is het nagaan in hoeverre het gecalibreerde model overeenstemt met waarnemingen die niet zijn gebruikt bij de calibratie. Het onderdeel parameter-optimalisatie is slechts één, zij het belangrijk, onderdeel in het calibratieproces. Ten minste zo belangrijk is de analyse van de verschillen tussen metingen en het model nadat de parameters zijn ingeregeld. Deze verschillen (de residuen) bevatten belangrijke informatie t.a.v. tekortkomingen van hetzij de meetresultaten (systematische fouten), het model (incorrecte of onvolledige procesbeschrijvingen), of de gebruikte vaste gegevens. Behalve een scala aan optimalisatietechnieken (klassieke gradiënt gebaseerde algoritmen en meer recent ontwikkelde technieken zoals genetische algoritmen) wordt aandacht besteed aan de analyse van de residuen. Het is niet uit te maken welke parameter-optimalisatietechniek de 'beste' is voor een gegeven geval; uit de behandelde voorbeelden blijkt dat een combinatie van genetische algoritmen en gradiënt gebaseerde technieken elkaar aanvullen. Een genetisch algoritme is in staat om betrekkelijk snel de globale ligging van de optimale parametercombinatie te vinden terwijl gradiënt gebaseerde methoden een schat aan informatie opleveren omtrent de identificeerbaarheid en onderlinge correlaties van de verkregen modelparameters. In de laatste twee paragrafen van dit hoofdstuk worden methoden beschreven waarmee de calibratie van een hydrodynamisch model met succes kan worden uitgevoerd. Deze methoden kunnen door eenieder die beschikt over een hydrodynamisch model en die toegang heeft tot eenvoudige hardware en een derde of vierde generatie programmeertaal eenvoudig worden uitgevoerd. Tevens wordt een methode geschetst waarmee het mogelijk is om aan de hand van een hydrodynamisch model de optimale meetlocaties binnen een rioolstelsel te kiezen als het meetdoel is 'het calibreren van een rekenmodel'. De geschetste methode is gebaseerd op een gevoeligheidsanalyse van het model gecombineerd met de informatieinhoud die de afzonderlijke meetpunten bevatten t.a.v. de te calibreren modelparameters.

7.2.7 Hoofdstuk 6: Case study

Hoofdstuk 6 beschrijft een voorbeeld waarin het proces van gegevensverzameling tot en met calibratie aan de hand van meetresultaten is beschreven. Het betreft een betrekkelijk klein bemalinggebied waarin een gemengd rioolstelsel aanwezig is. De onderliggende metingen zijn verricht in het najaar van 1998. Uiteindelijk blijkt dat het met in de voorgaande hoofdstukken beschreven werkmethoden mogelijk is om binnen vrij nauwe betrouwbaarheidsgrenzen voor waterstanden als functie van de tijd een hydrodynamisch model te calibreren. De gevonden waarden voor de modelparameters blijken echter per individuele bui te variëren Voor bepaalde parameters zoals de droogweer afvoer is dit deels te verklaren uit in- en exfiltratie van grondwater. Voor de overige modelparameters (met name de parameters die betrekking hebben op het zogenaamde inloopmodel) zijn de in werkelijkheid optredende processen gecompliceerder dan in het model is aangenomen. Gezien de informatieinhoud van de meetresultaten was een uitbreiding van de gemodelleerde processen niet mogelijk aangezien dit zou leiden tot een aanzienlijke toename van het aantal modelparameters. Als 'nevenproducten' van de calibratie zijn enkele fouten uit de geometrische database gehaald die anders nooit gevonden zouden zijn. Tevens is duidelijk naar voren gekomen dat de bergende inhoud van zogenaamde 'kleine objecten' (o.a. huisaansluitingen, straatkolken e.d.) een merkbare invloed heeft op het dynamisch gedrag van een rioolstelsel. Het correct incorporeren van de bergende inhoud van deze objecten bleek cruciaal om het model te kunnen calibreren, uiteindelijk is de bergende inhoud tot op circa 1.2 % nauwkeurig in het model ingebracht. Ook is aangetoond dat de resterende afwijking zich bevindt in een nauw gedefinieerd peilinterval van circa 15 cm lengte. Binnen de mogelijkheden van de database van de gebruikte software bleek het niet mogelijk de resterende afwijking weg te werken. Om in de toekomst te komen tot een verbetering van de resultaten van modelcalibratie worden de volgende aanbevelingen gedaan:

- Ontwerp het meetsysteem op een systematische wijze conform de in hoofdstuk 5 aangegeven methoden.
- Verbetering van de hydrodynamische software een aantal punten, het gaat hierbij om de volgende aspecten: het opheffen van het 'kleine waterdiepte probleem' (zie hoofdstuk 2). de mogelijkheid om invoergegevens met een aanzienlijk hogere nauwkeurigheid in te voeren en het variabel stellen van de machtsexponent in de overstortformule (zie hoofdstuk 4 en 6).
- Actueel houden van de gegevensbestanden; fouten in de geometrie van de objecten of de structuur in een stelsel hebben een grote invloed op de betrouwbaarheid van de modelresultaten.

REFERENCES

Abbot, M.B. & Basco, D.R. (1989)
Computational fluid dynamics (an introduction for engineers) John Wiley & Sons, New York. ISBN 0 470 21316 7

Ahyerre, M. , Chebbo, G. , Tassin, B. & Gaume, E. (1998)
Storm water quality modelling, an ambitious objective ? In: Wat. Sci. Tech., Vol 37, No 1, pp. 205-213.

Ando, Y., Takahasi, Y. & Kuan, M.F. (1984)
Relation between landsuse and final infiltration rate in urban areas.
In: proc of the 3rd ICUSD, Göteborg, Sweden. pp. 1029-1036.

Babovic, V. & Wu, Z. (1994)
Calibrating hydrodynamic models by means of simulated evolution. In: Hydroinformatics '94, p.p.193-200. ISBN 905410 5151.

Baffaut, C. & Delleur, J.W. (1989)
Expert system for calibrating SWWM. Jrn. Water Resources Planning and Management, ASCE 115(3), 278.

Baffaut, C. & Delleur, J.W. (1990)
Calibration of SWWM runoff quality model; with expert system. Jrn. Water Resources Planning and Management, ASCE 116(2), 147.

Batchelor, G.K. (1983)
An introduction to fluid dynamics.
Cambridge university press. ISBN 0-521-09817 3.

Bayley, G.V. & Hammersley, J.M. (1946)
The effective number of independent observations in auto-correlated time series. Journal of the Royal Statistical Society, volume 8, no. I-B, pp. 184-197.

Bebelaar, J.P. & Bakker, J.W. (1981)
Infiltratie van regenwater door verschillende wegdekken i.v.m. de watervoorziening van straatbomen. Instituut voor Cultuurtechniek en waterhuishouding, nota 1247.

Berlamont, J. (1998)
Riolering (In dutch), ISBN 90-334-3891-7.

Bernardo, J. M. & Smith A.F.M. (1994)
Bayesian theory, New York, Wiley.

Bielecki, R. & Schremmer, H. (1987)
Biogene Schwefelsäure-korrosion in teilgefüllten Abwasserkanälen. ISSN 0343-12223.

Bloemen, R. (1997)

Private communication on the topic of achievable accuracy in measures of height in an existing drainage system.

Boussinesq, J. (1897)
Theorie de l'ecoulement tourbillonnant et tumultueux des liquides dans les lits rectitilignes a grande section. Paris, Gautheir et fils, pp. 22-27.

De Bruin, H.A.R. (1973)
Data on precipitation and evaporation collected in Salland during the period 1970-1972 on behalf of the working group hydrological research Overijssel, KNMI scientific report WR 73-4. (In Dutch). Org. title: Gegevens betreffende neerslag en verdamping verzameld in Salland, gedurende 1970-1972 ten behoeve van de Werkgroep Hydrologisch Onderzoek Overijssel. KNMI Wetenschappelijk Rapport WR 73-4, De Bilt.

Brummelhuis, ten, P.G.J., Heemink, A.W. & Boogaard, van den, H.F.P. (1993)
Identification of sshallow sea models.
In: International journal for numerical methods in fluids, vol. 17, pp.637-664.

Buishand, T.A. & Velds, C.A. (1980)
The climate in the Netherlands: Rainfall and evotranspiration. (In Dutch with English summary). Org title: Klimaat van Nederland 1: Neerslag en verdamping. KNMI, 1980.

Cacucci, D. (1991)
Sensitivity theory for non linear systems. Part I: Non linear functional analysis

Cagiao, J., Diaz-Fierros, F. Suarez, J. & Puertas, J. (1998)
A numerical and experimental model of an urban catchment in the north of Spain: parameter fitting and an analysis of its behaviour. In: UDM '98, Fourth international conference on developments in urban drainage modelling, London, pp. 101-111.

Capelleveen, van, P.A. & Clemens, F.H.L.R. (1999)
3-D GIS applications in urban drainage modelling, more than a gimmick. (In Dutch). Org. title: 3-D GIS toepassing in rioleringsberekeningen, meer dan een gimmick. In: VI matrix, maart 1999.

Capelleveen, van, P.A. (1999):
Urban drainage network modelling better analysed using 3D-analist. ESRI User Conference 1999, San Diego.

Carrera, J. & Neumann, S.P. (1986[a])
Estimation of Aquifer Parameters Under Transient and Steady State Conditions: 1. Maximum Likelihood Method Incorporating Prior Information, In: Water resources research, vol. 22, no. 2 pp. 199-210.

Carrera, J. & Neumann, S.P. (1986[b])
Estimation of Aquifer Parameters Under Transient and Steady State Conditions: 2. Uniqueness, Stability and Solution Algorithms, In: Water resources research, vol. 22, no. 2 pp. 210-227.

Carrera, J. & Neumann, S.P. (1986[c])

Estimation of Aquifer Parameters Under Transient and Steady State Conditions: 3. Application to Synthetic and Field data, In: Water resources research, vol. 22, no. 2 pp. 228-242.

Chavent, G. (1971)
Analyse fonctionelle et identification de coefficients repartis des les equations aux derivees partielles. PHD Thesis, University of Paris, France.

Chow V.T. (1959)
Open channel hydraulics. McGraw-Hill, New York.

CIAD, Association for computer applications in engineering. (1985)
Engineering databases. Report of the CIAD project group on engineering databases. Edt: M.L. Boerstra. Elsevier, ISBN 0-444-42472-5

Clemens, F.H.L.R. & von der Heide, W. (1999)
Effect of geometrical data errors in hydrodynamic calculations in urban drainage. In: proc. Of the 8th ICUSD pp. 955-963, August 1999, Sydney, Australia.

Clemens, F.H.L.R., van Mameren H.J., & Kollen, J., (1993)
Optimising the efficiency of storm water settling basins by means of bypassing.
Wat. Sci. Tech. vol. 27, No. 5-6, pp. 105-110, 1993.

Clemens, F.H.L.R., Reydon, M.J.P. & Groosjohan, A. (1999)
Monitoring programme Deventer: calibration of a hydrodynamic model. (In Dutch). Org. title: Meetprogramma Deventer: de calibratie van een hydrodynamisch model.
In: Vakblad riolering, 1999

Clemens, F.H.L.R., Ganzevles, P.P.G., Boomgaard, M.E. Wixcey J. & van Luijtelaar, H. (1998)
Leidraad module C2100, berekening van het didactisch rioolstelsel: een toetsing. In: H_2O vol. 31-nr. 14, 1998, pp. 27-33.

Cock, de, W., Vaes, G. & Berlamont, J. (1998)
The efficiency of a storage sedimentation tank: numerical simulation and physical model experiments. In: Proc. of the fourth Int. Conf. on Developments in urban drainage, UDM '98, London, 1998. pp.417-424.

Conover, W.J. (1971)
Practical Non-parametric statistics. John Wiley, New York, 1971.

Cooley, R.L. & Naff, R.L. (1990)
Regression modelling of groundwater flow. chapter B4 in; Techniques of water resources investigations of the US geological survey. Appl. of hydraulics. US Doc., Washington.

Cunge J.A. & Mazaudou, B. (1984)
Mathematical modelling of complex surcharge systems: Difficulties in computation and simulation of physical situations. In: Proc. Of the third international conference on urban storm drainage, Göteborg, Sweden, 1984, pp. 363-374.

Cunge, J.A. & Wagner, M. (1964)
Numerical integration of the De Saint-Venant equations by an implicit finite difference scheme. Application for a case with partially surcharged and partially free surface flow. (In French), org. title: Intégration numérique des équations d'écoulement de Barré de Saint-Venant par un schéma implicite de différences finies. Application au cas d'une galerie tantôt en charge, tantôt à surface libre. La Houille Blanche, No. 1, pp. 33-39.

CUWVO werkgroep VI (1992)
Overstortingen uit rioolstelsels en regenwaterlozingen.Den Haag, 1992

Dam, van, V.H. & Schotkamp, J. (1983)
Infiltratie in bestratingen. RIJP-rapport 1983-38 Abw, Rijksdienst voor de IJsselmeerpolders, Lelystad.

Delleur, J.W. (1998)
Modelling quality of urban runoff
In: Nato ASI series, Kluwer academic Publishers, ISBN 0-7923-5097-9, Hydroinformatic tools for planing, design, operation and rehabilitation of sewer systems. (eds. Marsalek, J., Maksimovic,C., Zeman, E. & Price, R.K.) pp. 241-285.

Dicker, P. (1993)
Water in Pompeii (In Dutch). M.Sc. thesis Delft University of Technology, department of civil engineering, Delft.

Dijckmeester, P.N.M (1988)
Magnitude of hindrance due to groundwater in the urban area in the Netherlands; an estimate. (In Dutch). org. title: Omvang van de grondwateroverlast in het stedelijk gebied van Nederland; een inschatting. TU Delft/ RIZA, 1988

Diskin, M.H. (1963)
Temporary flow measurement in sewers and drains. In: Proc. of the Hydraulics division ASCE, July 1963, pp. 141-159.

Faber, W. (red) (1995, 1999)
'Guide line for urban drainage', Guidelines for urban drainage, module 'calculations in urban drainage, hydrodynamics' (In Dutch). Org. title: Leidraad riolering, module C2100 `rioleringsberekeningen, hydraulisch functioneren'.
Samsom Tjeenk Willink, Alphen aan de Rijn. ISBN 90-6501-636-8.

Farlow, S.J. (1982)
Partial differential equations for Scientist & Engineers. John Wiley and Sons Inc. New York. QA377.F37-515.3'53-81-12993-AACR2

Fernandez, E.B., Summers, R.C. & Wood, C. (1980)
Database Security and Integrity. Addison-Wesley publishing company, ISBN 0-201-14467-0

Field, R. (1974)
Design of combined sewer overflow regulator/separator. Journal water pollution control federation, 1974, No. 7 p. 1722.

Fletcher, R. (1987)
Practical methods of optimisation. New York, John Wiley and sons.

Fuchs, L., Măksimović, Č., Price, R.K. & Schilling, W. (1998)
Model development and application. In: Hydroinformatic tools for planning, Design, operation and rehabilitation of sewer systems. NATO ASI series. (2) volume. 44. Kluwer Academic Publishers.

Foundation RIONED (1992)
'Guidelines hydrodynamic calculations in urban drainage. Research Loenen (phase 1): Calculations and measurements.' (In Dutch). org title: Richtlijnen Rioleringsberekeningen. Onderzoek Loenen (fase 1): Berekeningen en metingen.

Foundation RIONED, STOWA, WRW & ARCADIS (1998)
Research into the effects of an improved system design on the pollutional load and the water quality. Measuring location Dorp-Oost at Stolwijk, Municipality of Vlist. (In Dutch). org. title: Onderzoek naar de effecten van een verbeterd stelselontwerp op de vuilemissie en de waterkwaliteit. Meetlocatie Dorp-Oost te Stolwijk, gemeente Vlist.

Foundation RIONED (1999)
Development of the cumulative length of urban drainage systems in the Netherlands (1925-1995), private communication.

Foundation RIONED (2000)
Urban drainage in numbers. (In Dutch), org. title: Het riool in cijfers 2000-2001, Ede, 1998

Ganzha, V.G. & Vorozhtsov, E.V. (1996)
Computer aided analysis of difference schemes for partial differential equations. New York, Wiley. ISBN 0-471-12946-1

Garb, N.K. & Send, F.J. (1994)
Determination of watershed features for surface runoff models. In: Jrn. hydr. engineering, vol 120., no. 4, pp. 427-447, April 1994.

Geer, van, F. (1987)
Application of Kalman filtering in the analysis of and design of groundwater monitoring networks. Ph. D. Thesis, Delft University of Technology. 1987.

Gelder, van, P.H.A.J.M. (2000)
Statistical methods for the risk-based design of civil structures. Ph.D. Thesis, Delft University of Technology. ISBN 90-9013452-2.

Giering, R. & Kaminski, T. (1989)
Recipes for adjoint code construction. In: Transactions on mathematical software, 1989, Vol. 24, nr. 4, pp. 437-474.

Godunov, S.K. & Ryaben'kii, V.S. (1964)
Theory of difference schemes, Publ. North-Holland Publishing Company, Amsterdam, 1964.

Golub, G.H. & Van Loan, C.F. (1996)

Matrix computations. Third edition, John Hopkins, Baltimore and London.

Griewank, A. (1989)
On automated differentiation. In: M. Iri & K. Tanabe (Ed's), Mathematical programming; recent developments and applications, pp. 83-108. Kluwer Academic Publishers.

Griffin, S., Bauwens, W. & Ahmad, K. (1994)
Urban drainage modelling intelligent assistant. Wat. Sci Techn. Vol 29 (1-2), pp. 427-436

Gujer W. & Krejci, V. (1987)
Urban storm drainage and receiving water ecology. Special lecture in Proc. of the 4th int. Conf. On urban storm drainage. p.p.1-20, Lausanne 1987.

Hamilton, J.D. (1994)
Time series analysis. Princeton university press. ISBN 0-691-04289-6.

Hassapis, G. (1991)
Wastewater flow monitoring with a personal computer. In: Computer in industry, 17, pp. 359-366.

Harremoës P. & Madsen H. (1998)
Fiction and reality in the modelling world, Balance between simplicity and complexity, calibration and identifiability, verification and falsification. pp.841-848, UDM '98

Hemker, C.J. (1996)
The reliability of parameters in automatic calibration (In Dutch). In: Modelkalibratie: Het automatisch ijken van grondwatermodellen, The Dutch hydrological association, NHV, special issue 2, 1996, pp 39-51.

Hillel, D. & Gardner, W.R. (1970)
Transient infiltration into crust-topped profiles, Soil science, vol. No. 109, no. 2, pp. 69-76.

Hillel, D. & Gardner, W.R. (1982)
Differences in water budgets for urban and rural areas. In: Three conference papers on urban and rural run-off, Report no.3058. Department of water resources Engineering. Lund institute of technology. University of Lund, Sweden.

Holland, J.H. (1975)
Adaptation in Natural and Artificial Systems. The University of Michigan Press, Ann Arbor, MI.

Horton, R.E.(1940)
An approach towards a physical interpretation of infiltration capacity.
In: Proceedings of the Soil Science of America, volume 5.

Houwen, van der, P.J. (1968)
Finite difference methods for solving partial differential equations. Mathematical centre Tracts, No. 20, Mathematical centre, Amsterdam, 1968.

Howart, D. & Saul, A. (1984)

Energy loss coefficients at manholes. In: Proc. Of the third international conference on urban storm drainage, Göteborg, Sweden, 1984, pp. 127-136.

Huber, W.C., Heany, J.P., Nix, S.J., Dickinson, R.E & Polmann, D.J. (1981)
Storm Water Management Model users manual, version III. University of Florida, Gainsville, Florida.

Ibrahim, Y. & Liong, S.Y. (1992)
Calibration Strategy for Urban Catchment parameters. Journal of Hydraulic engineering, Vol. 118, no 11 November 1992. pp. 1550-1569.

Idel'cik, I.E. (1963)
Memento des pertes de charge; coefficients de pertes de charge singulières et de pertes de charge par frottement.

Imhoff, K. (1932)
Taschenbuch der Stadentwasserung. Munchen und Berlin, 1932

ISO (1983[a])
ISO standards handbook 15; Measurement of fluid flow in closed conduits. Geneve, ISBN 92.67.10076.9

ISO (1983[b])
ISO standards handbook 16; Measurement of liquid flow in open channels. Geneve, ISBN 92.67.10077.7

Jennings, A. & Mckeon, J.J. (1992)
Matrix computation. Second edition, Wiley, Chichester.

Kamen, E. (1987)
Introduction to signals and systems. Macmillan Publishing Company, New York 1987. ISBN 0-02-362950-9

Kasabov, N.K. (1996)
Foundations of neural networks and knowledge engineering. MIT, ISBN 0-262-11212-4

Kidd, C.H.R. (1978)
Rainfall-runoff processes over urban surfaces.
In. Proc. of an Int. workshop, Report no. 53. Institute of hydrology, Wallingford, Oxon, England.

Kindsvater, C.E. & Carter, R.W. (1957)
Discharge characteristics of rectangular thin-plated weirs. Journal of the Hydraulics division ASCE 83(HY6), Paper 1453, pp. 1-36.

King, I.P. (1970)
An automated reordering scheme for simultaneous equations derived from network systems. Int. Journal numerical mathematical engineering (2) 1970, pp. 523-533.

Kleywegt, R.A. (1992)

On sediment transport in circular sewers with non-cohesive deposits. Ph.D thesis, DUT, Delft, The Netherlands.

Kleywegt, R.A. & Clemens, F.H.L.R. (1993)
Development of basic components for modelling sediment and pollutant transport in sewers. In: Proc. of the 6th Int. Conf. on Urban Storm Drainage, Niagara Falls, Canada, pp.610-618.

Kloet, van der, P. & van de Ven F.H.M (1981)
Laguerre series for linear and non-linear modelling of urban run-off. In: Proc. Of the 2nd Int. Conf. On Urban Storm Drainage. 1981, Urbana, Illinois.

Kluck, J. (1997)
The design of optimally functioning storm water settling tanks. Ph.D thesis DUT, Delft, The Netherlands. ISBN 90-407-1452-5.

KNMI, (Royal Dutch Meterological Institute),1993, Kwartiersommen van de neerslag te De Bilt, Tijdvak 1955-1979, De Bilt, 1993.

KNMI (1983)
Het beoordelen van regenmeters, met als voorbeelden de Thies regenmeter en de electrische KNMI-regenmeter, KNMI W.R. 83-16, De Bilt.

Kollen J. & Clemens F.H.L.R. (1991)
The Parallel Storage Sedimentation Sewer. In: proc. of the int conference sewage into 2000, Amsterdam 1991, pp. 362-365.

Koot, A.C.J. (1979)
Collection and transport of wastewater. (In Dutch), org title: Inzameling en transport van rioolwater. Waltman, Delft, 1979

Kop, J.H. (1972)
Vakantiecursus afvalwater. TU Delft

Koussis, A.D. & Bowen, J.D. (1984)
DEMOS: A diffusive model simulator. In: Proc of the third ICUSD, Göteborg, Sweden, June 1984, pp. 483-492.

Lei, J. & Schilling, W. (1993)
Requirements of spatial rain data resolution in urban rainfall runoff simulation. ICUSD Niagara Falls, Canada, pp. 447-452.

Lei, J., Li, J. & Schilling, W. (1999)
Procedure of urban runoff design model and an alternative of Monte-Carlo simulation. In: Proc 8th ICUSD, Sydney Australia, pp. 973-981.

Lettenmaier, D.P. (1976)
Detection of trends in water quality data from records with dependent observations. In: Water Resources Research, volume 12, no. 5, pp. 1037-1045.

Liong, S.Y., Weng, T.C. & Shreeram, J. (1995)

Peak flow forecasting with genetic algorithm and SWWM. Journal of hydraulic engineering, Vol 121, no. 8, August 1995, pp. 613-617.

Loke, E. , Arnbjerg-Nielsen, K. & Harremoës, P. (1999)
Artificial neural networks and grey-box modelling: A comparison. In: Proc. of the 8th ICUSD, Sydeney, Australia, pp 72-79.

Lovejoy, S. & Mandelbrot, B.B. (1985)
Fractal properties of rain and a fractal model. In: Tellus 37 A (1985), pp. 209-232.

Luenberger, D. (1973)
Introduction to linear and non-linear programming, Addison- Wesley, Reading, Mass., 1973

Luijtelaar, van, H. (1999)
Design criteria, Flooding of sewer systems in 'flat' areas. In: Proc. of the 8th ICUSD, Sydney, Australia, pp. 538-545.

Luijtelaar, van, H. & Rebergen, E. (1997)
Guidelines for hydrodynamic calculations in urban drainage in The Netherlands: backgrounds and examples.Paper in the proc. of the 7th ICUSD, Hannover, Germany, 1997.

Luijtelaar, van, H. & Ganzevles P.P.G. (1993)
Dynamic design of sewer detention tanks. Wat. Sci. Techn, Vol. 27, no's 5-6, pp. 133-143.

Máksimović, Č. (1998)
Application of GIS in urban drainage.
In: Nato ASI series, Kluwer academic Publishers, ISBN 0-7923-5097-9, Hydroinformatic tools for planing, design, operation and rehabilitation of sewer systems. (eds. Marsalek, J., Maksimovic,C., Zeman, E. & Price, R.K.) pp. 179-188.

Mameren, van , H.J. & Clemens F.H.L.R. (1993)
Het CUWVO VI referentiestelsel in de praktijk. H_2O (26) nr. 23 pp.692-695, 1993

Mameren, van, H.J. & Clemens, F.H.L.R. (1997)
Guidelines for hydrodynamic calculations on urban drainage: overview and principles.
In Wat. Sci. & Techn. Vol. 36.No 8-9, pp. 247-252.

Mandelbrot, B. B. (1984)
Fractals in physics: Squig clusters, Diffusion, Fractal measures, and the Unicity of fractal dimensionality. Journal of statistical Physics, Vol. 34, nos. 5/6, 1984, pp. 895-930.

Marquart, D.W. (1963)
An algorithm for the Least-Squares Estimation of Non linear Parameters. In: Journal of the society for industrial and Applied Mathematics., 11, 431-441.

Marsalek, J. Máksimović, Č., Zeman, E. & Price, R.K. (1998)
Hydroinformatic tools for planning, design, operation an rehabilitation of sewer systems. NATO ASI series: 2. Environment- Vol. 44. Kluwer, The Netherlands. ISBN 0-7923-5097-9

Mein, R.G. & Larson, C.L. (1973)

Modelling Infiltration during a steady rain. In: Water resources research. vol. 9, no. 2 pp. 384-394.

Meijer, E.V. (1998)
Adding stirm water to wastewater sewers, sensible or senseless ?
In: Proc of the Fourth int. conf. on Developments in Urban Drainage Modelling, UDM '98, London 1998, pp. 399-403.

Michalski, B. & Berger, W. (1984)
Fullstandmessung mittels Laufzeit von Ultraschall, Technisches Messen 51, Nr 9, 1984, pp. 306-311. (In German).

Michelbach, S., Wörhle, C. & Brombach, H.J. (1992)
Sedimentations- und Remobilisierungsvorgange im Abwasserkanal, Schlussbericht 1988-1991. Bonn 1992. (in German).

Miller, J.(1971)
On the location of zeros of certain classes of polynomials with applications to numerical analysis. In: Journal Inst. Math. Appl. Volume 8, pp 397-406.

Ministry of Housing, Land-use Planning and the Environment of the Netherlands (VROM) (1988)
Individuele behandeling van afvalwater bij verspreide bebouwing. Publicatiereeks milieubeheer 5. ISBN 90 346 1007 6. Den Haag, 1988

Monsma, S. (1999)
Fuzzy logic and neural networks in urban traffic control (In Dutch).
org .title: Fuzzy logic en neurale netwerken in de stedelijke verkeersbeheersing.
M.Sc. thesis Open University.

Monster N. & Leeflang M. (1996)
On the design of percolation facilities. Communication of the department of sanitary engineering and water management, Delft University of Technology, Faculty of civil engineering, Delft, 1996 ISSN 0169-6246.

Morton, K.W. & Mayers, D.F. (1994)
Numerical Solutions of partial differential equations. Cambridge University Press. ISBN 0 521 41855 0.

Mouthaan, E.E.A., Heemink, A.W. & Robaczewska, K.B. (1994)
Assimilation of ERS-1 altimeter data in a tidal model of the continental shelf.
In: German journal of hydrography, vol. 46, No. 4, pp.285-319.

Murray, J.B. (1987)
Infiltration rates for separated sewage collection. In: Wat. sci. Tech. vol. 19, pp. 589-601.

Nash, J.E. & Sutchiffe, J. (1970)
Riverflow forecasting through conceptual models. Journal of Hydrology, vol. 10 nr. 3, April 1970.

National Water Council (1982)
The Wallingford Procedure for the design and Analysis of Urban Storm Drainage in 5 volumes, NWC, UK 1982.

Nelder, J.A. & Mead, R.L. (1965)
A simplex method for function minimisation. Computer journal, vol. 7, pp. 308-313.

Neuman, S.P. (1980)
A statistical approach to the inverse problem of aquifer hydrology, 3, Improved solution and add perspective. Water Resources Res. 16(2), pp. 331-346.

Nikuradse, J. (1932)
Gesetzmässigkeiten der turbulenten Strömung in glätten Röhren. Ver. D. Ing. Forschung 356.

Nikuradse, J. (1933)
Gesetzmässigkeiten der turbulenten Strömung in rauhen Röhren. Ver. D. Ing. Forschung 361.

NWRW (1987)
Theme 8.1;Onderzoek naar het rendement van het bergbezinkbassin te Amersfoort. (In dutch).

NWRW (1989)
Nationale Werkgroep Riolering en Waterkwaliteit, thema 4.4; Regenmeters. ISBN-90.346.2397.X

Nijmeegse betonindustrie De Hamer (1998)
Private communication on the accuracy of profile dimensions.

Olsthoorn, T.N. (1995)
Effective parameter optimisation for Groundwater Model Calibration. In: Ground Water, Vol. No.33 nr.1, Jan-Feb. 1995, pp 42-48.

Olsthoorn, T.N. (1998)
Groundwater modelling: calibration and the use of spreadsheets, Ph.D. Thesis, University of Technology Delft, June 1998, ISBN 90-407-1702-8/CIP

Oomens, A.J. (1992)
Sewer management: structuring the management process on the basis of conditions for effective control. (*In Dutch), org. title*: Rioleringsbeheer: het structureren van het beheerproces aan de hand van de voorwaarden voor effectieve besturing. Ph.D. Thesis University of Technology, Delft, 1992.
ISBN: 90-90005687-4.

Oppenheim, A.V. & Schafer, R.W. (1989)
Discrete-time signal processing. Prentice-Hall inc., New Jersey. ISBN 0 13 216771 9

Oresken, N. Shrader-Frechette, Belitz, B. (1994)
Verification, validation and confirmation of numerical models in the earth sciences. Science, volme 263, 4 Frebruary 1994, pp. 641-646

Overgaard, S., El-Shaarawi A.H. & Arnbjerg-Nielsen K. (1998)
Calibration of tipping bucket rain gauges. In: Water Science and technology, vol. 37, No. 11 pp. 139-145.

Patel, R.V., Laub, A., Dooren, van, P.M. (eds.) (1993)
Numerical linear algebra techniques for systems an control. ISBN 0-7803-0443-8.

Pecher, R. (1969)
Der Abflussbeiwert und seine Abhängigkeit von der Regendauer. Berichte aus dem Institüt fur Wasserwirtschaft und Gesundheitsingenieurswesen, Technische Hochschule München, nr. 2.

Pedersen, F. & Mark, O. (1990)
Head losses in storm sewer manholes, submerged jet theory. Jrn. Hydraulic Division American Society of Civil Engineering. Vol. 116. No 16, pp. 1317-1328, 1990

Preissmann, A. (1961)
Propagation des intumescences dans les caneaux et rivieres. Proc. of the first congress of th French association for computation. Grenoble, France.

Press, W.H., Teukolsky, S.A., Saul, A., Vetterling, W.T. & Flannery, B.P. (1992)
Numerical recipes in C: The art of scientific computing. 2nd edition. New York, Cambridge University press.

Price, R.K. & Osborne, M.P. (1986)
Verification of sewer simulation models. In: Proceedings of the International Symposium on comparison of urban drainage models with real catchment data. UDM '86, Dubrovnik, Yugoslavia. Pp. 99-106.

Price, R.K. & Catterson G.J. (1997)
Monitoring and modelling in urban drainage. In: Water Science & Technology Vol. 36, No. 8-9, pp. 283-287.

Prodanović, D., Djondjević, S. & Măksimović, Č (1998)
GIS- assisted models for dual drainage simulation. In: proc. Int. conference Hydroinformatics 98, Copenhagen.

Province Utrecht (1991)
Onderzoek naar het rendement van de bergbezinkleiding Prinsenlaan te Maartensdijk.

Qualiarella D., Periaux, J., Poloni C. and Winter, G. (1998)
Genetic algorithms and evolutionary strategy in engineering and computer science. Recent advances and industrial applications. John Wiley & Sons, ISBN 0 471 97710 1

Rauch, W. & Harremoës, P. (1999)
On the potential of genetic algorithms in urban drainage modelling. Urban water 1, vol. 1 No1, pp. 79-89. ISSN 1462-0758.

Raudkivi, A.J. (1979)
Hydrology-An advanced introduction to hydrological processes and modelling.

Pergamon press, Paris.

Richtmyer, R.D. & Morton, K.W. (1967)
Difference methods for initial value problems
Interscience publishers, Wiley, New York, 1967.

Roo, de , H. (1982)
Richtlijnen voor veldwerkzaamheden bij de afdeling waterhuishouding.
RIJP rapport 1982-10. Rijksdienstvoor de Ijselmeerpolders, Lelystad.

Ruan, M. (1999)
Computational modelling of emission from combined sewer overflows. Ph.D. Thesis
University of Technology Delft. ISBN 90-9013023-3.

Rudolph, G. (1994)
An evolutionary algorithm for integer programming. In Parallel Problem Solving from Nature- PPSN III, International Conference on Evolutionary Computation, volume 866 of Lecture Notes in Computer Science, Springer, Berlin, pp. 139-148

Rustagi, J.S. (1994)
Optimization Techniques in Statistics. Academic Press, Harcourt Brace & Company.ISBN-0-12-604555-0.

Saint-Venant, de, B. (1871)
Théorie du movement non-permanent des eaux avec application aux crués des rivières et a l'introduction des marées dans leur lit. Acad. Sci. Paris Comptes rendues, volume 73, pp. 148-154, 237-240.

Sakakibara, T., Tanaka, S. & Imaida, T. (1996)
Energy losses at surcharged manholes-model experiment. In: Proc. Of the 7th Int. Conf. On urban Drainage, pp. 79-90.Hannover, 1996

Sangal, S.K. & Bonema, S.R. (1994)
A methodology for calibrating SWMM models. In: Current practices in modelling the management of storm water impact. Edt:.W. James, Lewis Publishers.

Schilperoort, T., Groot S., van de Wetering, B.G.M. & Dijkman, F. (1982)
Optimisation of the sampling frequency of water quality monitoring networks. Delft Hydraulics, publication no. 261, Delft.

Schilperoort, T. (1986)
'statistical aspects', in Design aspects of hydrological networks. In: Proceedings and information No. 35. of the TNO committee on hydrological research CHO, pp.35-55, ISBN 90-6743-082-X.

Schuurmans, W. (1991)
A model to study the hydraulic performance of controller irrigation canals.
Ph.D. Thesis, Delft University of Technology. ISBN 90-9004258-X

Schwefel, H.P. & Bäck, T. (1998)

Artificial evolution: how and why? In: Genetic algorithms and evolutionary strategy in engineering and computer science. Recent advances and industrial applications. Eds. Quagliarella et al. ISBN.0 471 97710 1

Schwefel, H.P. (1995)
Evolution and optimal seeking. Sixth-Generation Computer Technology Series. Wiley, New York.

Sieker, F. & Verworn H.R. (1997)
Selected proceedings of the 7th International conference on urban storm drainage, Water science & technology, volume 36, 8-9, 1997

Simon, H.A. (1996)
The sciences of the artificial. 3rd ed. The MIT press, Cambridge, Massachusetts. ISBN 0-262-19374-4.

Snow, J. (1854)
On the mode of communication of cholera, second edition 1854

Somer, De, M. (1984)
Flow instability occurred during the transition of full pipe flow to free surface flow and vice versa in a closed conduit. In: Proc. of the third ICUSD, Göteborg, Sweden, June, 1984, pp. 427-434.

Speed, D.E. & Ahlfeld, D.P. (1996)
Diagnosis of structural identifiability in groundwater flow and solute transport equations. In: Modelcare96, Proc. Of the golden Conference. IAHS. Publ. 237, pp. 287-298.

Stelling, G.S. & Booij, N. (1999)
Lecture notes CT wa4340; 'Computational modelling flow and transport'. Delft University of Technology. Department of Civil engineering and Geotechnical sciences.

Stelling, G.S. (1983)
On the construction of computational methods for shallow water flow problems. Ph.D thesis DUT, Delft, 1983.

Stevens, E.W. & Haan, C.T. (1993)
Effect of parameter variation and model non-linearity on first order analysis. In: Proceedings of the VIth International conference on Urban Storm Drainage, Niagara Falls, Canada, September 1993, pp. 610-618.

Stigter, L. (2000)
Measuring campaign drainage system in 'Deventer-Noord', time series analysis and area covering. (In Dutch) org. title: Meetprogramma riolering Deventer Noord: tijdreeks-analyse en gebiedsdekkende bepaling. M.Sc. thesis , Delft University of Technology.

STOWA (1996^{a})
Measuring and monitoring programs in urban drainage and surface waters (In Dutch). org. title: Metingen en meetprogramma's aan riolering en oppervlaktewater. STOWA rapport 96-09, ISBN 90.74476.48.1 , 1996.

STOWA (1996[b])
Measuring in urban drainage and surface waters. Simple techniques and observation methods (In Dutch). org. title: Metingen aan riolering en oppervlaktewater. Eenvoudige metingen en waarnemingen. STOWA rapport 96-10 ISBN 90.74476.49.X , 1996.

Streeter, H.W. & Phelps, E.B. (1925)
A study of the pollution and natural purification of the Ohio River. III. Factors concerned in the phenomena of oxidation an reaeration. U.S. Pub. Health Service, Public Health Bulletin No. 146, February 1925, Washington, Government Printing Office.

Taylor, J.R. (1982)
An introduction to error analysis. The study of uncertainties in physical measurements. University science books, Sausalito, California. ISBN 0-935702-75-X.

Terstriep, M.L. (1986)
Design of data collecting systems. In: NATO ASI series, Volume G10, Springer Verlag, Berlin. pp. 127-145.

Thenard, V., Ruban, G, Jaonnis, C. & Le Gal, C. (1999)
Assessment of level/flow formulae and Doppler velocimetry for the monitoring of storm overflows. In: Proc. of the 8th ICUSD Sydney, Australia, pp. 746-753.

Uhl, M.(1993)
Genauigkeit von Messungen, Grundlagen und Beispiele aus der Stadentwässerung. Schriftenreihe für Stadentwässerung und Gewässerschutz, nr. 7 1993. (In German). ISSN 0933-60 60.

Ummels T.P.M. & Clemens, F.H.L.R. (1998)
Propagation of errors in hydrodynamic calculations in urban drainage. In: UDM '98, Proc. of the 4th Int. Conf. on Dev. in Urban Drainage Modelling., 21-24 sept. 1998, pp. 221-228.

Vaes, G. & Berlamont, J. (1996)
Artificial design storms as input for hydrodynamic calculations in urban drainage (In Dutch), original title: Composietbuien als neerslaginvoer voor rioleringsberekeningen. Water, nr. 88, mei/juni 1996.

Vaes, G, Willems, P. & Berlamont, J. (1996)
IDF-relations and artificial design storms for the design of sewers. (in Dutch), original title: IDF-relaties en composietbuien voor het ontwerpen van rioleringen. Laboratorium voor Hydraulica, KU Leuven.

Vaterlaus, H.P. (1989)
Fullstandmessungen mit Schwingquarzen. Technisches Messen 56 Nr. 10, 1989, pp. 384-388. (In German).

Ven, van de , F.H.M. (1989)
Van neerslag tot rioolinloop in vlak gebied. Van zee tot land No. 57. Ministerie van rijkswaterstaat. ISBN 90-369-1060-9.

Veltri, P. & Pecora, S. (1999)
Genetic techniques to optimise urban drainage models. In: Proc of the 8th ICUSD, Sydney. pp. 1760-1767.

Verlaan, M. (1998)
Efficient Kalman Filtering Algorithms for Hydrodynamic models. Ph.D. Thesis, Delft University of Technology. ISBN 90-369-3452-4

Vermeer, P.A. & Verruijt, A. (1981)
An accuracy condition for consolidation by finite elements. In: International journal for numerical and analytical methods in geomechanics. Vol. 5, pp. 1-14.

Verwey, A., Minns, A.W., Babović, V & Măksimović, Č. (edt.) (1994)
Proc. of Hydroinformatics '94.

Vreugdenhil, C.B. (1989)
Computational hydaulics, an introduction. Springer verlag Berlin 1989

Watt, I.A. & Jefferies, C. (1996)
Portable sewage flow sensors: their calibration and accuracy. In: Water science and technology, Vol. 33, No. 1, pp. 127-137.

Wenzel, H.G. (1975)
Meter for sewer flow measurement. In: Journal of the Hydraulics division ASCE, HY1, January 1975, pp. 115-132.

Whitley, D., Rana, S. & Heckendorn, R. (1998)
Representation issues in search. In Genetic algorithms in engineering and computer science (Quagliarella et al (edt)) pp. 39-57. John Wiley & Sons, ISBN 0 471 97710 1

Willems, P. (1998)
Stochastic generation of spatial rainfall for urban drainage areas. In: Proc. of the fourth int. conf. on developments in urban drainage, London, 1994, pp.77-84.

Willems, P. & Berlamont, J. (1998)
Probabilistic modelling of sewer system overflow emission. UDM 98, pp. 213-220. London, 1998.

Willems, P. & Berlamont, J. (1998)
Stochastic modelling of spatial rain cells. In. Proc. of the Int. conf. on Hydrology in changing environment, Exeter 1998.

Willemsen, N. (2000)
Optimisation of a waste water system by application of a genetic algorithm. (In Dutch) Org. title: Optimalisatie van een afvalwatersysteem met behulp van een genetisch algoritme. M.Sc. thesis, Delft University of Technology.

Witteveen + Bos (1997)
Masterplan for the urban drainage systems in Boxtel and Liempde (In Dutch), org. title: Basisrioleringsplan Boxtel en Liempde.

Witteveen + Bos (1999)
Afvalwatersysteemoptimalisatie Apeldoorn.

Wixcey, J.R., Lewy, M., Price, R.K. (1992)
Computational Modelling of Highly Looped Networks of Storm Sewers, Fluid Flow Modelling (Eds. Blain, W.R. & Cabrera, E.), Computational Mechanics Publications, Southampton, Boston 1992.

Wolpert, D.H. & Macready, W.G. (1995)
No free lunch theorems for search. Technical report SFI-TR-95-02-010, Santa Fr institution.

World Meteorological Organisation
Guide to Meteorological Instruments and Methods of Observation. WMO No. 8, fifth edition, Geneva.

WRW (1999)
Tweede rioleringsbeleidnota., 1999

Yang, X. & Parent, E. (1996)
Reliability in hydrological modelling; concepts and applications to he G3 rainfall-runoff model. Revue des Sciences de l'eau/Journal of water science vol. 9 (1), 31, pp. 31-49. Paris and Quebec.

Yen, C.B. (1987)
Urban drainage hydraulics and hydrology; from art to science. In proc. of the 4th ICUSD, Lausanne, Switzerland.

Zeman, E., Vanecek, S. & Ingeduld P. (1998)
Tools for data archiving, visualisation and analysis applied in master planning.
In: Hydroinformatic tools for planning, design operation and rehabilitation of sewer systems. Edt: Jiri Marsalek, Cedo Măksimović, Evzen Zeman and Roland Price.

Zienkiewicz, O.C. & Taylor, R.L. (1991)
The finite element method, fourth edition, Volume 2, chapter 12 page 438-505. ISBN.0-07-084175-6 (v. 2).

Van Zon (1986)
A very nasty history, Study into the non- industrial pollution in the Netherlands, 1850-1920. (In Dutch), org. title: Een zeer onfrisse geschiedenis. Studies over niet-industriële vervuiling in Nederland, 1850-1920. Ph.D. thesis University of Groningen.

ANNEXE I: THE PROPERTIES OF CIRCULAR PIPES

I.1 Geometrical properties

Since circular pipes are the most frequently applied in urban drainage it is efficient to obtain analytical expressions for the geometrical properties used in solving several applications like hydrodynamic software or software used to calculate the storage capacity of a given drainage system.

The wetted area (A_s) as a function of water depth (a) is calculated in the following manner (see Figure I. 1: Definition sketch):

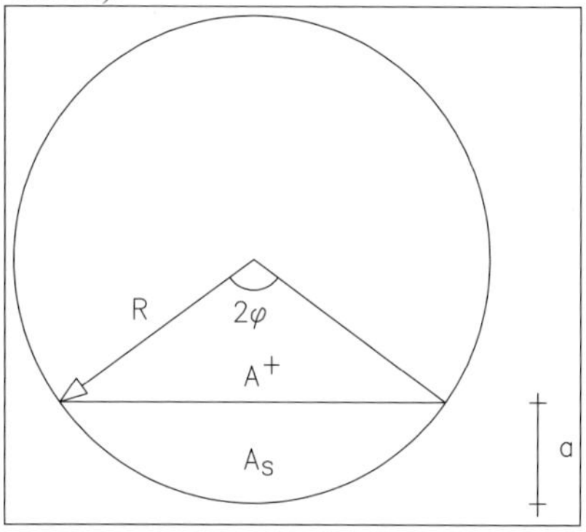

Figure I. 1: Definition sketch.

$$A^+ + A_s = \phi R^2$$

$$\phi = \arccos(1 - \frac{a}{R}), B = 2R \sin\phi$$

The width of the free water surface is $B = 2R\sin\phi = 2R\sin(\arccos(1-\frac{a}{R})) = 2\sqrt{2aR - a^2}$.

From this follows $A^+ = (R-a)\sqrt{2aR-a^2}$, resulting in :

$$A_s = R^2 \arccos(1-\frac{a}{R}) - (R-a)\sqrt{2aR-a^2}$$

The wetted perimeter is calculated from $P = 2R\phi = 2R\arccos(1-\frac{a}{R})$. Together with the wetted area this defines the hydraulic radius R_h:

$$R_h = 0.5R - \frac{(R-a)\sqrt{2aR-a^2}}{2R\arccos(1-\frac{a}{R})}$$ The hydraulic depth H_h is defined as

$$H_h = \frac{A_s}{B} = \frac{R^2\arccos(1-\frac{a}{R})}{2\sqrt{2aR-a^2}} - (R-a)$$

Furthermore the following expression proves to be useful when linearising the De Saint Venant equations: $\frac{dA_s}{da} = B(a) = 2\sqrt{2aR - a^2}$

The results are shown in Figure I. 2.

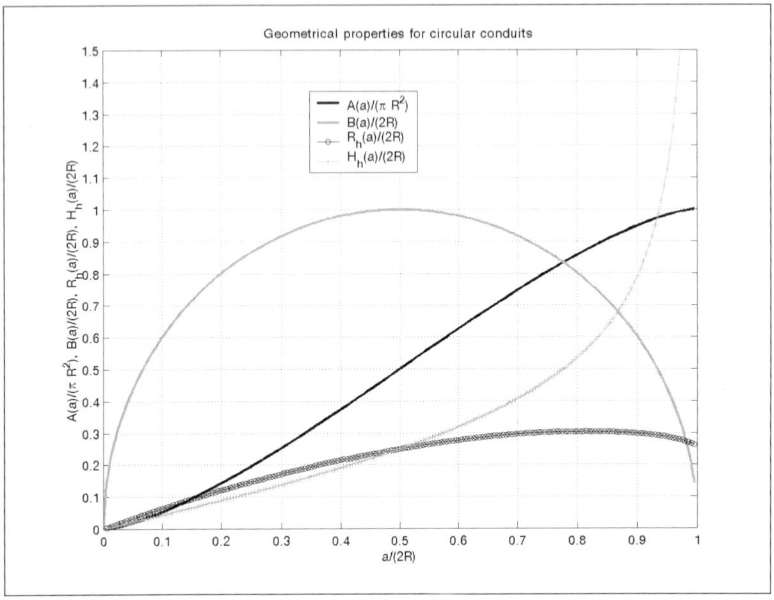

Figure I. 2: Geometrical properties of a circular conduit as function of the relative water depth.

I.2 The volume in partially filled circular pipes

In order to calculate the volume held by pipes in an urban drainage system as discussed in chapter 3 of this thesis the volume as a function of level should be calculated. For pipes being filled completely the calculation of the volume is trivial. When a sloping pipe is only partly filled the calculation of the volume is less obvious. This may be done conveniently using numerical integration techniques for irregular profile shapes. For circular profiles however, an analytical method is available. In Figure I. 3 a definition sketch is shown.

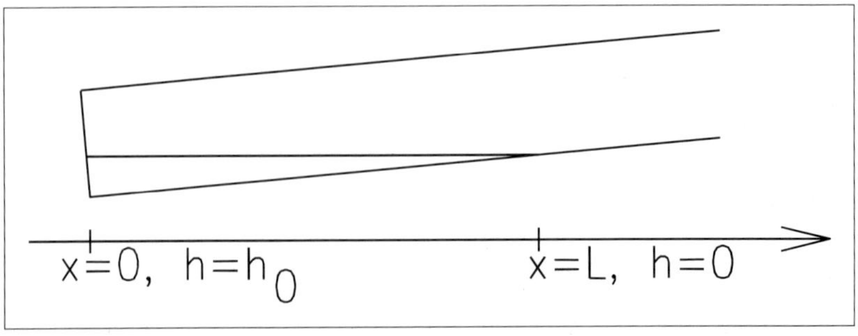

Figure I. 3: Definition sketch.

The volume in the shaded part of the pipe is:

$$V = \int_0^L A(x)dx \qquad \text{(eq. I.1)}$$

In which h(x) is the water depth h as a function of the length, and A(x) is the wetted area as a function of length. In fact A(x) can be expressed to be:

$$\left.\begin{aligned} &A(x) = A(h(x)) \\ &h(x) = h_0 - \frac{h_0}{L}x \\ &A(x) = R^2 \arccos(1 - \frac{1}{R}(h_0 - \frac{h_0}{L}x)) - (R - h_0 + \frac{h_0}{L}x)\sqrt{2R(h_0 - \frac{h_0}{L}x) - (h_0 - \frac{h_0}{L}x)^2} \end{aligned}\right\} \text{(eq. I.2)}$$

Using equations I.1 and I.2 the expression for the volume becomes:

$$V = \int_0^L R^2 \arccos(1 - \frac{1}{R}(h_0 - \frac{h_0}{L}x)) - (R - h_0 + \frac{h_0}{L}x)\sqrt{2R(h_0 - \frac{h_0}{L}x) - (h_0 - \frac{h_0}{L}x)^2}\,dx \qquad \text{(eq. I.3)}$$

Or:

$$\left.\begin{aligned} V = {} & R^2 \int_0^L \arccos(1 - \frac{1}{R}(h_0 - \frac{h_0}{L}x))dx - (R - h_0)\int_0^L \sqrt{2R(h_0 - \frac{h_0}{L}x) - (h_0 - \frac{h_0}{L}x)^2}\,dx - \\ & + \frac{h_0}{L}\int_0^L x\sqrt{2R(h_0 - \frac{h_0}{L}x) - (h_0 - \frac{h_0}{L}x)^2}\,dx \end{aligned}\right\} \text{(eq. I.4)}$$

Each term in equations I.4 has an analytical solution, using the standard primitives as summarised in the Handbook of chemistry and Physics ed. '82-'83, section A, page A-46 the following solution is found:
The first term is written as:

$$V_1 = R^2 \int_0^L \arccos(1 - \frac{h_0}{R} + \frac{h_0}{LR}x)dx \qquad \text{(eq. I.5)}$$

Using a transformation in the principal unknown x: $t = 1 - \frac{h_0}{R} + \frac{h_0}{LR}x$, the primitive of the integrand is:

$$V_1 = \frac{LR^3}{h_0}\left[\left(1 - \frac{h_0}{R} + \frac{h_0}{LR}x\right)\arccos\left(1 - \frac{h_0}{R} + \frac{h_0}{LR}x\right) - \sqrt{1 - \left(1 - \frac{h_0}{R} + \frac{h_0}{LR}x\right)^2}\right]_0^L \qquad \text{(eq. I.6)}$$

After some algebra, this is:

$$V_1 = \frac{LR^3}{h_0}\left[\sqrt{2\frac{h_0}{R} - \left(\frac{h_0}{R}\right)^2} - \left(1 - \frac{h_0}{R}\right)\arccos\left(1 - \frac{h_0}{R}\right)\right] \qquad \text{(eq. I.7)}$$

The second term is written as:

$$V_2 = -(R - h_0)\int \sqrt{h_0(2R - h_0) + 2\frac{h_0}{L}(h_0 - R)x - \left(\frac{h_0}{L}\right)^2 x^2}\,dx \qquad \text{(eq. I.8)}$$

The primitive of the integrand is known to be (see Handbook of Chemistry and Physics, section A, page A-46, formula nr. 242 and 246)

$$V_2 = d\left[\frac{(2cL+b)\sqrt{a+bx+cx^2}}{4c} - \frac{4ac-b^2}{8c}\left(\frac{1}{\sqrt{-c}}\arcsin\left(\frac{2cx+b}{\sqrt{b^2-4ac}}\right)\right)\right]_0^L \quad \text{(eq. I.9)}$$

Which is:

$$\left.\begin{aligned} V_2 = d\left[\frac{(2cL+b)\sqrt{a+bL+cL^2}}{4c} - \frac{4ac-b^2}{8c}\left(\frac{1}{\sqrt{-c}}\arcsin\left(\frac{2cL+b}{\sqrt{b^2-4ac}}\right)\right)\right] - \\ d\left[\frac{b\sqrt{a}}{4c} - \frac{4ac-b^2}{8c}\left(\frac{1}{\sqrt{-c}}\arcsin\left(\frac{b}{\sqrt{b^2-4ac}}\right)\right)\right] \end{aligned}\right\} \text{(eq. I.10)}$$

In which:

$$a = h_0(2R-h_0), b = 2\frac{h_0}{L}(h_0-R), c = -\left(\frac{h_0}{L}\right)^2, d = (h_0-R)$$

This may be further simplified to:

$$\left.\begin{aligned} V_2 = d\left[\frac{(2cL+b)\sqrt{a+bL+cL^2}}{4c} + \frac{1}{2}R^2\left(\frac{L}{h_0}\frac{1}{2}\pi\right)\right] - \\ d\left[\frac{b\sqrt{a}}{4c} - \frac{1}{2}R^2\left(\frac{L}{h_0}\arcsin\left(\frac{h_0-R}{R}\right)\right)\right] \end{aligned}\right\} \quad \text{(eq. I.11)}$$

The third term is written as:

$$V_3 = \frac{h_0}{L}\int_0^L x\sqrt{h_0(2R-h_o)+2\frac{h_0}{L}x(h_0-R)-\left(\frac{h_0}{L}\right)^2 x^2}\,dx \quad \text{(eq. I.12)}$$

or:

$$V_3 = d\int_0^L x\sqrt{a+bx+cx^2}\,dx \quad \text{(eq. I.13)}$$

In which $a = h_0(2R-h_0), b = 2\frac{h_0}{L}(h_0-R), c = -\left(\frac{h_0}{L}\right)^2, d = \frac{h_0}{L}$

$$\left.\begin{aligned} V_3 = d\left[\frac{\sqrt{a+bx+cx^2}\left(a+bx+cx^2\right)}{3c} - \frac{b(2cx+b)}{8c^2}\sqrt{a+bx+cx^2}\right]_0^L + \\ + d\left[+\frac{b\left(4ac-b^2\right)}{16c^2}\left(\frac{1}{\sqrt{-c}}\arcsin\left(\frac{2cx+b}{\sqrt{b^2-4ac}}\right)\right)\right]_0^L \end{aligned}\right\} \quad \text{(eq. I.14)}$$

The total volume becomes:

$$V = \sum_{i=1}^{i=3} V_i \quad \text{(eq. I.15)}$$

Using this analytical solution proves to be faster and more accurate than a numerical integration method normally applied. Since the volume has to be evaluated many times in a storage calculation the analytical method is to be preferred. In Table 84 and Table 85 a comparison has been made between a numerical estimate (using a second order integration method) and the analytical solution with respect to accuracy, number of floating point operations and relative CPU time. The accuracy of the analytical method is limited only by the accuracy with which the computer operates, while the accuracy of numerical methods depends strongly on the step size applied in the integration method. (R=0.5 m, i_b=0.002, h_0=0.2 m----> L=100 m).

method	$(V-V_{analytic})/V_{analytic}$	CPU (anal.=1)	volume (m^3)
dx=10m	$2.7129*10^{-3}$	7.25	4.57213392012838
dx=1m	$2.8437*10^{-5}$	51.5	4.55989305048317
dx=0.1 m	$1.5239*10^{-7}$	518.0	4.55976408084878
analytical	0	1.0	4.55976338598068

Table 84: Comparison between numerical solutions and the presented analytical solution.

method	multiplications	divisions	sqrt	acos
dx=10m	200	82	4	2
dx=1 m	2000	820	40	20
dx=0.1m	20000	8200	400	200
analytical	51	18	5	-

Table 85 : Number and type of arithmetic operations.

I.3 An algorithm for locating and quantifying lost storage in a drainage network

A possible algorithm to identify and quantify unused storage in a drainage network is shown in the schemes I through IV. The following symbols are used:

VARIABLE	TYPE	DESCRIPTION
H1	Real	
H2	Real	
HH()	Real array	
HMIN()	Real array	
B1	Real	
B2	Real	
BOB1()	Real array	Invert level node 1
BOB2()	Real array	Inver level node 2
I	Integer	Counter
J	Integer	Counter
NC	Integer	Number of conduits
STL()	Real array	Street level for nodes
SLOSS	Logical array	Storage loss present (T) else (F)
IK1	Integer	Counter
IK2	Integer	Counter
NN	Integer	Number of nodes
BOTTOM	Real	Bottom level manhole

The algorithm is initialised by setting SLOSS(NN)=.FALSE., indicating no storage loss is present at the outset. The arrays BOB1(NC),BOB2(NC), STL(NN),KN(NC) are obtained from the structural/geometrical database. In scheme I the vector HH(NN) is set to street level for all nodes, if HH(i) for node 1 (IK1) is smaller than the invert level at node 2 for a conduit j, HH(i) is set to BOB2(j). This procedure is repeated for all conduits. In Scheme II for all conduits it is checked whether HH(IK1)=HH(IK2), if not scheme III is entered to find out if storage loss is present for each node, in that case (SLOSS(I)=.T.). The corresponding minimum water level at the nodes is stored in the array HH(NN). Using this value the amount of unused storage is easily calculated (scheme IV) using the analytical solution presented in I.2.

SCHEME I

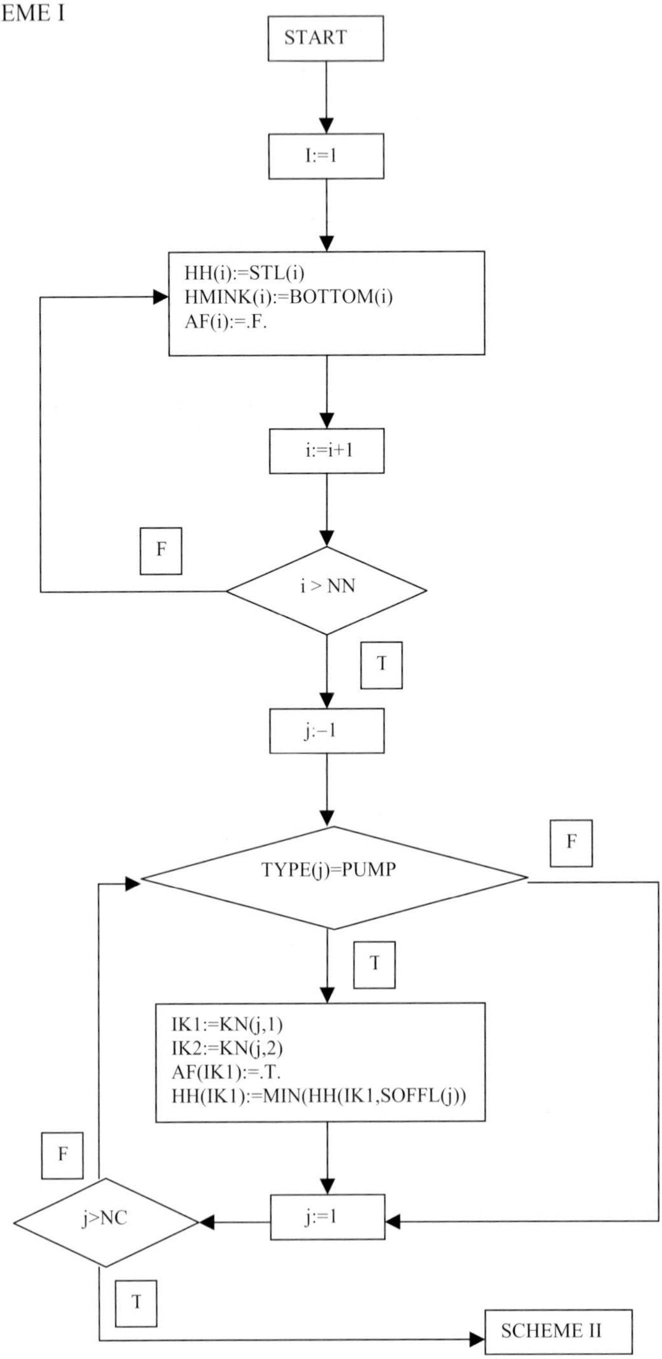

SCHEME II

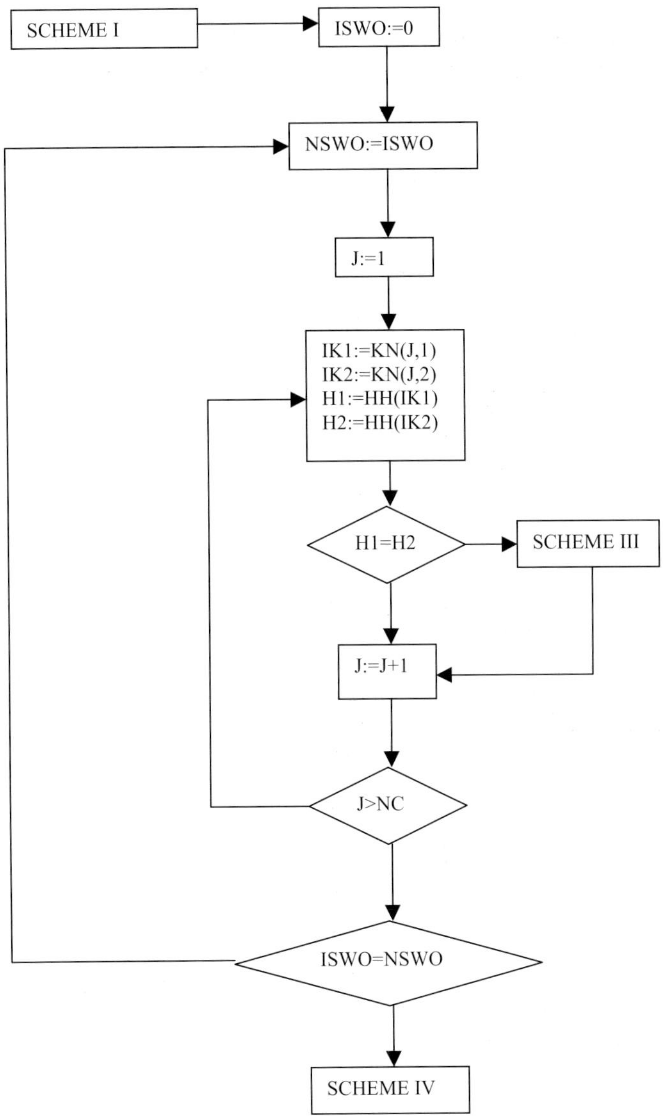
SCHEME I
ISWO:=0
NSWO:=ISWO
J:=1
IK1:=KN(J,1)
IK2:=KN(J,2)
H1:=HH(IK1)
H2:=HH(IK2)
H1=H2
SCHEME III
J:=J+1
J>NC
ISWO=NSWO
SCHEME IV

scheme III

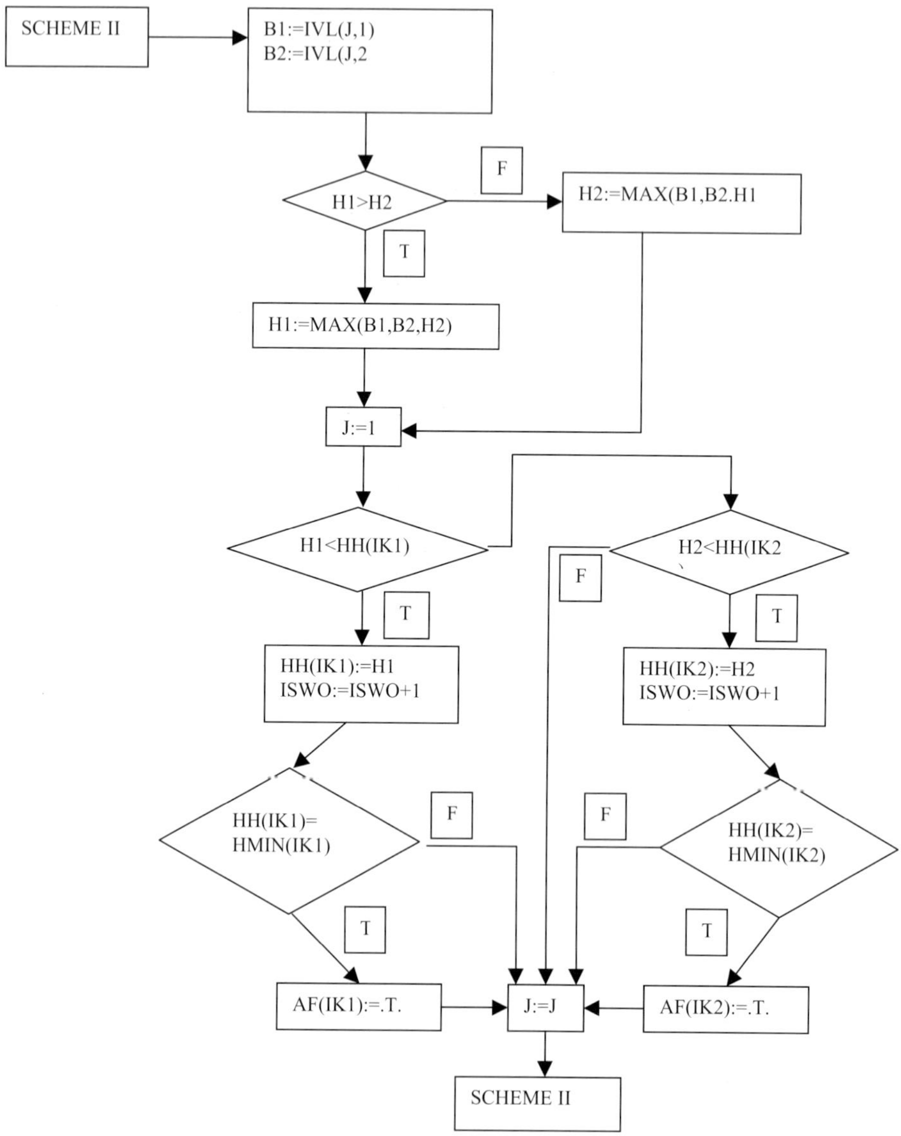
SCHEME II
B1:=IVL(J,1)
B2:=IVL(J,2
H1>H2
F
H2:=MAX(B1,B2.H1
T
H1:=MAX(B1,B2,H2)
J:=1
H1<HH(IK1)
H2<HH(IK2
F
T
HH(IK1):=H1
ISWO:=ISWO+1
T
HH(IK2):=H2
ISWO:=ISWO+1
HH(IK1)=
HMIN(IK1)
F
F
HH(IK2)=
HMIN(IK2)
T
T
AF(IK1):=.T.
J:=J
AF(IK2):=.T.
SCHEME II

SCHEME IV

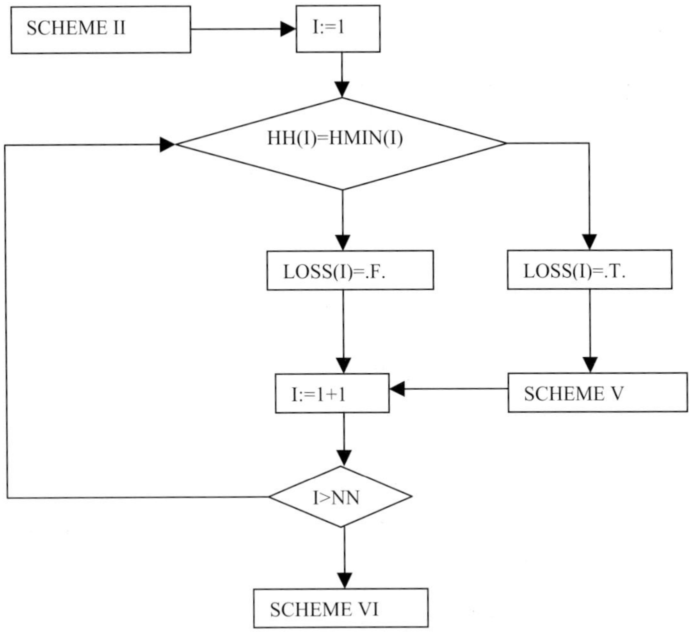
SCHEME II
I:=1
HH(I)=HMIN(I)
LOSS(I)=.F.
LOSS(I)=.T.
I:=1+1
SCHEME V
I>NN
SCHEME VI

ANNEXE II: NUMERICAL PROCEDURE.

In order to solve the De Saint-Venant equation the Preissmann scheme is applied. Since this set equations is non-linear a multi-dimensional Newton-Rhapson method was used to solve the non-linear difference equations.

The de Saint-Venant equations read as:

$$\left.\begin{aligned} & B\frac{\partial h}{\partial t}+\frac{\partial Q}{\partial x}=0 \\ & \frac{\partial Q}{\partial t}+\frac{\partial\left(\frac{Q^2}{A}\right)}{\partial x}+gA\frac{\partial h}{\partial x}+g\frac{Q|Q|}{RAC^2}=0 \end{aligned}\right\} \qquad \text{(eq. II.1)}$$

Taking the water level h and the discharge Q as principal parameters the following non linear difference equations are obtained:

Mass-balance:

$$\left.\begin{aligned} & \theta\frac{Q_{j+1}^{n+1}-Q_j^{n+1}}{\Delta x}+(1-\theta)\frac{Q_{j+1}^{n}-Q_j^{n}}{\Delta x}+ \\ & \left[\frac{\theta}{2}\left(B_{j+1}^{n+1}+B_j^{n+1}\right)+\frac{1-\theta}{2}\left(B_{j+1}^{n}+B_j^{n}\right)\right]\frac{h_{j+1}^{n+1}-h_{j+1}^{n}+h_j^{n+1}-h_j^{n}}{2\Delta t}=0 \end{aligned}\right\} \qquad \text{(eq. II.2)}$$

Momentum-balance:

$$\left.\begin{aligned} & \frac{Q_{j+1}^{n+1}-Q_{j+1}^{n}+Q_j^{n+1}-Q_j^{n}}{2\Delta t}+\theta\frac{\frac{\left(Q_{j+1}^{n+1}\right)^2}{A_{j+1}^{n+1}}-\frac{\left(Q_j^{n+1}\right)^2}{A_j^{n+1}}}{\Delta x}+(1-\theta)\frac{\frac{\left(Q_{j+1}^{n}\right)^2}{A_{j+1}^{n}}-\frac{\left(Q_j^{n}\right)^2}{A_j^{n}}}{\Delta x}+ \\ & g\left[\frac{\theta}{2}\left(A_{j+1}^{n+1}+A_j^{n+1}\right)+\frac{1-\theta}{2}\left(A_{j+1}^{n}+A_j^{n}\right)\right]*\left[\frac{\theta}{2}\frac{h_{j+1}^{n+1}-h_j^{n+1}}{\Delta x}+(1-\theta)\frac{h_{j+1}^{n}-h_j^{n}}{\Delta x}\right]+ \\ & \frac{g\theta}{2}\left[\frac{Q_{j+1}^{n+1}\left|Q_{j+1}^{n+1}\right|}{R_{j+1}^{n+1}A_{j+1}^{n+1}\left(C_{j+1}^{n+1}\right)^2}+\frac{Q_j^{n+1}\left|Q_j^{n+1}\right|}{R_j^{n+1}A_j^{n+1}\left(C_j^{n+1}\right)^2}\right]+ \\ & \frac{g(1-\theta)}{2}\left[\frac{Q_{j+1}^{n}\left|Q_{j+1}^{n}\right|}{R_{j+1}^{n}A_{j+1}^{n}\left(C_{j+1}^{n}\right)^2}+\frac{Q_j^{n}\left|Q_j^{n}\right|}{R_j^{n}A_j^{n}\left(C_j^{n}\right)^2}\right]=0 \end{aligned}\right\} \qquad \text{(eq. (II.3)}$$

In which:

j	index for discretisation in space dimension	(-)
n	index for discretisation in the time dimension	(-)
h	water level	(m)
Q	discharge	(m^3/s)
A	cross sectional area	(m^2)
B	width of the free water level	(m)
C	Chezy's coefficient	($m^{0.5}/s$)
g	gravitational acceleration (9.813)	(m/s^2)

R	hydraulic radius	(m)
θ	weighing factor $(0.50<\theta<1.0)$	(-)
Δt	time step	(s)
Δx	space step	(m)

The Newton process reads as (in vector notation):

$$\underline{\Delta u} = -\underline{\underline{D}}^{-1}\underline{f} \qquad \text{(eq. II.4)}$$

In which

$\underline{\Delta u}$	vector containing the increments for the unknown parameters (water level and discharge at time level n+1)
$\underline{\underline{D}}$	matrix containing the first derivatives of the discretised mass and momentum balance to the unknown parameters
$\underline{f}$	vector containing the values for mass and momentum balance

The initial estimate for the parameter values at time level n+1 are the values they have at time level n. Then an iterative process is followed until mass and momentum conservation are obtained to within a prescribed accuracy, which is limited by the calculation accuracy of the computer applied.

The elements in $\underline{\underline{D}}$ for the mass balance equation are:

$$\left.\begin{aligned}
\frac{\partial f}{\partial Q_j^{n+1}} &= \frac{-\theta}{\Delta x} \\
\frac{\partial f}{\partial Q_{j+1}^{n+1}} &= \frac{\theta}{\Delta x} \\
\frac{\partial f}{\partial h_j^{n+1}} &= \frac{\theta}{4\Delta t}\left[B_{j+1}^{n+1} + B_j^{n+1} + h_j^{n+1}\frac{\partial B_j^{n+1}}{\partial h_j^{n+1}}\right] + \frac{1-\theta}{4\Delta t}\left(B_{j+1}^{n} + B_j^{n}\right) \\
\frac{\partial f}{\partial h_{j+1}^{n+1}} &= \frac{\theta}{4\Delta t}\left[B_{j+1}^{n+1} + B_j^{n+1} + h_{j+1}^{n+1}\frac{\partial B_{j+1}^{n+1}}{\partial h_{j+1}^{n+1}}\right] + \frac{1-\theta}{4\Delta t}\left(B_{j+1}^{n} + B_j^{n}\right)
\end{aligned}\right\} \qquad \text{(eq. II.5)}$$

The elements in $\underline{\underline{D}}$ for the momentum balance equation are:

$$\left.\begin{array}{l}
\dfrac{\partial f}{\partial Q_j^{n+1}} = \dfrac{1}{2\Delta t} - \dfrac{\theta Q_j^{n+1}}{\Delta x A_j^{n+1}} + \dfrac{g\theta}{R_j^{n+1} A_j^{n+1} \left(C_j^{n+1}\right)^2} \\
\dfrac{\partial f}{\partial Q_{j+1}^{n+1}} = \dfrac{1}{2\Delta t} - \dfrac{\theta Q_{j+1}^{n+1}}{\Delta x A_{j+1}^{n+1}} + \dfrac{g\theta}{R_{j+1}^{n+1} A_{j+1}^{n+1} \left(C_{j+1}^{n+1}\right)^2} \\
\dfrac{\partial f}{\partial h_j^{n+1}} = \dfrac{\theta}{\Delta x}\left(\dfrac{Q_j^{n+1}}{A_j^{n+1}}\right)^2 \dfrac{\partial A_j^{n+1}}{\partial h_j^{n+1}} + \dfrac{g\theta^2}{2\Delta x}\left[-A_{j+1}^{n+1} + h_{j+1}^{n+1}\dfrac{\partial A_j^{n+1}}{\partial h_j^{n+1}} - A_j^{n+1} - h_j^{n+1}\dfrac{\partial A_j^{n+1}}{\partial h_j^{n+1}}\right] + \\
\dfrac{g\theta(1-\theta)}{2\Delta x}\left[h_{j+1}^{n}\dfrac{\partial A_j^{n+1}}{\partial h_j^{n+1}} - h_j^{n}\dfrac{\partial A_j^{n+1}}{\partial h_j^{n+1}}\right] - \dfrac{g\theta(1-\theta)}{2\Delta x}\left(A_{j+1}^{n} + A_j^{n}\right) - \\
\dfrac{g\theta Q_j^{n+1}\left|Q_j^{n+1}\right|}{2\Delta x}\left[\dfrac{2A_j^{n+1}R_j^{n+1}C_j^{n+1}\dfrac{\partial C_j^{n+1}}{\partial h_j^{n+1}} + \left(C_j^{n+1}\right)^2 R_j^{n+1}\dfrac{\partial A_j^{n+1}}{\partial h_j^{n+1}} + A_j^{n+1}C_j^{n+1}\dfrac{\partial R_j^{n+1}}{\partial h_j^{n+1}}}{\left(R_j^{n+1}A_j^{n+1}\left(Q_j^{n+1}\right)^2\right)^2}\right] \\
\dfrac{\partial f}{\partial h_{j+1}^{n+1}} = \dfrac{-\theta}{\Delta x}\left(\dfrac{Q_{j+1}^{n+1}}{A_{j+1}^{n+1}}\right)^2 \dfrac{\partial A_{j+1}^{n+1}}{\partial h_{j+1}^{n+1}} + \dfrac{g\theta^2}{2\Delta x}\left[A_{j+1}^{n+1} + h_{j+1}^{n+1}\dfrac{\partial A_{j+1}^{n+1}}{\partial h_{j+1}^{n+1}} + A_j^{n+1} - h_j^{n+1}\dfrac{\partial A_{j+1}^{n+1}}{\partial h_{j+1}^{n+1}}\right] + \\
\dfrac{g\theta(1-\theta)}{2\Delta x}\left[h_{j+1}^{n}\dfrac{\partial A_{j+1}^{n+1}}{\partial h_{j+1}^{n+1}} - h_j^{n}\dfrac{\partial A_{j+1}^{n+1}}{\partial h_{j+1}^{n+1}}\right] + \dfrac{g\theta(1-\theta)}{2\Delta x}\left(A_{j+1}^{n} + A_j^{n}\right) - \\
\dfrac{g\theta Q_{j+1}^{n+1}\left|Q_{j+1}^{n+1}\right|}{2\Delta x}\left[\dfrac{2A_{j+1}^{n+1}R_{j+1}^{n+1}C_{j+1}^{n+1}\dfrac{\partial C_{j+1}^{n+1}}{\partial h_{j+1}^{n+1}} + \left(C_{j+1}^{n+1}\right)^2 R_{j+1}^{n+1}\dfrac{\partial A_{j+1}^{n+1}}{\partial h_{j+1}^{n+1}} + A_{j+1}^{n+1}C_{j+1}^{n+1}\dfrac{\partial R_{j+1}^{n+1}}{\partial h_{j+1}^{n+1}}}{\left(R_{j+1}^{n+1}A_{j+1}^{n+1}\left(Q_{j+1}^{n+1}\right)^2\right)^2}\right]
\end{array}\right\} \quad \text{(eq II.6)}$$

Implementing eq II.5 and eq. II.6 in eq. II.4 results in a system of equations looking like:

$$\begin{bmatrix} \Delta h_1^{n+1} \\ \Delta Q_1^{n+1} \\ \Delta h_2^{n+1} \\ \Delta Q_2^{n+1} \\ . \\ . \\ \Delta h_j^{n+1} \\ \Delta Q_j^{n+1} \end{bmatrix} = -\begin{bmatrix}
HR & QR & 0 & 0 & 0 & 0 & 0 & 0 \\
\dfrac{\partial mass}{\partial h_1^{n+1}} & \dfrac{\partial mass}{\partial Q_1^{n+1}} & \dfrac{\partial mass}{\partial h_2^{n+1}} & \dfrac{\partial mass}{\partial Q_2^{n+1}} & 0 & 0 & 0 & 0 \\
\dfrac{\partial mom}{\partial h_1^{n+1}} & \dfrac{\partial mom}{\partial Q_1^{n+1}} & \dfrac{\partial mom}{\partial h_2^{n+1}} & \dfrac{\partial mom}{\partial Q_2^{n+1}} & 0 & 0 & 0 & 0 \\
0 & 0 & x & x & x & x & 0 & 0 \\
0 & 0 & x & x & x & x & 0 & 0 \\
0 & 0 & 0 & 0 & \dfrac{\partial mass}{\partial h_{j-1}^{n+1}} & \dfrac{\partial mass}{\partial h_{j-1}^{n+1}} & \dfrac{\partial mass}{\partial h_j^{n+1}} & \dfrac{\partial mass}{\partial h_j^{n+1}} \\
0 & 0 & 0 & 0 & \dfrac{\partial mom}{\partial h_{j-1}^{n+1}} & \dfrac{\partial mom}{\partial h_{j-1}^{n+1}} & \dfrac{\partial mom}{\partial h_j^{n+1}} & \dfrac{\partial mom}{\partial h_j^{n+1}} \\
0 & 0 & 0 & 0 & 0 & 0 & BH_j & BQ_j
\end{bmatrix}^{-1} \begin{bmatrix} Bound \\ mass_1 \\ mom_1 \\ y \\ y \\ mass_{j-1} \\ mom_{j-1} \\ Bound \end{bmatrix}$$

(eq. II.7)

In which Bound indicates a boundary condition mass indicates mass-balance equation and mom indicates momentum balance equation. HR, QR, BH_j and BQ_j are matrix elements following from the boundary conditions theta are implemented.

After each iteration the right hand vector is split into two vectors; one containing all elements for the mass balance equation (m) and one vector containing the elements for the momentum balance equation (im). The Newton process is stopped when:

$$\begin{aligned} &|m_i| < \varepsilon \wedge |im_i| < \varepsilon \\ &i = 1, \ldots, j \end{aligned} \qquad \text{(eq. II.8)}$$

In which j is the number of calculation points in the space domain and ε is a small real positive number.

Boundary conditions should be treated in the same manner. For example the boundary conditions as applied in chapter 2 are discussed.
Since the flow is sub critical an upstream (i=1) and a downstream (i=j) boundary condition are to be specified.
The upstream boundary condition is:

$$Q_1^{time} = q(t) \text{ or } Q_1^{time} - q(t) = 0 \qquad \text{(eq II.9)}$$

Differentiating eq. II.10 to Q_1^{time} results in $HR = 1, QR = 0$, while Bound is given by eq. II.9.

At the downstream end a free overflow is assumed, implying the flow is critical:

$$(Q_j)^2 - g\frac{(A_j)^3}{B_j} = 0 \qquad \text{(eq. II.10)}$$

Differentiating eq. II.11 to Q and h results in:

$$\left.\begin{aligned} &BH_j = -g\left(3(A_j^{n+1})^2 - \frac{(A_j^{n+1})^3 \sqrt{Da_{j+1}^{n+1} - (a_{j+1}^{n+1})^2}}{D - a_{j+1}^{n+1}}\right) \\ &BQ_j = 2Q_j^{n+1} \end{aligned}\right\} \qquad \text{(eq. II.11)}$$

In which a_{j+1}^{n+1} is the water depth (i.e. the water level relative to some reference level minus the local bottom level relative to the same reference) and D is the diameter of the pipe.

Following the method outlined, basically any boundary condition can be implemented.

ANNEXE III : STABILITY CONDITIONS FOR THE PREISSMANN SCHEME

In this annexe the results of the cases discussed in Section 2.4.4.4 are graphicaly shown.

The cases considered are:

	I	II	III	IV	V	VI	VII	VIII
θ	0.55	0.55	0.55	0.55	0.6	0.6	0.6	0.6
ā	0.01	1.00	0.01	1.00	0.01	1.00	0.01	1.00
w	0	0	10	10	0	0	10	10

In which:

a	water depth	(m)
w	linearised resistance factor	(s^{-1})
θ	weighting factor	(-)

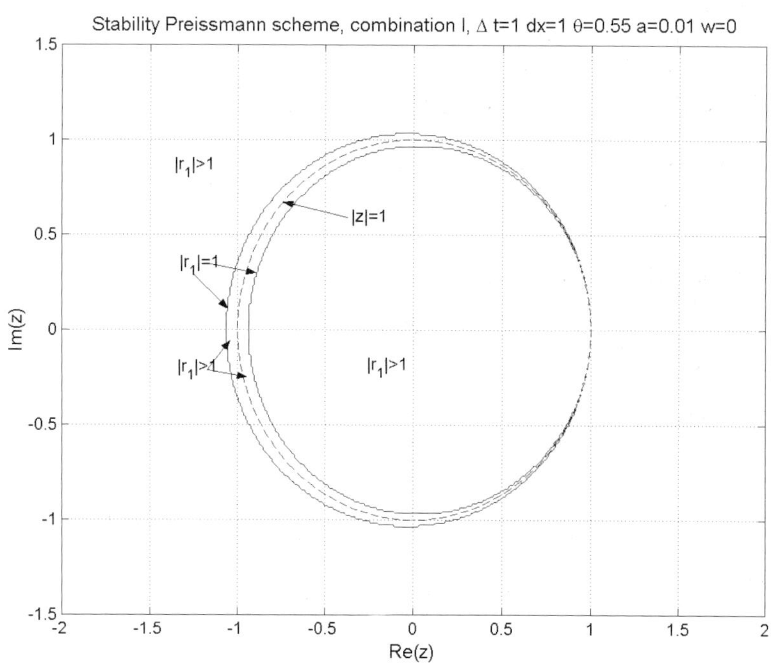

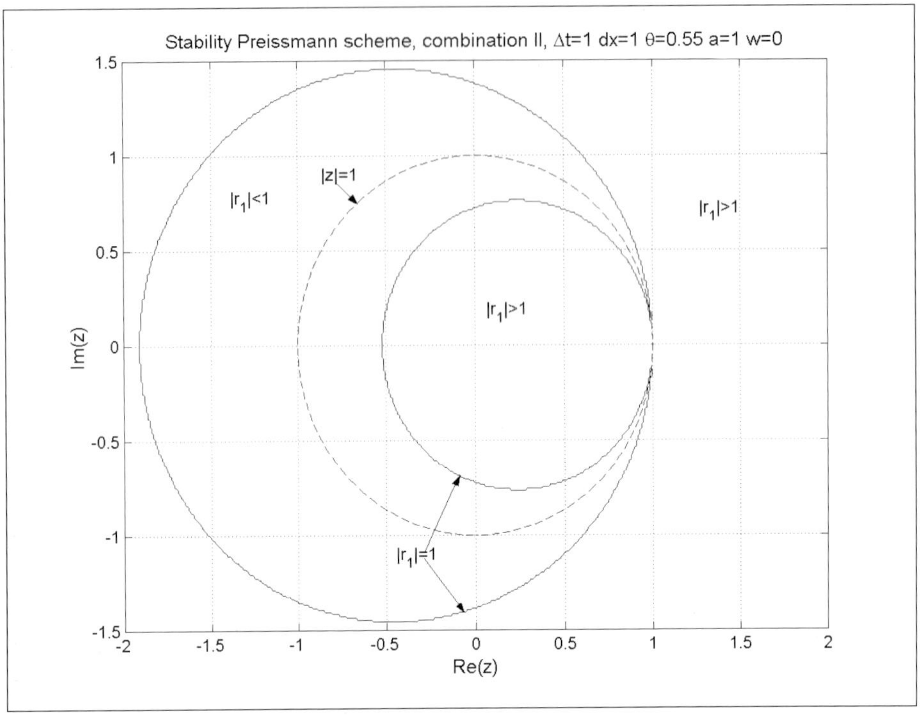
Stability Preissmann scheme, combination II, Δt=1 dx=1 θ=0.55 a=1 w=0
|z|=1
|r1|<1
|r1|>1
|r1|>1
|r1|=1
Im(z)
Re(z)

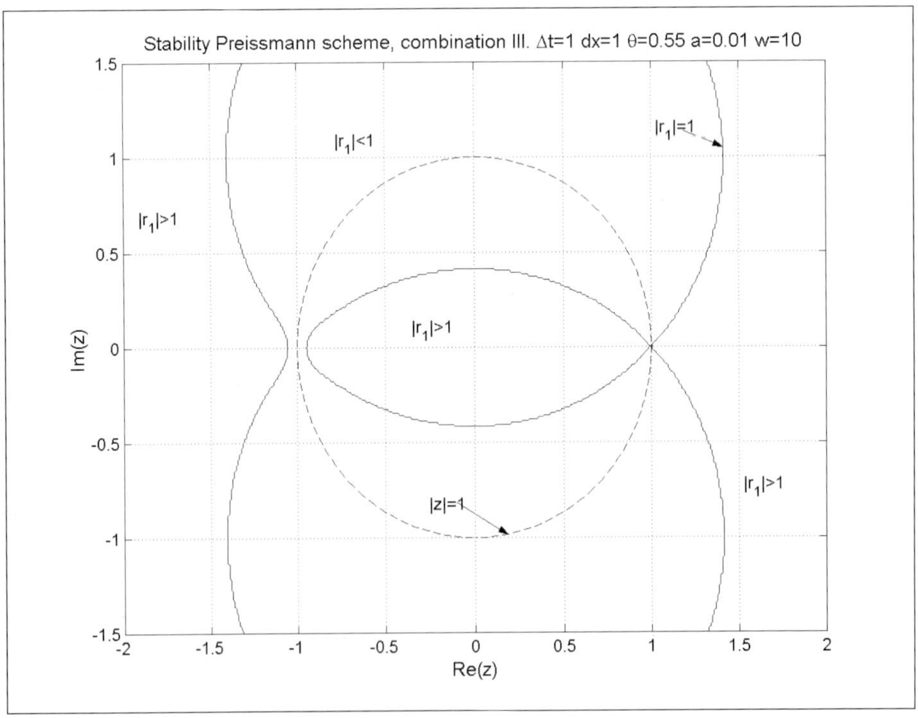
Stability Preissmann scheme, combination III. Δt=1 dx=1 θ=0.55 a=0.01 w=10
|r1|<1
|r1|=1
|r1|>1
|r1|>1
|r1|>1
|z|=1
Im(z)
Re(z)

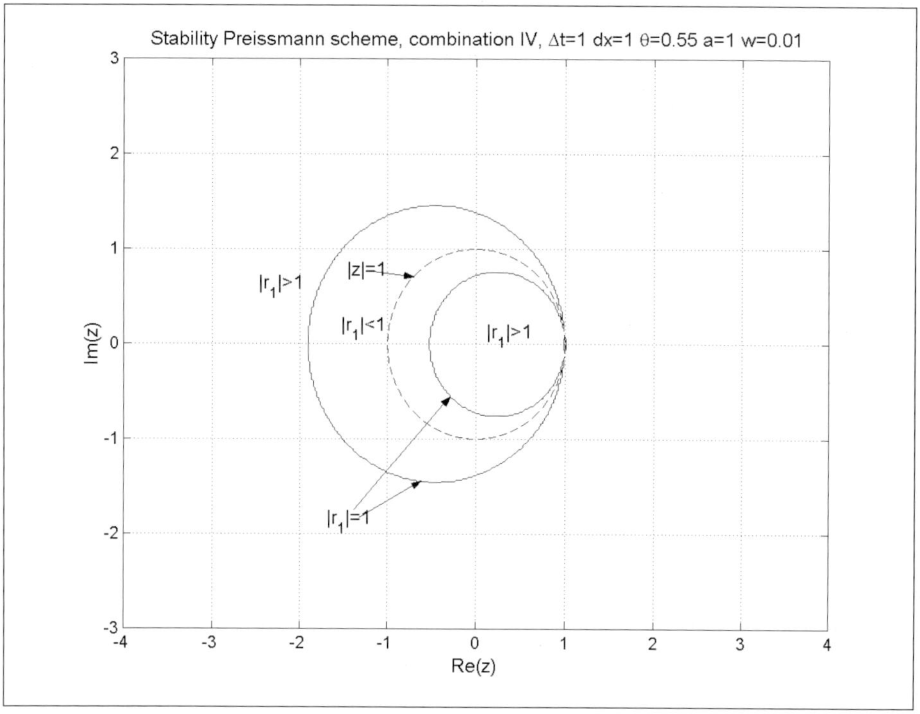

Stability Preissmann scheme, combination IV, Δt=1 dx=1 θ=0.55 a=1 w=0.01
|z|=1
|r1|>1
|r1|<1
|r1|>1
|r1|=1
Im(z)
Re(z)

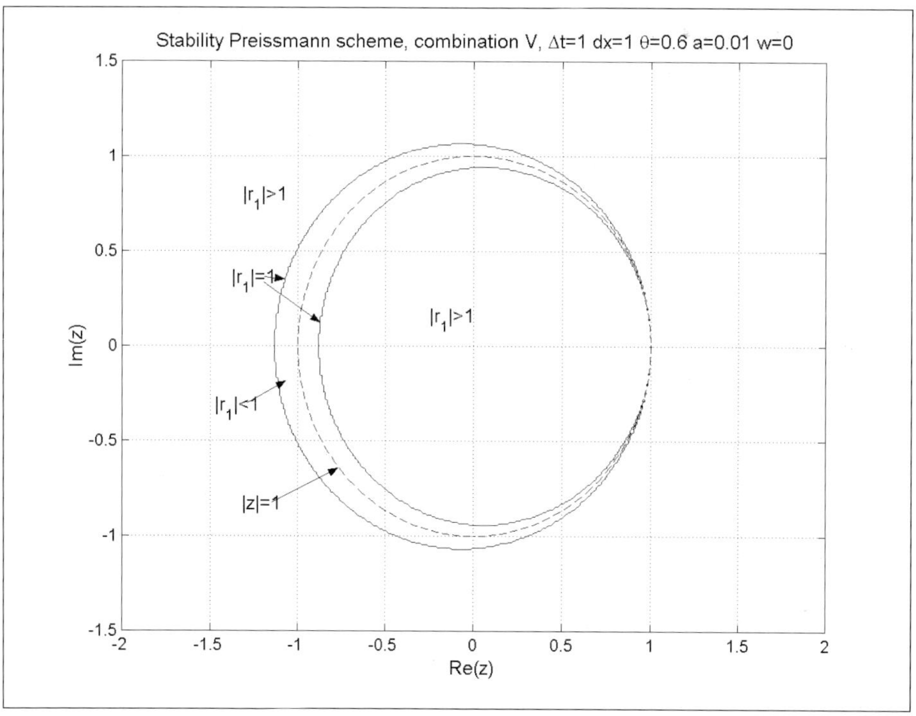

Stability Preissmann scheme, combination V, Δt=1 dx=1 θ=0.6 a=0.01 w=0
|r1|>1
|r1|=1
|r1|>1
|r1|<1
|z|=1
Im(z)
Re(z)

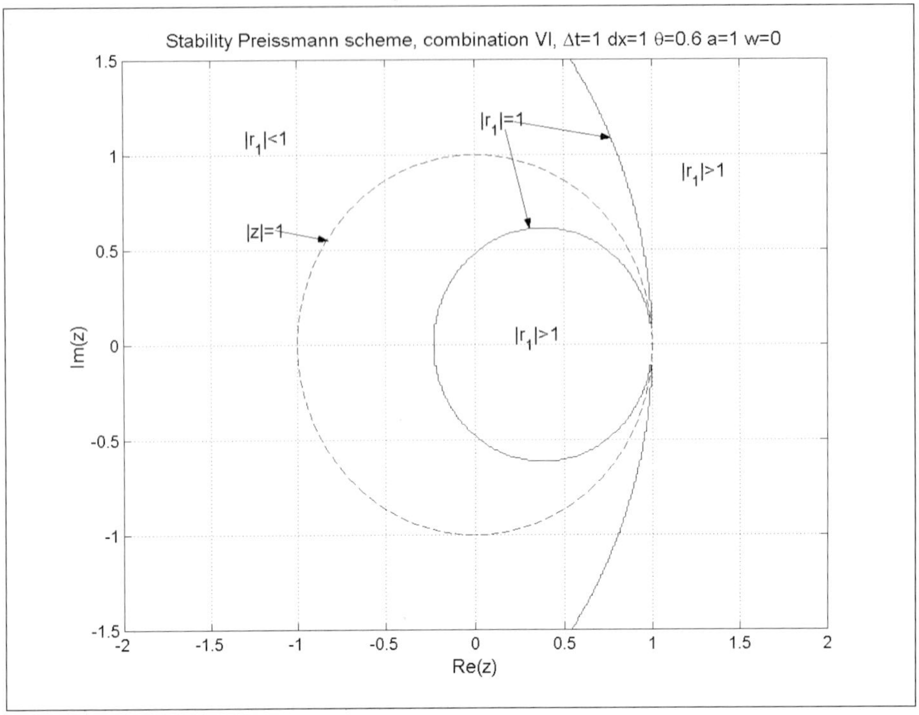
Stability Preissmann scheme, combination VI, Δt=1 dx=1 θ=0.6 a=1 w=0
|r1|<1
|r1|=1
|r1|>1
|z|=1
|r1|>1
Im(z)
Re(z)

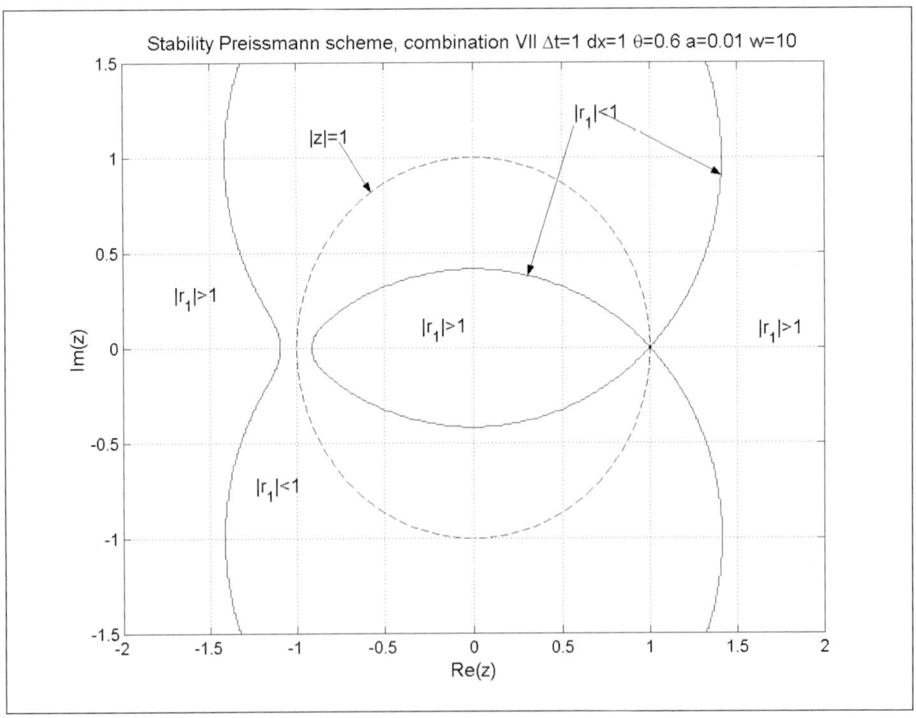
Stability Preissmann scheme, combination VII Δt=1 dx=1 θ=0.6 a=0.01 w=10
|r1|<1
|z|=1
|r1|>1
|r1|>1
|r1|>1
|r1|<1
Im(z)
Re(z)

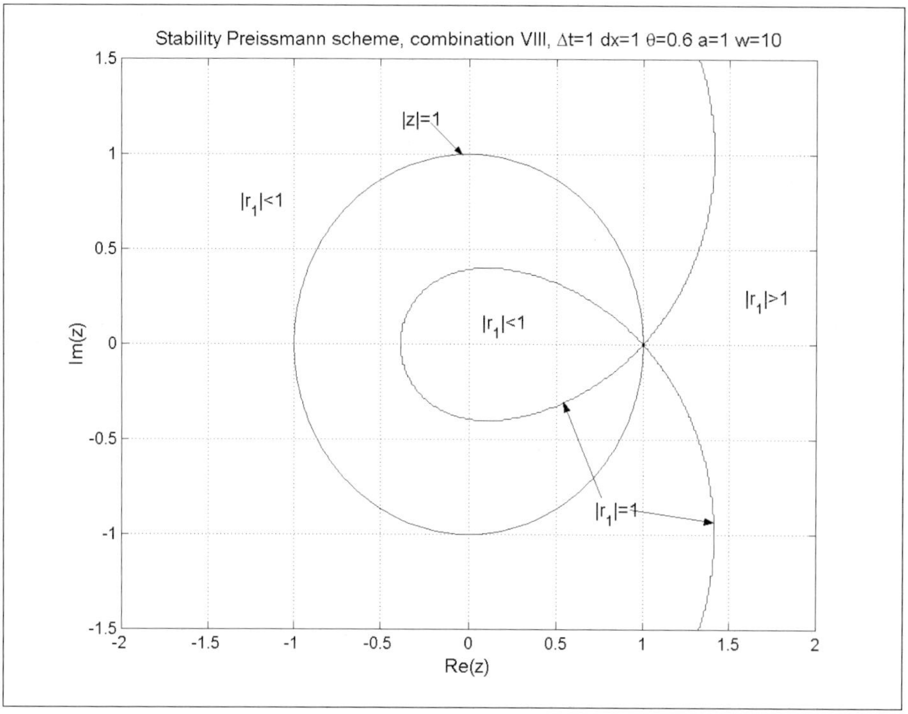
Stability Preissmann scheme, combination VIII, Δt=1 dx=1 θ=0.6 a=1 w=10
|z|=1
|r1|<1
|r1|<1
|r1|>1
|r1|=1
Im(z)
Re(z)

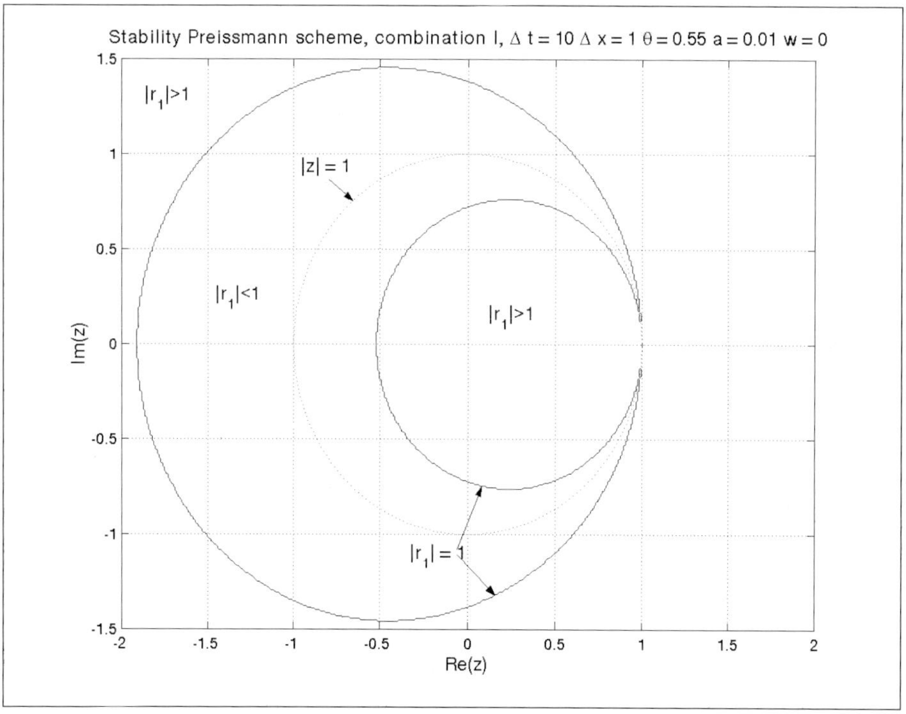
Stability Preissmann scheme, combination I, Δ t = 10 Δ x = 1 θ = 0.55 a = 0.01 w = 0
|r1|>1
|z| = 1
|r1|<1
|r1|>1
|r1| = 1
Im(z)
Re(z)

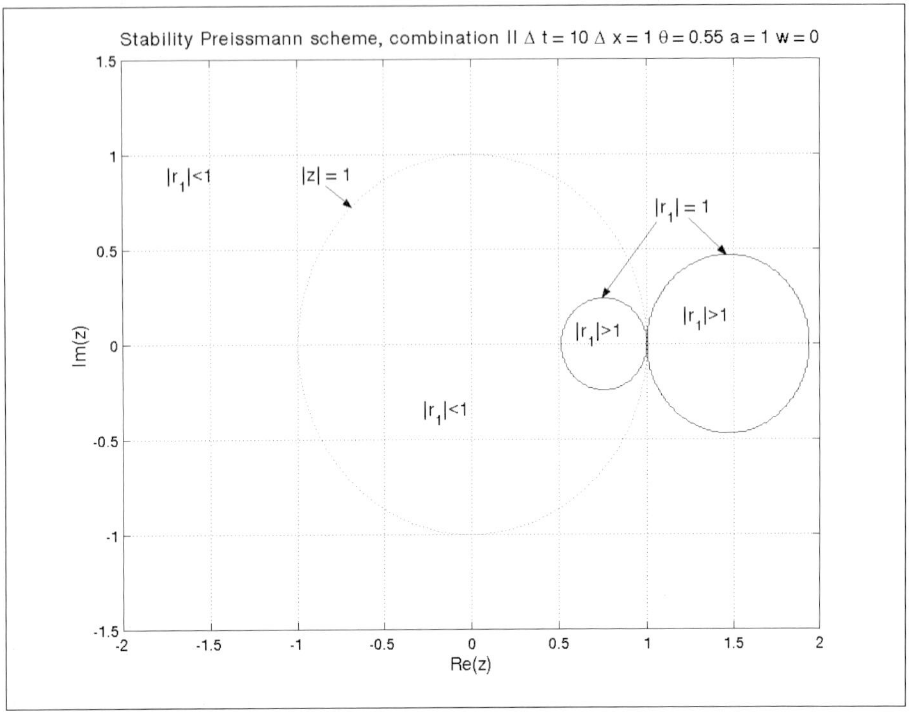
Stability Preissmann scheme, combination II Δ t = 10 Δ x = 1 θ = 0.55 a = 1 w = 0
|r1|<1
|z| = 1
|r1| = 1
|r1|>1
|r1|>1
|r1|<1
Im(z)
Re(z)
1.5
1
0.5
0
-0.5
-1
-1.5
-2
-1.5
-1
-0.5
0
0.5
1
1.5
2

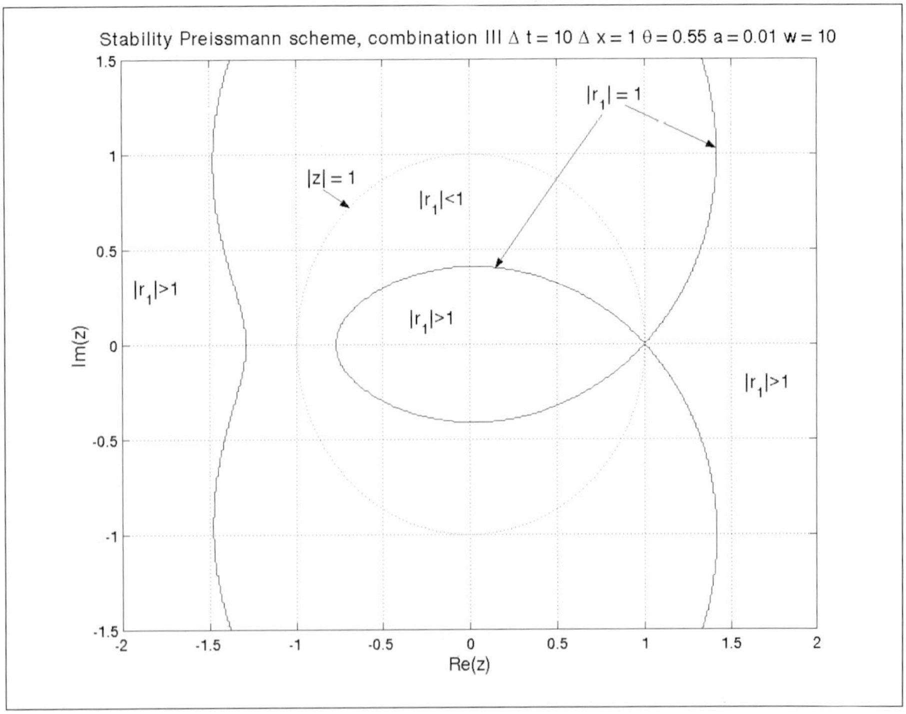
Stability Preissmann scheme, combination III Δ t = 10 Δ x = 1 θ = 0.55 a = 0.01 w = 10
|r1| = 1
|z| = 1
|r1|<1
|r1|>1
|r1|>1
|r1|>1
Im(z)
Re(z)
1.5
1
0.5
0
-0.5
-1
-1.5
-2
-1.5
-1
-0.5
0
0.5
1
1.5
2

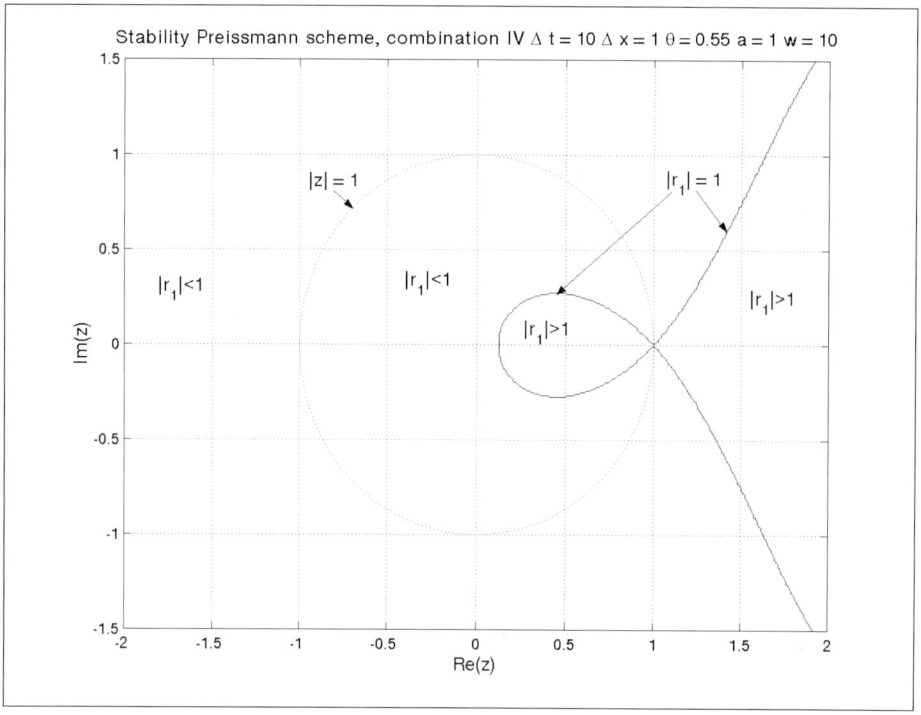

Stability Preissmann scheme, combination IV Δ t = 10 Δ x = 1 θ = 0.55 a = 1 w = 10
Im(z)
Re(z)
|z| = 1
|r1| = 1
|r1|<1
|r1|<1
|r1|>1
|r1|>1

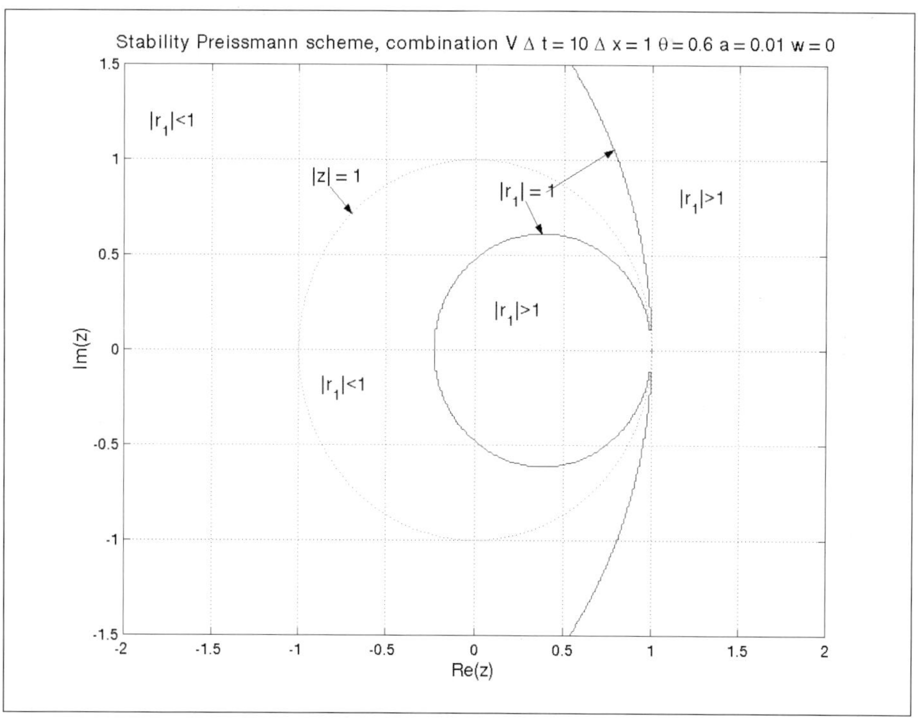

Stability Preissmann scheme, combination V Δ t = 10 Δ x = 1 θ = 0.6 a = 0.01 w = 0
Im(z)
Re(z)
|r1|<1
|z| = 1
|r1| = 1
|r1|>1
|r1|>1
|r1|<1

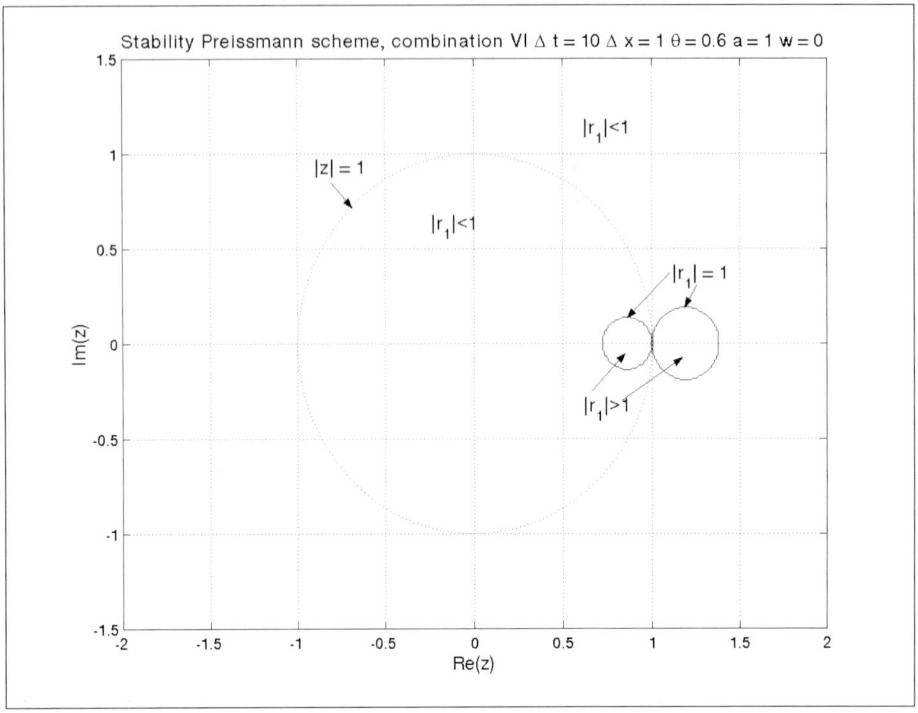
Stability Preissmann scheme, combination VI Δ t = 10 Δ x = 1 θ = 0.6 a = 1 w = 0
|r1|<1
|z| = 1
|r1|<1
|r1| = 1
|r1|>1
Im(z)
Re(z)

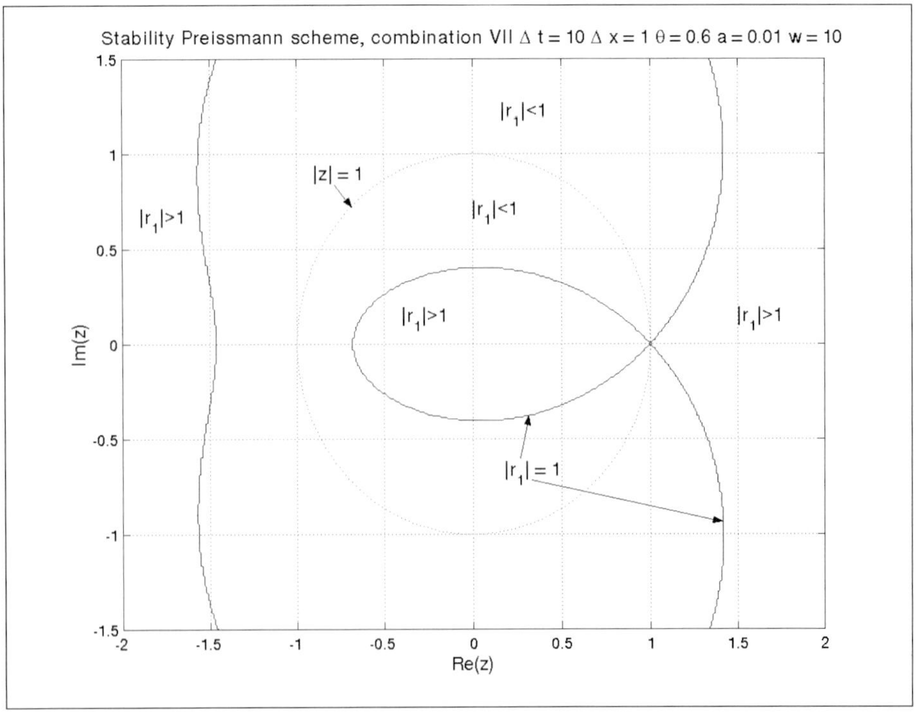
Stability Preissmann scheme, combination VII Δ t = 10 Δ x = 1 θ = 0.6 a = 0.01 w = 10
|r1|<1
|z| = 1
|r1|>1
|r1|<1
|r1|>1
|r1|>1
|r1| = 1
Im(z)
Re(z)

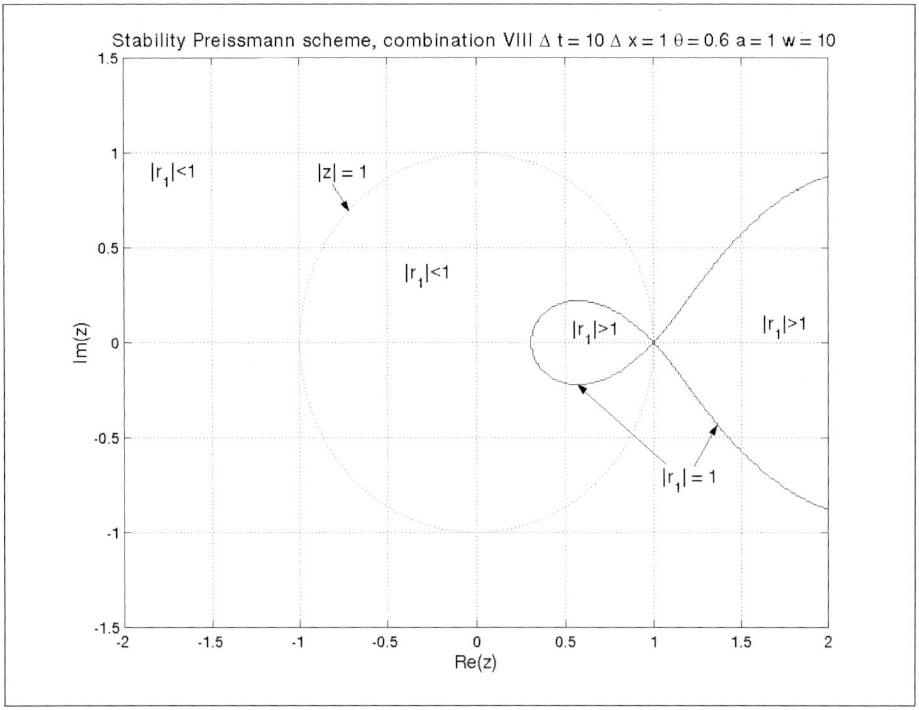
Stability Preissmann scheme, combination VIII Δ t = 10 Δ x = 1 θ = 0.6 a = 1 w = 10
|r_1|<1
|z| = 1
|r_1|<1
|r_1|>1
|r_1|>1
|r_1| = 1
Im(z)
Re(z)
1.5
1
0.5
0
-0.5
-1
-1.5
-2
-1.5
-1
-0.5
0
0.5
1
1.5
2

ANNEXE IV: RESULTS MONTE CARLO SIMULATIONS

IV.1 PROBABILITY DISTRIBUTIONS

In order to characterise the probability distribution of the results obtained, the following analyses were made on the results:

- A χ^2-test([57]).
- The 'fit' of the results to several analytical distributions, the fit was defined as: $$s = \sqrt{\frac{\sum_{i=1}^{i=n}\left(F(h_i) - p_i\right)^2}{n-2}}$$
- The Skewness, defined as $Sk = \frac{n}{(n-1)(n-2)} i = n \sum_{i=1}\left(\frac{x_i - x}{s}\right)^3$. A positive value points at a relatively long 'tail' for values above the average, whereas a negative value points at a relatively long tail at values smaller than average.
- The Kurtosis, defined as $K = \left[\frac{n(n+1)}{n(n-1)(n-2)(n-3)}\sum_{i=1}^{i=n}\left(\frac{x_i - x}{s}\right)^4\right] - \frac{3(n-1)^2}{(n-2)(n-3)}$. A positive value for K indicates a relatively high maximum in the distribution, whereas a negative value point at a relatively flat peak in the distribution ([58]).

The figures presented for Skewness and Kurtosis should not be regarded as exact estimates, due to the fact that the number of Monte Carlo simulations was to little, but are merely indications for the characterisation of the probability functions.

The probability distributions analysed are:

The Normal distribution (or Gauss-distribution), indicated as 'N' and defined by:

$$f(x) = \frac{1}{\sigma\sqrt{2\pi}} e^{-\frac{(x-\mu)^2}{2\sigma^2}}$$, in which μ is the mean and $\sigma$is the standard deviation.

The Weibull distribution (indicated as 'W'), is defined by;

$$f(x) = \frac{k}{u}\left(\frac{x}{u}\right)^{k-1} e^{-\left(\frac{x}{u}\right)^k}$$

The cumulative Weibull distribution is:

$$F(x) = 1 - e^{-\left(\frac{x}{u}\right)^k}$$

in which:

([57]) The results of the χ^2-test are not very consistent, since they depend to a certain extent on the partition made in the original data set, therefore in the analysis the emphasis is on the 's'-value obtained.

([58]) Both Kurtosis and Skewness are relative to the Normal distribution, for a perfect fit with this distribution both parameter are zero.

$$u = \frac{\mu}{\Gamma\left(1+\frac{1}{k}\right)}, \text{and } u^2 = \frac{\sigma^2+\mu^2}{\Gamma\left(1+\frac{2}{k}\right)}.$$

The Gumbel distribution, indicated as 'G' and defined by:

$$f(x) = \alpha e^{\left(\alpha(x-u)-e^{\alpha(x-u)}\right)}$$

The cumulative 's' distribution is defined as:

$$F(x) = 1 - e^{\alpha(x-u)}, \text{ in which } \alpha = \frac{\pi}{\sigma\sqrt{6}}, \text{ and } u = \mu + \frac{0.5772}{\alpha}$$

(0.5772 is known as Eulers constant).

If none of the distribution types mentioned fit the results, this is marked in the tables by an 'X'.

IV.2 Results system I

The results are presented in Table 86 through Table 88 and Figure 52 through Figure 54. In these graphs some characteristic results with respect to probability distributions found and the comparison between 250 Monte Carlo runs and 1000 Monte Carlo runs are shown. From these results it was concluded that using 250 runs gives enough information for further analysis, therefore the results presented in the tables are based upon 250 simulations.

set	$\mu(m^3)$	$\sigma(m^3)$	Type	parameters	kurtosis	Skewness	χ^2	s
I	1354.28	132.12	N	-	-0.0983	-0.1403	36.66	$1.1555*10^{-4}$
II	1368.81	27.97	N	-	-0.0312	-0.2444	87.82	$1.0264*10^{-3}$
III	1376.17	124.11	N	-	-0.3102	-0.1519	82.06	$3.3804*10^{-4}$
IV	1370.07	4.09	G	U=1372.60 α=0.31302	1.2411	-1.0552	67.08	$1.5036*10^{-4}$
V	1348.15	11.75	N	-	0.1143	-0.1377	48.72	$2.5672*10^{-4}$
VI	1369.00	5.2	X					
VII	1324.52	29.2	X	peaks at 1303 m^3 and 1370 m^3				
VIII	1322.11	32.25	X	peaks at 1305 m^3, 1310 m^3 and 1370 m^3				

Table 86: Overflow volume, system I.

set	$\mu(m^3/s)$	$\sigma(m^3/s)$	Type	parameters	Kurtosis	Skewness	χ^2	s
I	0.905	0.073	N	-	0.1687	0.1818	45.14	$3.4688*10^{-5}$
II	0.899	0.026	N	-	0.0312	-0.0414	60.92	$4.6841*10^{-4}$
III	0.903	0.025	N	-	-0.1127	-0.3713	87.98	$7.7085*10^{-4}$
IV	0.903	0.045	N	-	0.8645	0.5762	59.13	$8.4528*10^{-4}$
V	0.899	0.007	N	-	0.0476	0.0700	137.66	$1.0950*10^{-3}$
VI	0.903	0.002	X					
VII	0.871	0.022	X	peaks at 0.854 m^3/s and 0.904 m^3/s				
VIII	0.876	0.024	X	peaks at 0.856 m^3/s and 0.904 m^3/s				

Table 87: Maximum overflow discharge, system I.

Set	$\mu(-)$	$\sigma(-)$	Type	Parameters	kurtosis	Skewness	χ^2	s
I	6.78	1.066	N	Min 2 Max 13	4.74569	-0.4046	7454.24 (nkl=50) 6733.01 (nkl=11)	$3.6063*10^{-2}$
II	6.97	0.188	X					
III	6.90	0.568	X					
IV	6.92	0.389	X					
V	6.83	0.039	X					
VI	7.00	No variation						
VII	6.66	0.473	X					
VIII	6.66	0.587	X					

Table 88: Number of flooded manholes, system I.

set	Location	μ(m)	σ(m)	Type	Parameters	kurtosis	Skewness	χ^2	s
I	A	0.656	0.114	N	-	1.2492	0.21196	27.97	$7.3776*10^{-5}$
	B	0.737	0.099	N	-	$2.9677*10^{-2}$	$-2.8163*10^{-2}$	58.36	
	C	0.069	0.029	W	K=0.9671 u=0.0271 h_{min}=0.029 m	8.4585	2.45852	311.62	$4.0521*10^{-3}$
	D	2.278	0.135	N	-	2.60533	0.47311	20.01	$1.0064*10^{-4}$
	E	0.067	0.044	W	K=0.8776 U=0.0362 H_{min}=0.029	6.11828	2.06138	60.07	$1.1386*10^{-3}$
II	A	0.651	0.035	N	-	-0.18512	-0.26311	69.80	$6.4651*10^{-4}$
	B	0.761	0.032	N		-0.26505	0.375851	86.09	$8.7614*10^{-4}$
	C	0.055	No variation						
	D	2.256	0.007	N	-	0.191804	-0.15603	114.23	1.7874810^{-3}
	E	0.038	No variation						
III	A	0.653	0.066	N	-	-0.11061	-0.54225	77.93	$1.5032*10^{-3}$
	B	0.775	0.045	N		-0.63755	0.06665	94.56	$3.6267*10^{-4}$
	C			X	0.054 (12%) 0.055(28%) 0.056(42%) 0.057(18%)				
	D	2.257	0.021	N	-	-0.06138	-0.28855	106.22	$7.6528*10^{-4}$
	E	0.039	0.0044	W	K=0.4662 U=$7.825*10^{-4}$ h_{min}=0.037 m	56.28515	7.27072	907.27	$5.1300*10^{-2}$
IV	A	0.664	0.080	N	-	21.1592	2.46035	70.10	$2.8432*10^{-3}$
	B	0.770	0.057	N	-	-0.00998	-0.30502	74.17	$5.8742*10^{-4}$
	C	0.056	No variation						
	D	2.277	0.108	N	-	22.97901	3.0086	25.86	
	E	0.042	0.017	W	K=0.3452 u=$7.911*10^{-4}$ h_{min}=0.032 m	164.0186	12.2499	836.85	$2.7206*10^{-4}$
V	A	0.646	0.060	N	-	0.64552	-0.176061	107.98	$5.3081*10^{-4}$
	B	0.740	0.074	N		-0.24875	-0.012564	50.67	$2.9799*10^{-4}$
	C	0.067	0.022	W	K=0.9151 u=0.0191 h_{min}=0.047 m	5.91502	2.08817	69.93	$7.2033*10^{-4}$
	D	2.248	0.098	N		-		52.75	$2.0680*10^{-4}$
	E	0.062	0.034	W	K=0.8778 u=0.0362 h_{min}=0.029 m	2.00327	1.495235	73.64	$2.5588*10^{-2}$
VI	A	0.656	0.010	X					
	B	0.766	0.024	X					
	C	0.057	0.000	X					
	D	2.255	0.019	X					
	E	0.038	0.000	X					
VII	A	0.607	0.037	X					
	B	0.748	0.015	X					
	C	0.056	0.000	X					
	D	2.245	0.022	X					
	E	0.867	0.575	X	peaks at 0.038 m and 1.260 m				
VIII	A	0.604	0.044	X	peaks at 0.585 m and 0.656 m				
	B	0.769	0.036	X					
	C	0.056	0.001	X					
	D	2.245	0.031	X					
	E	0.841	0.589	X	peaks at 0.038 m and 1.275 m				

Table 89: Results Monte Carlo simulations: system I, maximum water levels.

	$H_{max,A}$	$H_{max,B}$	$H_{max,C}$	$H_{max,D}$	$H_{max,E}$	$Q_{max\ I}$	V
$H_{max,A}$	1	0.044654	0.0481	0.2694	0.38055	0.14667	0.5331
$H_{max,B}$		1	0.082097	-0.28138	0.079643	0.06292	0.4000025
$H_{max,C}$			1	-0.04813	0.046077	0.048562	0.055828
$H_{max,D}$				1	0.254188	0.178617	0.102282
$H_{max,E}$					1	-0.06308	0.112909
Q_{max}						1	0.750319
V							1

Table 90 : Correlation matrix for system I, set I. As can be seen only the correlation between the volume spilled over the weir and the maximum overflow discharge show a mutual correlation of any significance. The other results prove to be almost independent of each other.

The sole effect of structural and geometrical errors is seen when studying the results from sets VI, VII and VIII, having respectively 1%, 10% and 20 % randomly introduced error. For brevity, only the most informative graphs of the density distributions obtained are shown here.

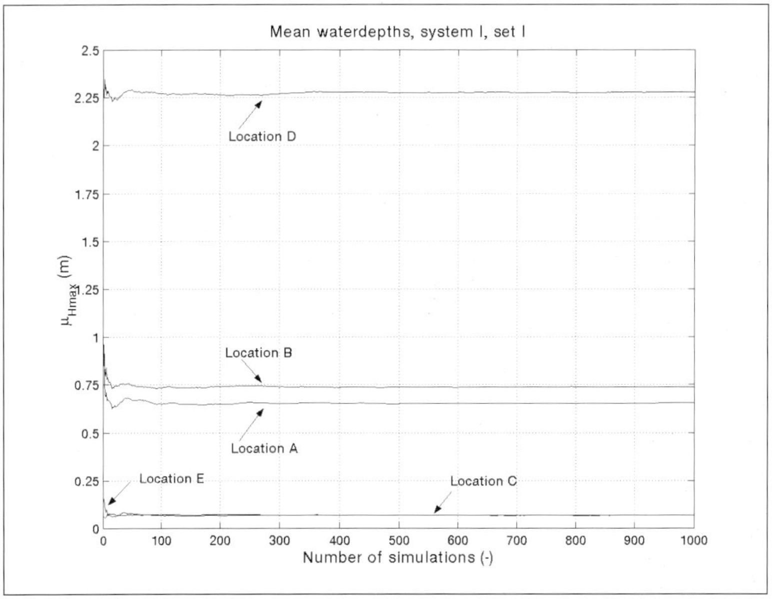

Figure 176 : Mean value for the maximum water-depths as function of the number of Monte Carlo simulations.

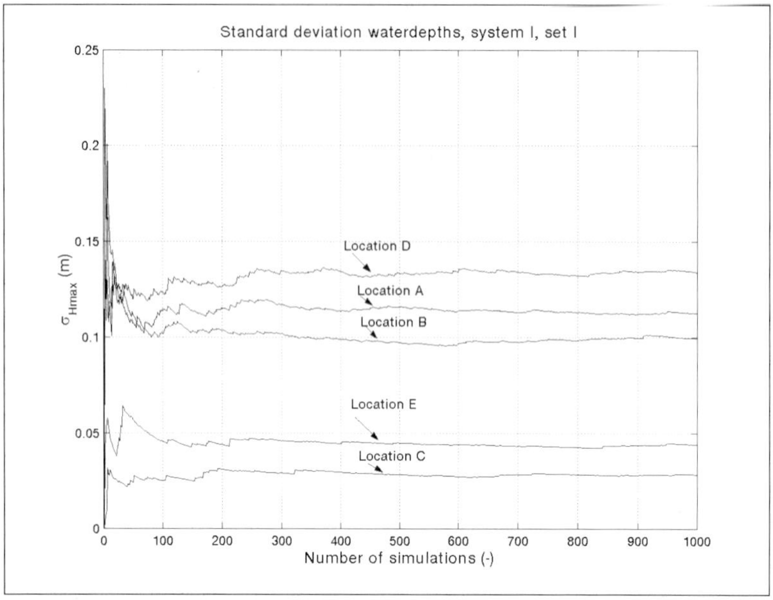

Figure 177: Standard deviation in maximum water depths as function of the number of Monte Carlo simulations. After circa 250 simulations the results are almost stable.

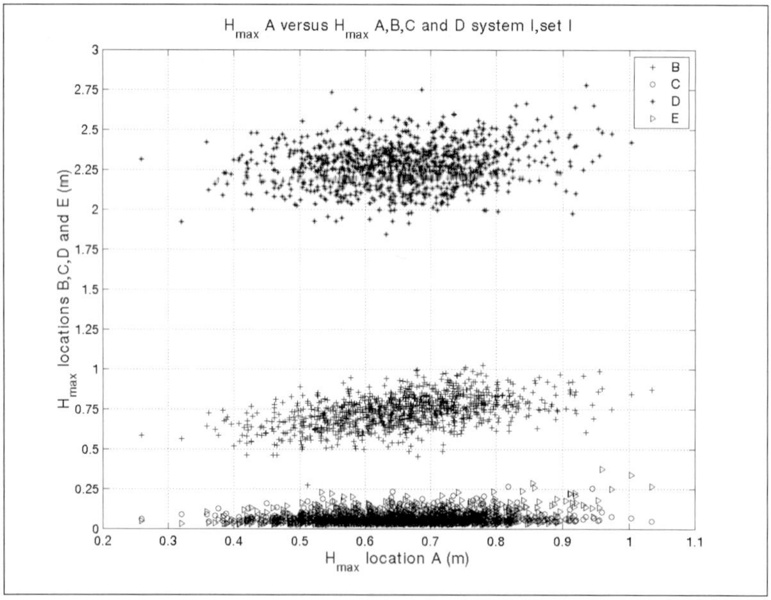

Figure 178 : Hmax location A versus H max at locations B,C,D and E. This Figure illustrates some of the correlations presented in Table 90.

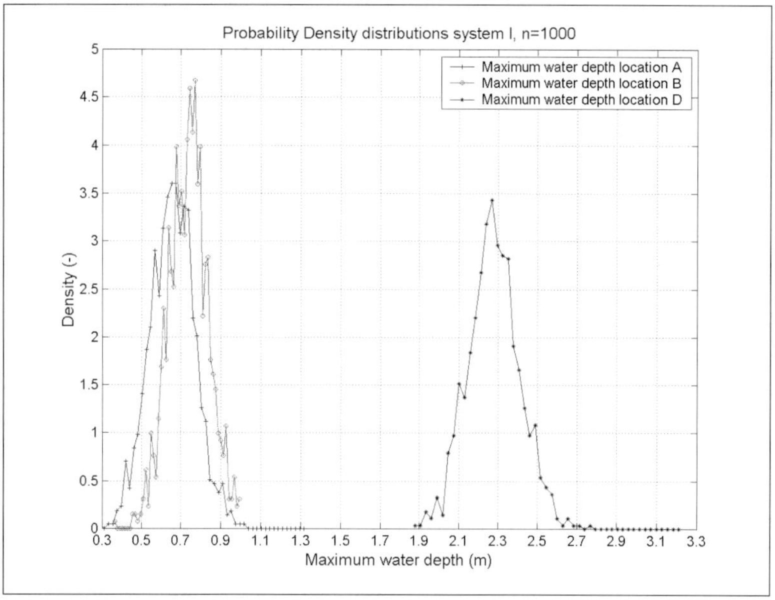

Figure 179: Density distributions maximum water depth location A, B and D, set I, system I.

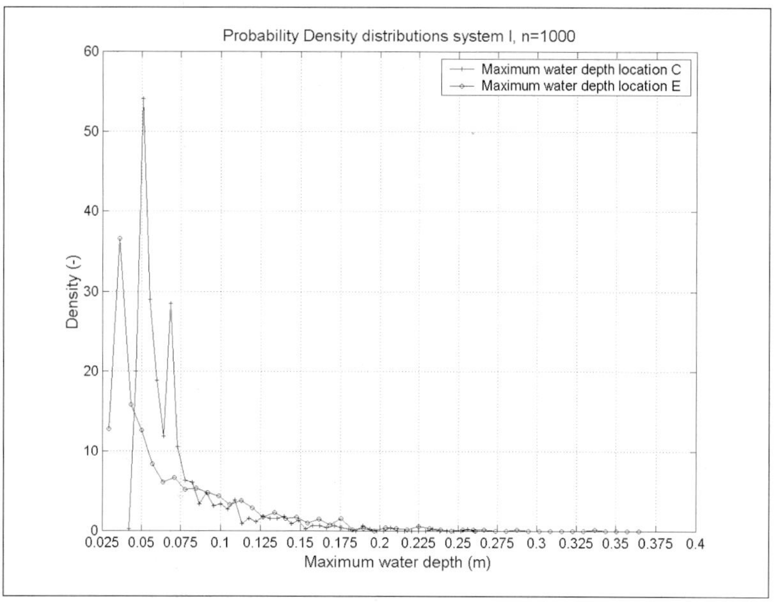

Figure 180: Density distribution maximum water depth location C and E, system I, set I

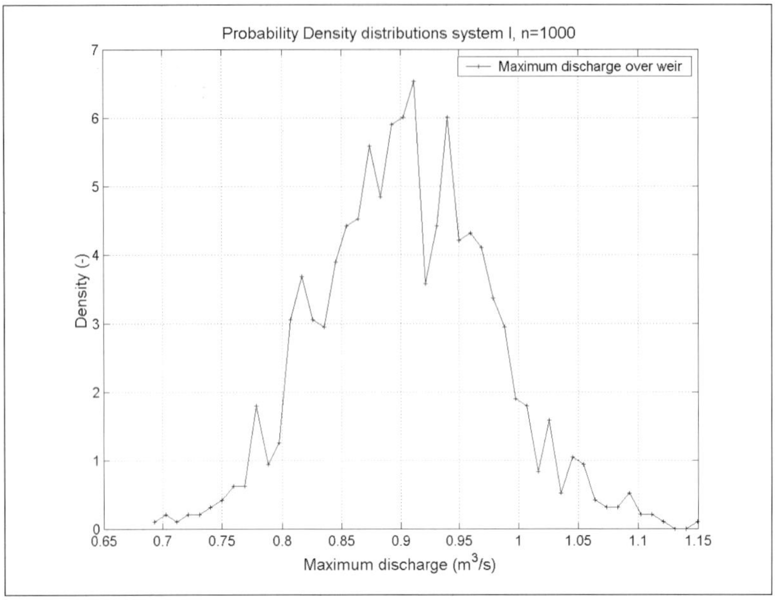

Figure 181: Density distribution maximum discharge, set I, system I.

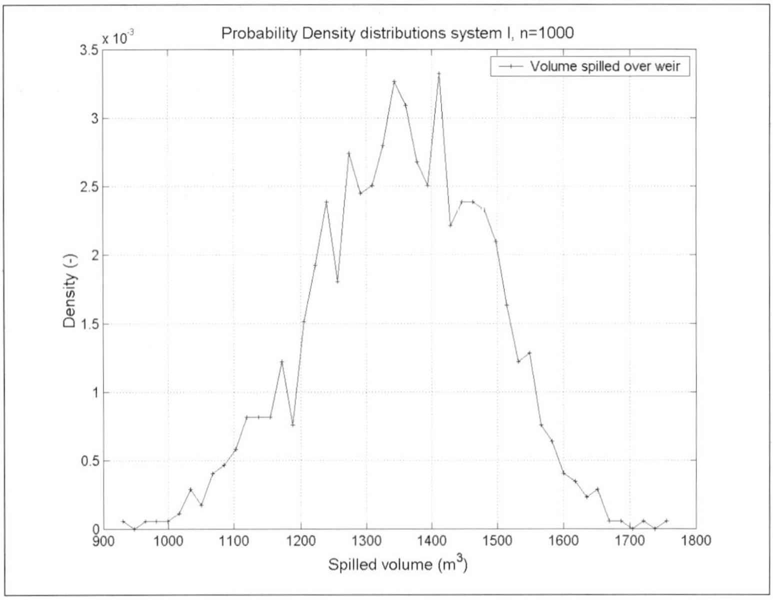

Figure 182 : Density distribution overflow volume, set I, system, I.

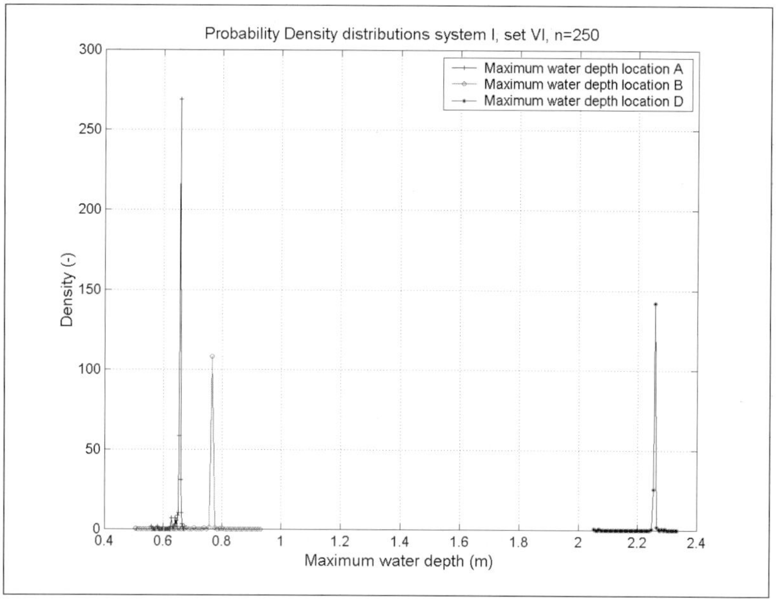

Figure 183: Density distribution maximum water depth, location A, B and D set VI (1% structural errors), system I. The density distribution is concentrated at sharply defined peaks, this implies that hardly no variance is present.

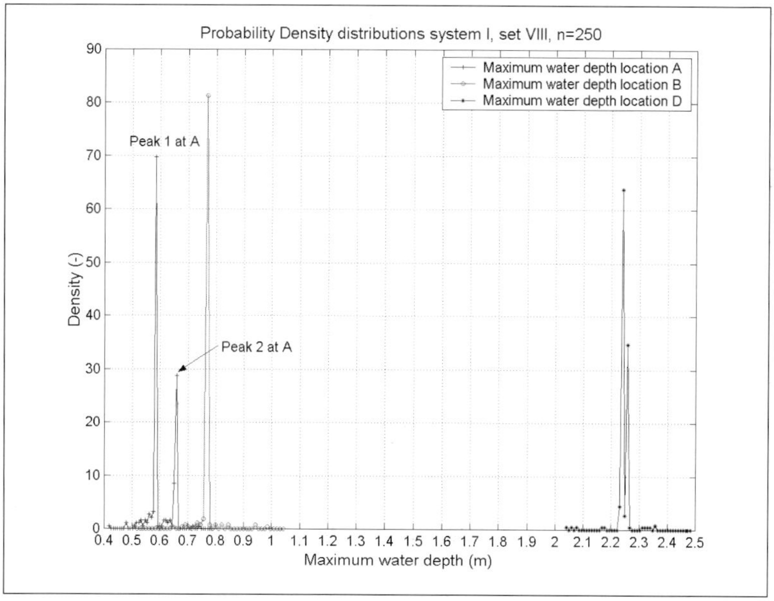

Figure 184: Density distribution maximum water depth, locations A, B and D, set VIII (20% structural errors), system I. As can be seen at location A two distinct peaks evolve, though at a relative small mutual distance.

IV.3 Results system II

set	Location	$\mu(m)$	$\sigma(m)$	type	Parameters	kurtosis	Skewness	χ^2	S
I	A	1.206	0.122	N	-	$-6.412*10^{-2}$	$-8.098*10^{-2}$	46.54	$5.54431*10^{-5}$
	B	2.431	0.074	N	-	0.141132	-0.26271	119.6	$2.89354*10^{-4}$
	C	2.147	0.074	N	-	0.279047	$-7.565*10^{-2}$	211.14	$1.01665*10^{-4}$
	D	1.585	0.115	G	U=1.6366 α=11.099	2.960650	-1.21718	4499.28	$6.40150*10^{-4}$
	E	1.803	0.067	N	-	-0.11820	-0.10145	47.26	$1.42663*10^{-3}$
II	A	1.209	0.032	N	-	0.176592	$7.4609*10^{-2}$	74.29	$1.76375*10^{-4}$
	B	2.438	0.016	N	-	-0.179886	0.1541743	82.38	$2.11833*10^{-4}$
	C	2.159	0.016	N	-	0.165194	$5.8099*10^{-2}$	79.15	$3.06227*10^{-4}$
	D	1.589	0.016	G	U=1.5956 α=81.081	0.108324	-0.71240	284.83	$1.69213*10^{-3}$
	E	1.809	0.011	N	-	-0.286882	-0.10431	116.31	$5.57377*10^{-4}$
III	A	1.182	0.086	N	-	0.23387	-0.30778	107.55	$4.88274*10^{-4}$
	B	2.418	0.043	N	-	0.64964	-0.52658	218.05	$7.67217*10^{-4}$
	C	2.143	0.038	N	-	0.37719	-0.50551	153.80	$1.11304*10^{-3}$
	D	1.549	0.108	G	U=1.5978 α=11.8752	3.23775	-1.61415	243.07	$2.11526*10^{-3}$
	E	1.791	0.035	N	-	0.73479	-0.58749	212.55	$9.36574*10^{-4}$
IV	A	1.185	0.060	N		-0.22822	$-8.8864*10^{-3}$	45.06	$1.37766*10^{-4}$
	B	2.419	0.030	N		-0.32101	-0.22345	63.22	$6.17075*10^{-4}$
	C	2.147	0.047	N		-0.18702	-0.21141	71.28	$8.08988*10^{-4}$
	D	1.563	0.016	N		0.36475	$6.3328*10^{-2}$	65.59	$6.22905*10^{-4}$
	E	1.792	0.019	N		-0.22696	-0.30703	70.23	$9.18069*10^{-4}$
V	A	1.204	0.073	W	K=2.3401 U=0.1815 H_{min}=1.043	2.63396	1.03990	225.45	$7.60530*10^{-4}$
	B	2.421	0.051	N	-	0.65497	0.30687	43.44	$1.68159*10^{-4}$
	C	2.133	0.057	N	-	-0.17837	$2.3677*10^{-2}$	46.60	$1.09335*10^{-4}$
	D	1.611	0.119	W	K=1,8348 U=0.2378 H_{min}=1.400	5.39249	1.92077	276.29	$3.66523*10^{-3}$
	E	1.794	0.049	N	-	-0.3109	$1.8125*10^{-2}$	60.11	$2.35423*10^{-4}$
VI	A	1.184	0.073	X					
	B	2.421	0.007	X					
	C	2.146	0.018	X					
	D	1.577	0.006	X					
	E	1.794	0.004	X					
VII	A	1.139	0.088	X	peaks at 0.91 m, 1.169 m and 1.183 m				
	B	2.473	0.091	X	peaks at 2.425 m, 2.455 m, 2.62 m and 1.675 m				
	C	2.224	0.220	X	peaks at 2.07 m, 2.15 m and 2.825 m				
	D	1.579	0.063	X	peaks 1.58 m				
	E	1.831	0.059	X	peaks at 1.765 m, 1.800 m, a,865 m and 1.97 m.				
VIII	A	1.161	0.097	X	peaks at 0.91 m, 0.95 m 1.865 m and 1.20 m				
	B	2.496	0.093	X	peaks at 2,44 m, 2.625 m, and 2.692 m				
	C	2.235	0.223	X	peaks at 2.08 m, 2.16 m and 2.84 m				
	D	1.589	0.012	X	peak at 1.58 m				
	E	1.897	0.046	X	peaks at 1.83 m, 1,.88 m, 1.938 m and 1.997 m				

Table 91: Maximum water depths, system II.

set	$\mu(m^3)$	$\sigma(m^3)$	Type	parameters	kurtosis	skewness	χ^2	S
I	14.51	9.00	N	-	$7.3358*10^{-2}$	0.591064	142.62	$4.92951*10^{-3}$
II	10.11	1.66	N	-	-0.25997	$-8.90905*10^{-2}$	109.07	$3.64144*10^{-4}$
III	10.80	7.73	W	K=1.3886 U=11.6108 V_{min}=0.200	0.278083	0.791389	220.08	$1.31342*10^{-3}$
IV	8.09	1.63	N	-	-0.35628	$6.61656*10^{-2}$	66.02	$3.90339*10^{-4}$
V	15.93	20.19	W	K=0.7505 U=12.5501 V_{min}=1.00	7.85789	2.82490	229.18	$6.70555*10^{-3}$
VI	8,07	1.47	X	peak at 8.03 m^3				
VII	8.21	0.60	X	peak at 8.30 m^3				
VIII	10.10	1.31	X	peak at 10.40 m^3				

Table 92: Overflow volume, weir I, system II.

set	$\mu(m^3/s)$	$\sigma(m^3/s)$	Type	Parameters	Kurtosis	Skewness	χ^2	S
I	0.0388	0.0177	N	-	-0.38304	0.160483	156.12	$7.04320*10^{-4}$
II	0.0333	0.0054	G	U=0.03576 α=236.66	-0.50339	-0.48212	722.09	$3.53554*10^{-3}$
III	0.0318	0.0018	N	-	-0.87551	$-1.27739*10^{-2}$	490.77	$2.97689*10^{-3}$
IV	0.0284	0.0044	N	-	0.31046	-0.37152	431.95	$4.64315*10^{-3}$
V	0.0365	0.0220	W	K=1.4033 U=0.0335 Q_{min}=0.006	3.4504	1.67942	248.49	$1.46085*10^{-3}$
VI	0.0282	0.004	X	peak at 0.029 m^3/s				
VII	0.0291	0.002	X	peak at 0.029 m^3/s				
VIII	0.0343	0.004	X	peak at 0.035 m^3/s				

Table 93: Maximum overflow discharge, weir I, system II.

set	$\mu(m^3)$	$\sigma(m^3)$	Type	parameters	kurtosis	skewness	χ^2	S
I	908.36	75.83	N	-	$-7.4625*10^{-2}$	$-6.07934*10^{-2}$	52.07	$3.83102*10^{-5}$
II	856.97	8.81	N	-	$3.8189*10^{-2}$	-0.31210	92.13	$3.66214*10^{-4}$
III	809.71	79.48	N	-	$-3.8946*10^{-2}$	$4.3346*10^{-2}$	103.65	$2.68368*10^{-4}$
IV	812.54	14.05	G	U=818.87 α=0.0913	2.825	-0.75503	6005.90	$9.88096*10^{-4}$
V	831.98	70.90	W	K=1.2027 U=90.3994 V_{min}=747.0	1.63522	1.480327	89.68	$3.35462*10^{-3}$
VI	808.81	10.90	X	peak at 810 m^3				
VII	782.53	131.82	X	peaks at 482 m^3, 702 m^3, 774.5 m^3, 791 m^3 and 1055 m^3				
VIII	852.82	137.75	X	peaks at 615 m^3, 778 m^3 ,828 m^3 ,870 m3 and 1157 m^3				

Table 94: Overflow volume, weir II, system II.

set	$\mu(m^3/s)$	$\sigma(m^3/s)$	Type	Parameters	Kurtosis	Skewness	χ^2	S
I	0.537	0.035	N	-	0.679901	0.349862	166.31	$1.17277*10^{-4}$
II	0.535	0.010	N	-	$-1.3427*10^{-2}$	$-9.72712*10^{-2}$	160.27	$9.07283*10^{-4}$
III	0.527	0.025	N	-	0.34726	-0.48097	136.26	$9.54440*10^{-4}$
IV	0.532	0.022	N	-	-0.23113	0.12963	71.92	$1.74476*10^{-4}$
V	0.532	0.015	W	K=1.1681 U=0.0179 Q_{min}=0.515	6.64793	2.41247	237.60	$7.03680*10^{-3}$
VI	0.529	0.007	X	peak at 0.529 m^3/s				
VII	0.511	0.030	X	peaks at 0.42 m^3/s and 0.52 m^3/s				
VIII	0.521	0.032	X	peaks at 0.44 m^3/s, 0.51 m^3/s and 0.54 m^3/s				

Table 95: Maximum overflow discharge, weir II system II.

set	$\mu(m^3)$	$\sigma(m^3)$	Type	Parameters	kurtosis	skewness	χ^2	S
I	743.22	77.98	N	-	-0.113638	$-5.15717*10^{-2}$	49.52	$3.76765*10^{-5}$
II	746.17	8.04	N	-	$5.21345*10^{-2}$	-0.313219	69.28	$3.01603*10^{-4}$
III	700.51	71.84	N	-	$-3.11729*10^{-2}$	$9.6265*10^{-5}$	57.91	$2.41226*10^{-4}$
IV	699.25	13.51	N	-	1.10110	-0.281028	69.33	$9.65295*10^{-4}$
V	709.53	31.32	W	K=1.553 U=52.976 V_{min}=661.9	0.92430	1.148566	61.13	$1.03066*10^{-3}$
VI	700.00	7.21	X	peak at 702 m^3				
VII	670.01	233.23	X	peaks at 60 m^3 , 711 m^3 , 857 m^3 and 952 m^3				
VIII	693.54	239.11	X	peaks at 72 m^3 , 732 m^3 ,871 m^3 and 938 m^3				

Table 96: Overflow volume, weir III, system II.

set	$\mu(m^3/s)$	$\sigma(m^3/s)$	Type	Parameters	Kurtosis	Skewness	χ^2	S
I	0.505	0.031	N	-	0.269229	0.22559	54.73	$1.49295*10^{-4}$
II	0.500	0.015	N	-	$-7.93381*10^{-2}$	0.90325	287.96	$3.64144*10^{-3}$
III	0.500	0.021	N	-	$-1.52356*10^{-2}$	$-2.53169*10^{-2}$	312.75	$1.63002*10^{-3}$
IV	0.500	0.021	N	-	0.386244	0.29272	61.82	$5.26098*10^{-4}$
V	0.505	0.022	N	-	-0.33559	0.85729	373.44	$1.17422*10^{-2}$
VI	0.493	0.006	X	peak 0.492 m^3/s				
VII	0.443	0.140	X	peaks at 0.056 and m^3/s and 0.500 m^3/s				
VIII	0.456	0.142	X	peaks at 0.070 m^3/s and 0.507 m^3/s				

Table 97: Maximum overflow discharge, weir III, system II.

Set	$\mu(-)$	$\sigma(-)$	Type	Parameters	kurtosis	Skewness	χ^2	s
I	147.15	7.79	N	Min=117 Max=173	0.87328	0.13622	226.62 (nkl=50) 134.91 (nkl=56)	$8.6782*10^{-4}$
II	146.83	2.12	N	Min=143 Max=153	-0.33461	0.25873	2335.07(nkl=50) 156.60 (nkl=10)	$7.9843*10^{-3}$
III	144.83	5.13	N	Min=126 Max=158	0.04698	-0.20797	500.89 (nkl=50) 156.37 (nkl=28)	$3.5116*10^{-3}$
IV	145.30	4.43	N	Min=134 Max=165	0.98438	0.33747	512.49 (nkl=40) 167.37 (nkl=31)	$2.2930*10^{-3}$
V	146.19	4.37	W	K=1.4402 U=6.8231 N_{min}=140 N_{max}=170	9.09772	2.59149	586.08 (nkl=50) 235.76 (nkl=30)	$6.4352*10^{-3}$
VI	144.96	0.84						
VII	146.12	4.00						
VIII	149.03	3.33	W	K=1.5426 U=5.5842 N_{min}=144 N_{max}=158	2.16626	1.84662	3504.03 (nkl=50) 792.76 (nkl=14)	$2.2532*10^{-3}$

Table 98: Number of flooded manholes, system II.

	$H_{max,A}$	$H_{max,B}$	$H_{max,C}$	$H_{max,D}$	$H_{max,E}$	$Q_{max\ I}$	$Q_{max,II}$	$Q_{max,III}$	V_I	V_{II}	V_{III}
$H_{max,A}$	1	0.6367	0.6071	0.6928	0.5622	0.1111	-0.0800	-0.1078	0.6086	0.6512	0.6511
$H_{max,B}$		1	0.5304	0.5531	0.5002	0.1073	-0.0903	-0.1126	0.4918	0.4770	0.4748
$H_{max,C}$			1	0.4802	0.4742	0.0063	-0.2699	-0.2862	0.4408	0.3441	0.3422
$H_{max,D}$				1	0.5336	0.4062	0.1556	0.1771	0.7438	0.7397	0.7402
$H_{max,E}$					1	0.1614	-0.0390	-0.0609	0.5014	0.4685	0.4691
$Q_{max\ I}$						1	0.6886	0.8495	0.9708	0.6236	0.6205
$Q_{max\ II}$							1	0.8116	0.3185	0.7082	0.4333
$Q_{max\ III}$								1	0.4006	0.5079	0.4321
V_I									1	0.8889	0.8738
V_{II}										1	0.6236
V_{III}											1

Table 99: Correlation matrix for system II, set I. As can be seen the correlation between water levels, maximum discharges and spilled volumes are one order of magnitude larger than for system I (see Table 90) although the values found indicate that (except for the shaded relations) no prediction for e.g. an overflow volume can be based on a maximum water level. The larger correlation values in system II compared to system I are caused by the characteristics of this system. System II is relatively flat, implying that the systems functions more or less uniformly over the separate elements of which it is composed.

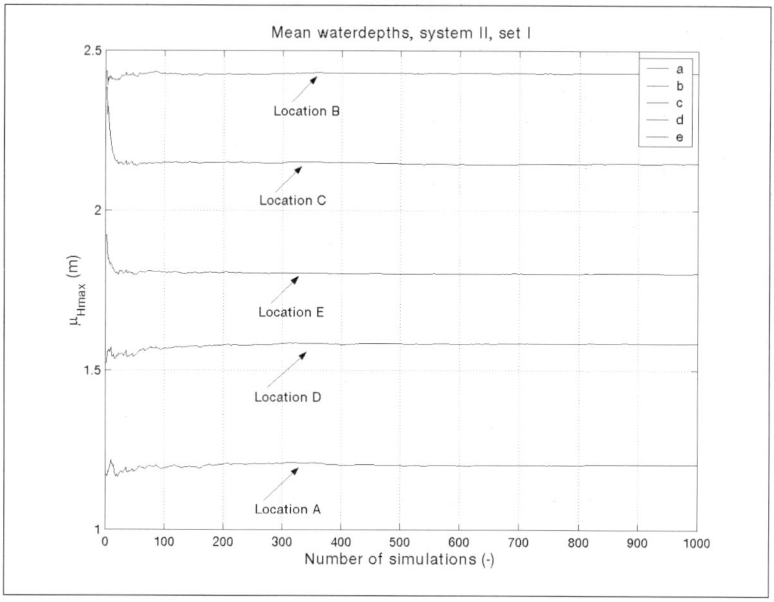

Figure 185: Mean values for H_{max} versus the number of simulations.

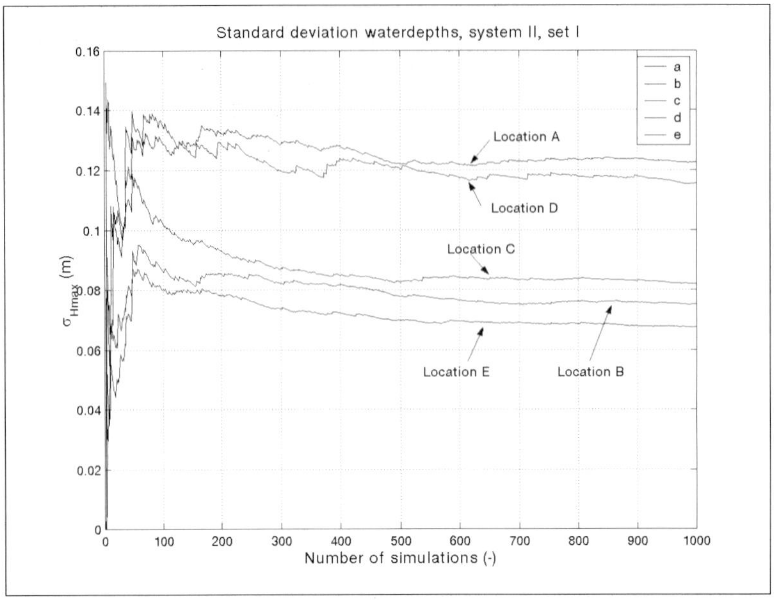

Figure 186: The standard deviation in H_{max} versus the number of simulations.

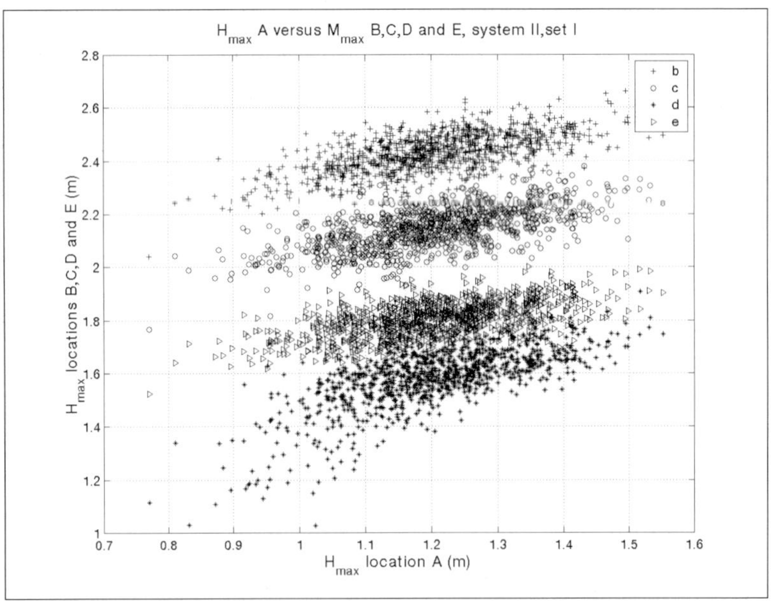

Figure 187: H_{max} at location A versus H_{max} at locations B, C, D and E. This Figure illustrates some of the mutual correlation presented in Table 99, as can be seen there is a larger correlation between water depths in system II than found for system I.

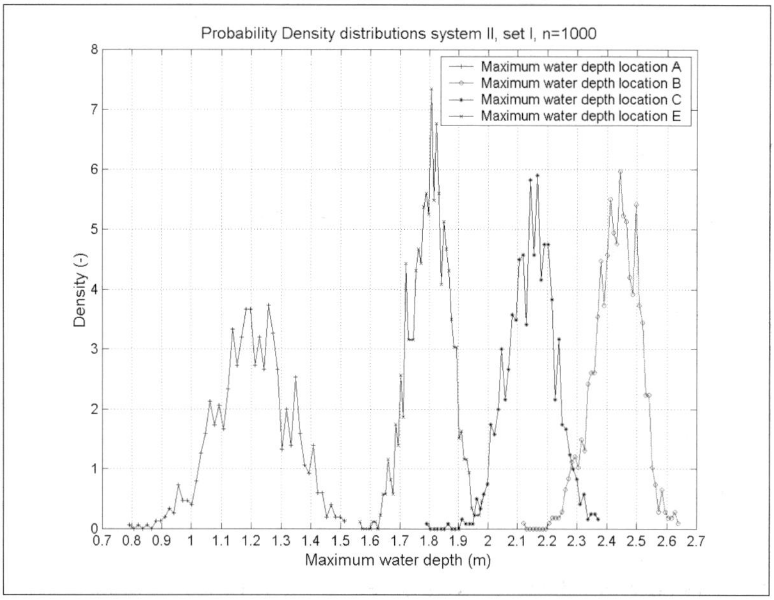

Figure 188: Density distributions for maximum water depths, locations A, B, C and E, set I, system II.

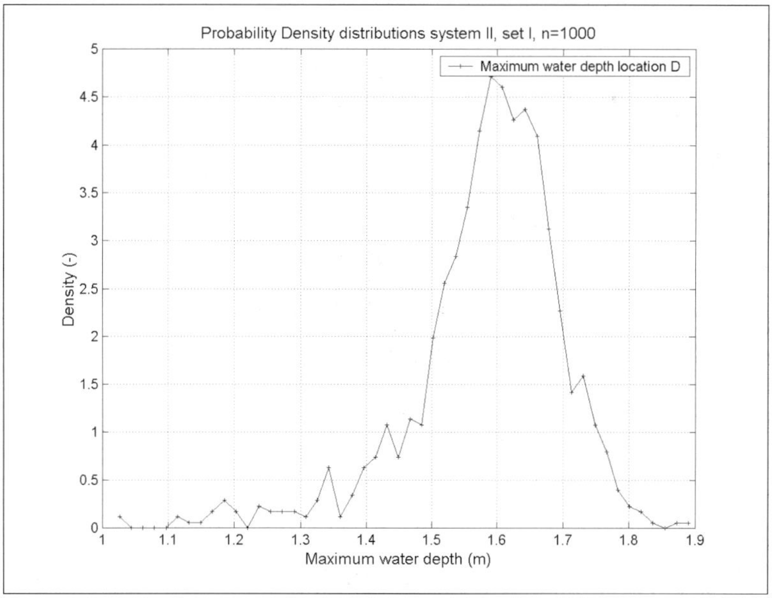

Figure 189 : Density distribution maximum water depth location D, set I, system II.

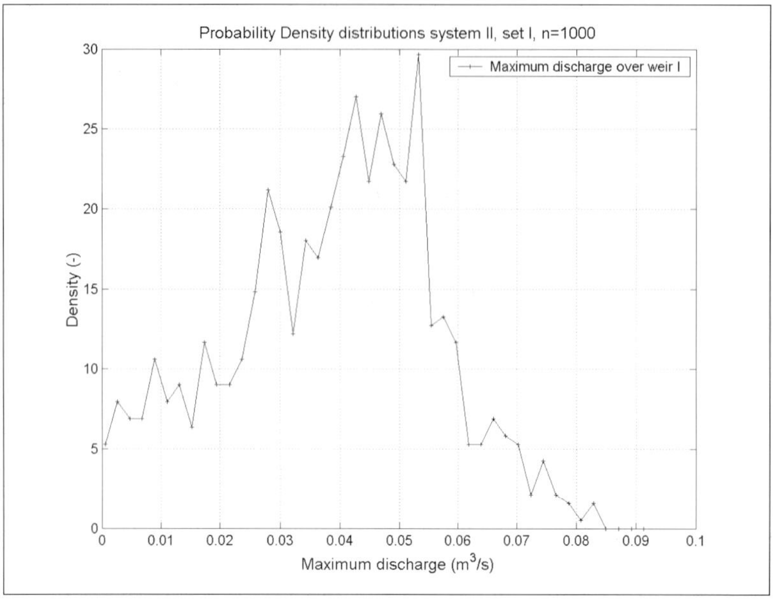

Figure 190: Density distribution maximum discharge weir I, set I, system II. As can be seen there is a substantial density at Q=0, this implies that there is realistic chance that no volume is spilled at all.

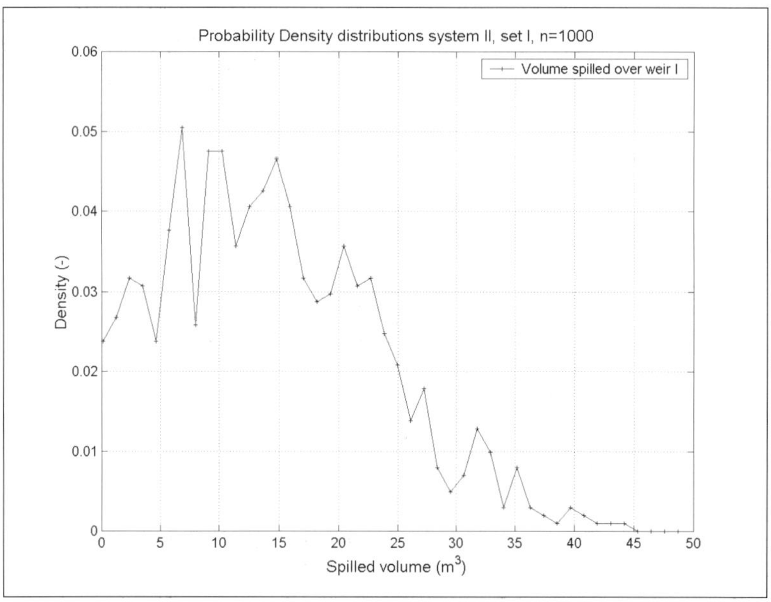

Figure 191: Density distribution spilled volume, weir I, set I, system II.

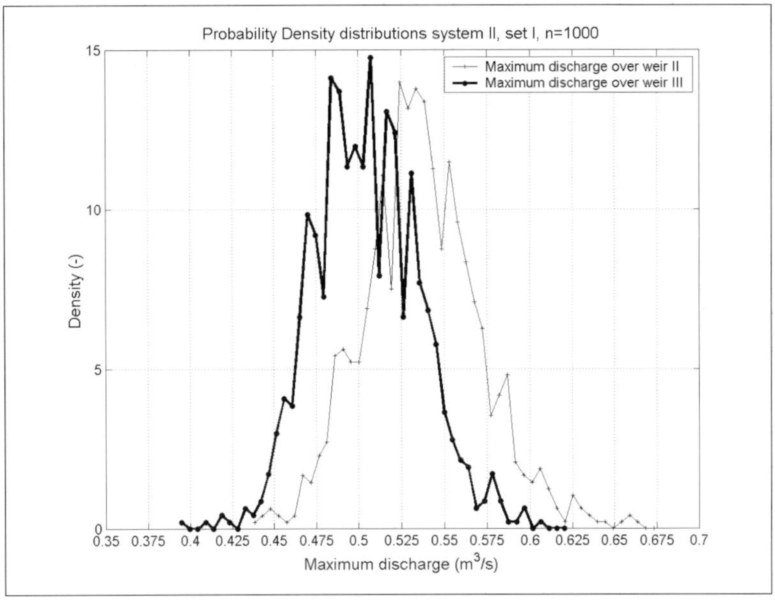

Figure 192 : Density distributions for maximum discharge over weirs II and III , set I, system II.

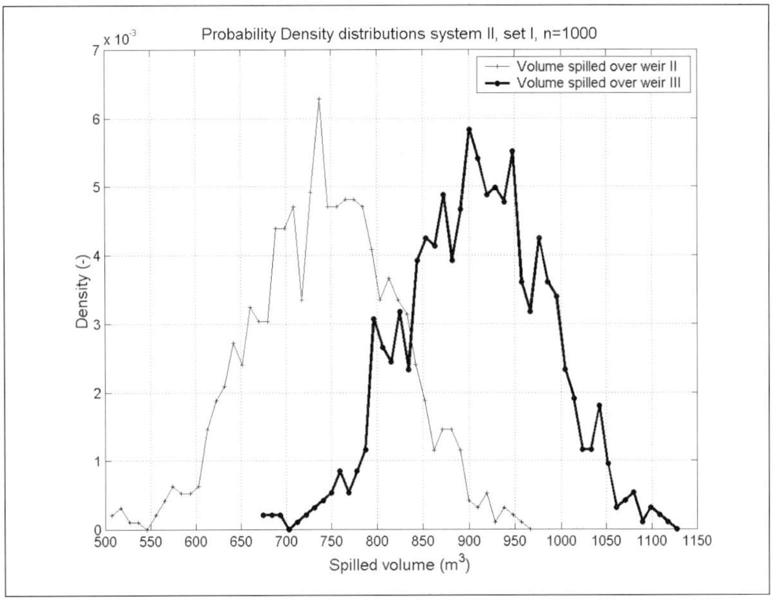

Figure 193 : density distributions for spilled volumes over weirs II and III, set I, system II.

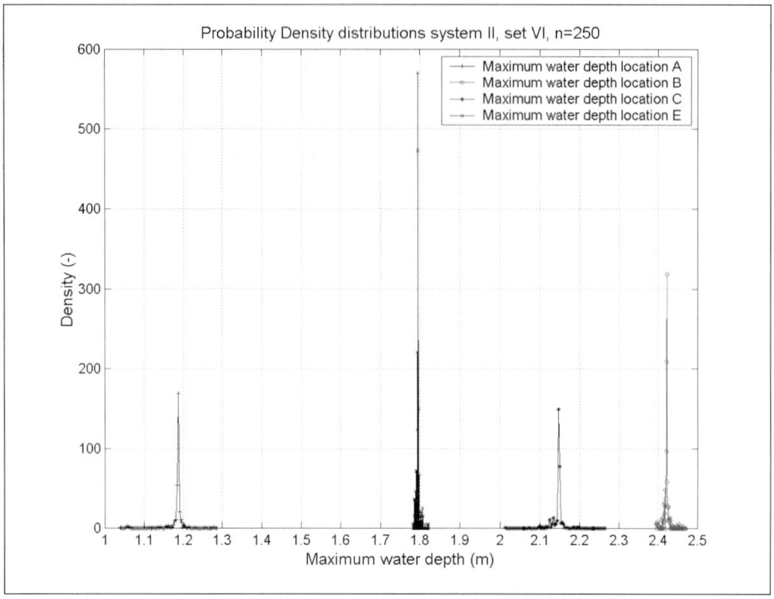

Figure 194: Density distributions for maximum water depths locations A, B, C and E set VI (1% structural errors), system II. The density is concentrated in sharply defined peaks, implying only a small variance in maximum water depths is caused by a 1% errors level with respect to structural errors.

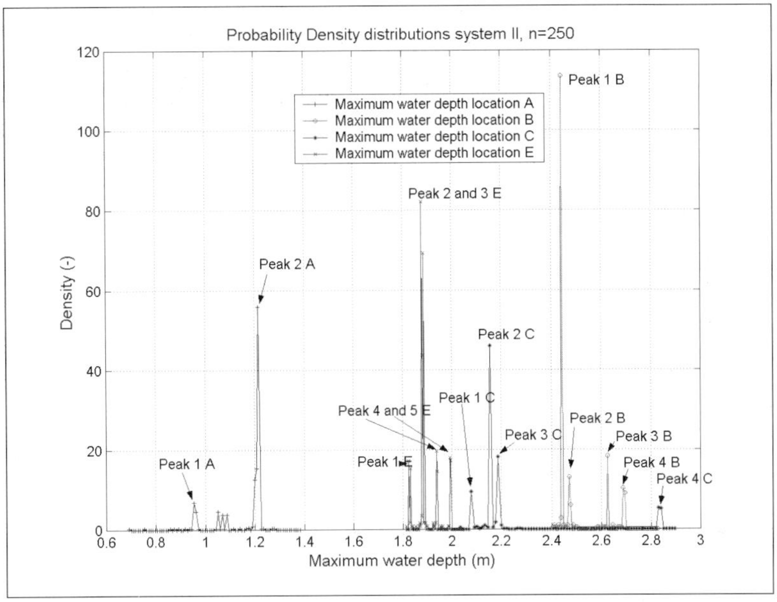

Figure 195: Density distributions for maximum water depths locations A, B, C and E set VIII (20% structural errors), system II. When compared with the density distributions for 1% (Figure 194) several distinct peaks evolve.

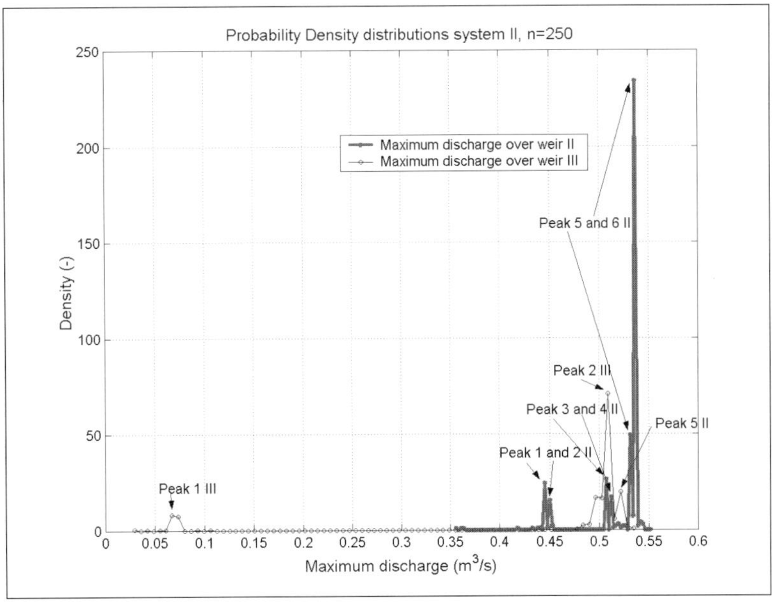

Figure 196: Density distributions for maximum discharges over weirs II and III, set VIII (20% structural errors), system II.

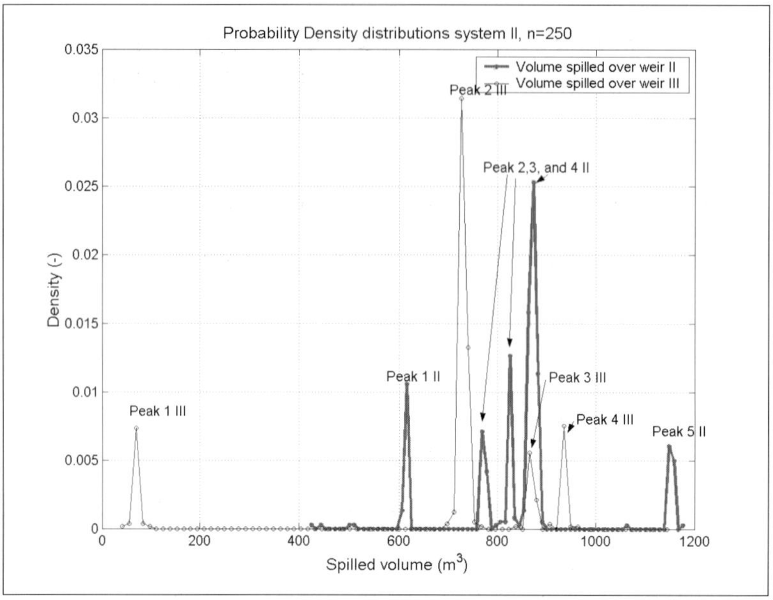

Figure 197: Density distributions for spilled volumes over weirs II and III, set VIII (20% structural errors), system II.

ANNEXE V: LABORATORY EXPERIMENTS INTO THE Q-H RELATIONSHIP OF A VENTURI FOR CIRCULAR CHANNELS.

V.1 general

Discharge measurement using a venturi is based on the principle of a constriction in a channel causing a measurable drop in water level in the constriction.
If the geometry of the construction is know, the discharge can be calculated using measured water levels:

$$h_1 + \alpha_1 \frac{Q^2}{2gA_1^2(h_1)} + \Delta z - \Delta H_f = h_2 + \alpha_2 \frac{Q^2}{2gA_2^2(h_2)} \qquad \text{(eq. V.1)}$$

In which:

h_1	water depth upstream of the constriction	(m)
h_2	water depth in the constriction	(m)
g	gravitational acceleration	(m/s^2)
A_1	cross-sectional area at section 1	(m^2)
A_2	cross-sectional area at section 2	(m^2)
Q	discharge	(m^3/s)
ΔH_f	head loss between section 1 and 2	(m)
Δz	difference in bottom level between section 1 and 2 relative to datum level	(m)
α_1	Boussinesq number section 1	(-)
α_2	Boussinesq number section 2	(-)

When assuming h_1 and h_2 to be known from measurement and the values for the Boussinesq values α_1 and α_2 to be known, this equation can be solved for Q. When the assumption is made that in the constriction the flow becomes critical a different equation may be applied:

$$Fr = 1 = \frac{Q^2}{g\frac{A(h_2)}{B_{wl}(h_2)}} \Rightarrow Q = \sqrt{g\frac{A(h_2)}{B_{wl}(h_2)}} \Rightarrow h_2 = f(Q, geometry). \qquad \text{(eq. V.2)}$$

In which:

h_2	water depth in the constriction	(m)
g	gravitational acceleration	(m/s^2)
$A(h_2)$	cross-sectional area at section 2	(m^2)
Q	discharge	(m^3/s)
B_{wl}	Width free water surface at section 2	(m)
Fr	Froude number	(-)

This would imply that, at known geometry the discharge is directly known from measurement of h_2. Because the location at which the flow is critical is not known exactly in advance, another strategy is to be followed. Implementing the known relation between Q and h_2 (eq V.2) in eq. V.1 reduces the problem to solving the resulting equation for Q assuming h_1 to be obtained by measurement. The resulting equation to be solved in this case is:

$$h_1 + \alpha_1 \frac{Q^2}{2gA_1^2(h_1)} + \Delta z - \Delta H_f = f(Q, geometry) + \alpha_2 \frac{Q^2}{2gA_2^2(f(Q, geometry)} \qquad \text{(eq. V.3)}$$

For simple geometries this equation can be solved directly, resulting in the well known formulas for venturis as applied in rectangular prismatic channels:

$Q = mB\frac{2}{3}\sqrt{\frac{2}{3}g}h^{\frac{3}{2}}$, in which h is the water level measured as some distance upstream from the constriction and m is some construction specific value compensating for frictional losses. In the case considered here however, the geometry does not allow for a simple analytic solution, since eq. V.3 is recursive and is to be solved by iterative methods. In Figure V. 1 the geometry is shown.

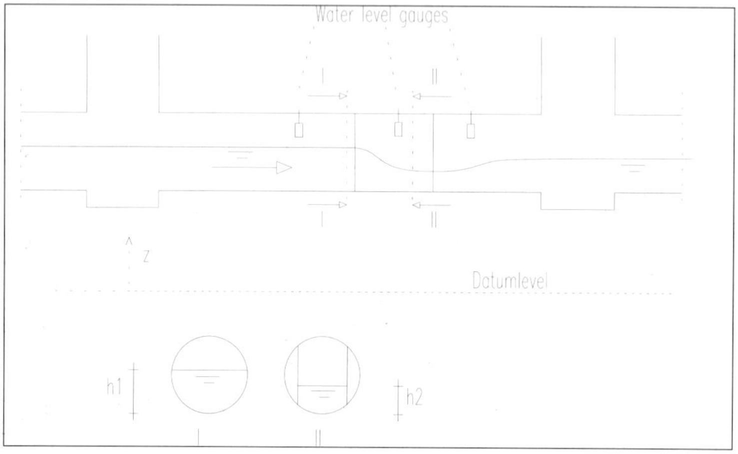

Figure V. 1: definition sketch venturi.

The value for A_1 is calculated from the formula derived in annexe I. A_2 (h_2) is calculated as follows:

$$A_2(h_2) = B(h_2 - a_t) + A_t \qquad \text{(eq. V.4)}$$

$$a_t = \frac{D - \sqrt{D^2 - B^2}}{2} \qquad \text{(eq. V.5)}$$

In which:

D	diameter	(m)
B	width free water surface at section 2	(m)

and finally

$$A_t = \frac{1}{4}D^2\left(\pi - a\cos\left(-1 + \frac{2a_t}{D}\right)\right) + \frac{1}{2}D^2\sqrt{\frac{a_t}{D}\left(1 - \frac{a_t}{D}\right)}\left(-1 + \frac{a_t}{D}\right) \qquad \text{(eq. V.6)}$$

Head losses due to wall friction can be neglected for all practical purposes, since the distance between measuring point and the point at which the flow becomes critical is limited (in practice in the order of magnitude of meters). The local losses due to curvature of the streamlines however, can generally not be neglected when striving for accurate measurement. Wenzel (1975) derived the following first approximation for these losses for venturis:

$$\Delta H_{fr} = \xi \frac{u_1^2}{2g} \qquad \text{(eq. V.7)}$$

with

$$\xi \approx 0.225 \text{ for } \frac{Q}{\sqrt{gD^5}} \geq 0.1 \text{ and } \xi \approx -3.75\log(2.5\frac{Q}{\sqrt{gD^5}} \text{ for } \frac{Q}{\sqrt{gD^5}} < 0.1 \qquad \text{(eq. V.8)}$$

This approximation is valid for curved inlet geometries only. Therefore, when implementing another geometry, as the one used in the experiments presented here, an adjustment is needed. These will be extracted from the experimental data.

V.2 experimental set up and measuring methods

The laboratory experiments([59]) were done using a circular PVC pipe with an inner diameter of 188.2 mm and a total length of 6 m. Figure V. 2 shows the measuring setup. Water was extracted from a basin (in fact a spare bath tub) by a pump and transported into a constant head tank. From this tank the water was led to one extreme of the pipe. After passing a diffusor, the water was allowed 4 m (circa 20 pipe diameters) of prismatic channel before entering the venturi. At the downstream end, the water was collected in a measuring flume in which at the downstream end a V-notch was inserted. Both measuring flume and 90° V-notch were constructed in accordance with the requirements set in ISO 1438/1-1980 E (ISO handbook of standards no. 16. pp 238-239). The discharge was directly calculated from the water level readings in the measuring flume by applying:

$$Q = C_e \frac{8}{15} \tan\left(\frac{\alpha}{2}\right)\sqrt{2g}h_e^{\frac{5}{2}}, \text{ since } \alpha\text{=}90^0 \text{ this is } Q = 0.4555C_e\sqrt{2g}h_e^{\frac{5}{2}} \qquad \text{(eq. V.9)}$$

In which h_e is the recorded water level in the flume compensated for surface tension by: $h_e = h + h_k = h + 0.85 * 10^{-3}m$, in which h is the original reading. The value for C_e depends on the ratio between water depth above lowest weir level and the distance between the bottom of the measuring flume and the lowest weir level (p). For this specific flume the ratio between width and p is 0.5. According to the ISO the value for C_e may be interpolated using Table V. 1.

h/p	C_e	h/p	C_e
0.1	0.5779	0.6	0.5841
0.2	0.5781	0.7	0.5889
0.3	0.5783	0.8	0.5950
0.4	0,5785	0.9	0.6017
0.5	0.5806	1.0	0.6100

Table V. 1: relation between h/p and C_e.

Using a point gauge, the water level in the measuring flume was measured (reading ± 0.1 mm) allowing for accurate discharge measurement. During the experiments the temperature was regularly measured, since the Q-h relationship of the V-notch applied is valid only in a temperature range between 10 and 30°C.

The geometry of the venturi applied is shown in detail in Figure V. 2 and Figure V. 3 along with the location of the measuring point for the water level in the construction.

([59]) It must be mentioned that just a 'poor mans' laboratory was available; a non-heated hall, normally used by a construction firm. Equipment was either lent from Delft Hydraulics (point gauges) or obtained from the authors private workshop. In this workshop the venturi as well as the other facilities applied were constructed.

Measuring point D1 is the principal measuring point, which is used in the Q-h relationship for the venturi when the flow is critical at some point in the constriction. Measuring point D5 was monitored to identify the occurrence of backwater effects while D2, D3 and D4 were used to follow the curvature of the water level in the venturi.

The water level was measured by point gauges in stilling pipes, these were connected by rubber hoses to nipples inserted in the bottom of the pipe.

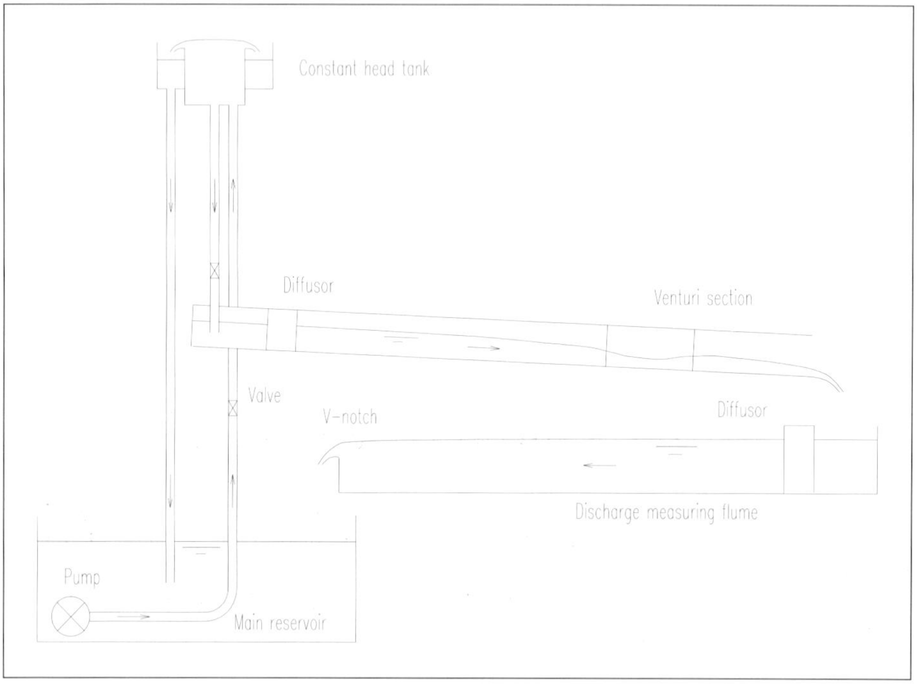

Figure V. 2: Experimental set-up.

During each experiment readings were done only after all levels were stable, so as to cancel out any dynamic effects due to damping in the whole setup (caused by level variations in the measuring gauge, the reservoir and the venturi pipe.

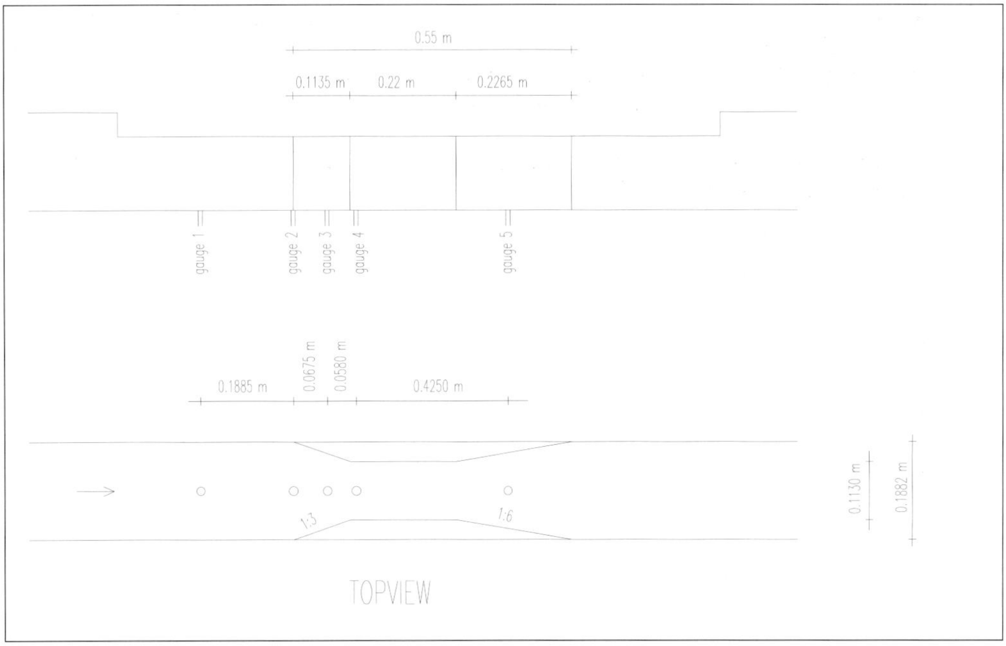

Figure V. 3: Experimental layout, including measures.

In total 93 experiments were done at 2 inclination levels and in a range of discharges (set by means of the valve at the downstream and of the constant head tank). The bottom inclination for series A was negative:
The results of the experiments are presented in Table V. 2: Experimental results series A. through Table V. 5.

experiment	time	temperature (°C)	depth 1 (m)	depht 2 (m)	depht 3 (m)	depht 4 (m)	depht 5 (m)	reading v-notch (m)	discharge (m^3/s)
1		13	0.0534	0.0546	0.0527	0.0487	0.0352	0.0777	0.00237514
2		-	0.0533	0.0542	0.0527	0.0487	0.0352	0.0777	0.00237514
3	13:15	13.5	0.0474	0.0487	0.0462	0.0433	0.0385	0.0714	0.00192346
4		-	0.072	0.0722	0.0708	0.065	0.0406	0.0933	0.00376043
5		-	0.072	0.0724	0.0707	0.0651	0.0406	0.0936	0.00379108
6		-	0.0721	0.0725	0.0708	0.0651	0.0407	0.0932	0.00375024
7		-	0.0722	0.0727	0.0708	0.0651	0.0406	0.0934	0.00377063
8		-	0.0722	0.0727	0.0707	0.065	0.0407	0.0934	0.00377063
9	14:13	15	0.0723	0.0727	0.0708	0.0651	0.0406	0.0934	0.00377063
10		-	0.0663	0.0674	0.0651	0.0603	0.0396	0.0887	0.00331014
11		-	0.0664	0.0673	0.0653	0.0603	0.0396	0.0888	0.00331952
12		-	0.0664	0.0674	0.0654	0.0603	0.0395	0.0887	0.00331014
13		-	0.0665	0.0673	0.0654	0.0603	0.0395	0.0887	0.00331014
14		-	0.0666	0.0674	0.0654	0.0603	0.0396	0.0886	0.00330078
15	14:59	16	0.0666	0.0672	0.0655	0.0603	0.0395	0.0883	0.00327279
16		-	0.0814	0.0817	0.0794	0.0735	0.0429	0.1005	0.00453892
17		-	0.0814	0.0817	0.0795	0.0735	0.0428	0.1007	0.00456184
18	15:35	16.5	0.0815	0.0819	0.0796	0.0735	0.0428	0.1002	0.00450467
19		-	0.0798	0.0811	0.0771	0.0711	0.0422	0.0987	0.00433581
20		-	0.0796	0.0808	0.0771	0.0711	0.0421	0.0987	0.00433581
21		-	0.0793	0.0805	0.0771	0.071	0.0422	0.0987	0.00433581
22	16:03	17	0.0793	0.0802	0.0771	0.071	0.0422	0.0988	0.00434694
23		-	0.0606	0.0621	0.0592	0.0547	0.0379	0.0836	0.00285313
24	16:33	17.5	0.0603	0.0618	0.0592	0.0547	0.038	0.0837	0.00286169
25	16:54	18	0.0436	0.0451	0.0435	0.0408	0.0403	0.069	0.00176666

Table V. 2: Experimental results series A.

experiment	time	temperature (°C)	depth 1 (m)	depht 2 (m)	depht 3 (m)	depht 4 (m)	depht 5 (m)	reading v-notch (m)	discharge (m^3/s)
26		11.0	0.08094	0.08158	0.07926	0.07328	0.0431	0.1021	0.0047243
27		-	0.08104	0.08158	0.07936	0.07338	0.0431	0.1018	0.00468919
28		-	0.08104	0.08158	0.07936	0.07328	0.0431	0.1022	0.00473604
29	10:37	11.2	0.08104	0.08168	0.07936	0.07338	0.0431	0.1021	0.0047243
30		-	0.07694	0.07768	0.07526	0.06948	0.0421	0.0987	0.00433581
31		-	0.07694	0.07758	0.07526	0.06948	0.0421	0.0987	0.00433581
32		-	0.07694	0.07758	0.07526	0.06938	0.0422	0.0988	0.00434694
33	11:40	13.5	0.07694	0.07768	0.07526	0.06948	0.0421	0.0987	0.00433581
34	12:50	15.1	0.06934	0.07008	0.06816	0.06288	0.0407	0.0926	0.00368951
35		-	0.06934	0.07008	0.06806	0.06298	0.0407	0.0928	0.00370968
36		-	0.06934	0.07018	0.06816	0.06298	0.0407	0.0928	0.00370968
37		-	0.06934	0.07018	0.06816	0.06298	0.0406	0.0928	0.00370968
38	13:40	16	0.06934	0.07018	0.06826	0.06298	0.0406	0.0927	0.00369959
39	14:00	16.5	0.06614	0.06698	0.06526	0.06028	0.0399	0.0903	0.0034623
40		-	0.06624	0.06708	0.06526	0.06028	0.0399	0.0904	0.003472
41		-	0.06614	0.06708	0.06536	0.06028	0.04	0.0903	0.0034623
42		-	0.06614	0.06708	0.06526	0.06038	0.0399	0.0903	0.0034623
43	14:25	17	0.06614	0.06708	0.06526	0.06028	0.0399	0.0903	0.0034623
44		-	0.06294	0.06388	0.06186	0.05728	0.0389	0.0874	0.0031897
45		-	0.06284	0.06368	0.06186	0.05718	0.0389	0.0874	0.0031897
46		-	0.06274	0.06358	0.06196	0.05718	0.0389	0.0873	0.00318055
47		-	0.06274	0.06348	0.06186	0.05708	0.0389	0.0873	0.00318055
48		-	0.06264	0.06548	0.06176	0.05708	0.0389	0.0874	0.0031897
49		-	0.06264	0.06438	0.06186	0.05708	0.0390	0.0872	0.00317142
50		-	0.06264	0.06408	0.06176	0.05718	0.0389	0.0873	0.00318055
51		17.8	0.06254	0.06378	0.06186	0.05708	0.0390	0.0874	0.0031897

Table V. 3: Experimental results series B.

experiment	time	temperature (°C)	depth 1 (m)	depht 2 (m)	depht 3 (m)	depht 4 (m)	depht 5 (m)	reading v-notch (m)	discharge (m^3/s)
52	15:17	17.8	0.06254	0.06378	0.06186	0.05708	0.0389	0.0874	0.0031897
53		-	0.05884	0.05968	0.05796	0.05358	0.0378	0.084	0.00288748
54		-	0.05874	0.05958	0.05796	0.05358	0.0378	0.0839	0.00287887
55	15:50	18	0.05854	0.05948	0.05806	0.05358	0.0379	0.0841	0.00289611
56	16:00	18	0.05364	0.05468	0.05336	0.04928	0.0376	0.0798	0.00253916
57		-	0.05364	0.05458	0.05336	0.04938	0.0376	0.0799	0.00254714
58		-	0.05364	0.05458	0.05336	0.04948	0.0374	0.0798	0.00253916
59		-	0.05364	0.05458	0.05336	0.04948	0.0375	0.0799	0.00254714
60	16:20	18.3	0.05364	0.05458	0.05336	0.04948	0.0374	0.0798	0.00253916
61		-	0.05004	0.05108	0.04996	0.04638	0.0429	0.0765	0.0022844
62		-	0.04994	0.05108	0.04996	0.04638	0.0429	0.0765	0.0022844
63	16:48	19	0.04994	0.05088	0.04986	0.04638	0.0429	0.0765	0.0022844

Table V. 4: Experimental results series B (continued from Table V. 3).

experiment	time	temperature (°C)	depth 1 (m)	depth 2 (m)	depth 3 (m)	depth 4 (m)	depth 5 (m)	reading v-notch (m)	discharge (m^3/s)
64	10:50	12	0.04532	0.04534	0.04538	0.043	0.0396	0.07308	0.002038061
65			0.04542	0.04514	0.04528	0.0429	0.0395	0.07318	0.002045009
66			0.04542	0.04524	0.04528	0.0429	0.0393	0.07298	0.002031127
67			0.04532	0.04524	0.04528	0.0428	0.0393	0.07288	0.002024207
68	11:15	13	0.04532	0.04514	0.04528	0.0428	0.0393	0.07278	0.002017302
69	11:25	13.3	0.04192	0.04194	0.04178	0.0398	0.039	0.06968	0.001810271
70			0.04182	0.04174	0.04178	0.0398	0.039	0.06968	0.001810271
71			0.04182	0.04164	0.04178	0.0398	0.039	0.06968	0.001810271
72			0.04172	0.04164	0.04178	0.0398	0.039	0.06968	0.001810271
73	11:47	14	0.04172	0.04154	0.04178	0.0398	0.039	0.06968	0.001810271
74	12:03	14.3	0.03922	0.03914	0.03948	0.0379	0.0375	0.06668	0.001622693
75			0.03922	0.03924	0.03948	0.038	0.0377	0.06678	0.001628746
76			0.03922	0.03914	0.03958	0.038	0.0377	0.06668	0.001622693
77			0.03922	0.03924	0.03948	0.038	0.0377	0.06678	0.001628746
78	12:30	14.9	0.03922	0.03924	0.03948	0.038	0.0377	0.06688	0.001634813
79	12:40	15.1	0.03712	0.03704	0.03738	0.0363	0.0358	0.06398	0.001464371
80			0.03712	0.03704	0.03738	0.0363	0.0358	0.06398	0.001464371
81			0.03712	0.03704	0.03738	0.0364	0.0358	0.06428	0.001481478
82			0.03712	0.03704	0.03748	0.0363	0.0358	0.06418	0.001475762
83	13:02	15.5	0.03712	0.03704	0.03748	0.0363	0.0358	0.06398	0.001464371
84	13:15	16	0.03522	0.03514	0.03558	0.0349	0.0343	0.06178	0.001342567
85			0.03522	0.03514	0.03558	0.0348	0.0344	0.06178	0.001342567
86			0.03522	0.03514	0.03558	0.0348	0.0344	0.06168	0.001337182
87			0.03522	0.03514	0.03558	0.0348	0.0344	0.06168	0.001337182
88	13:40	16.2	0.03522	0.03514	0.03558	0.0348	0.0344	0.06168	0.001337182
89	13:55	16.5	0.03392	0.03384	0.03418	0.0338	0.0333	0.05988	0.001242509
90			0.03392	0.03384	0.03428	0.0337	0.0333	0.05978	0.001237396
91			0.03392	0.03374	0.03428	0.0337	0.0333	0.05978	0.001237396
92			0.03392	0.03384	0.03428	0.0338	0.0333	0.05988	0.001242509
93	14:11	17	0.03392	0.03374	0.03428	0.0338	0.0333	0.05988	0.001242509

Table V. 5: Experimental results series C.

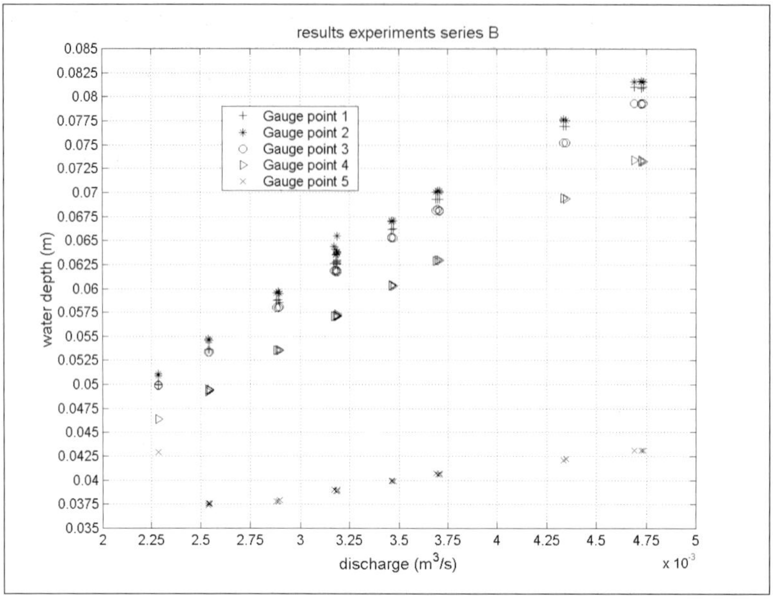

Figure V. 4: Experimental results series B.

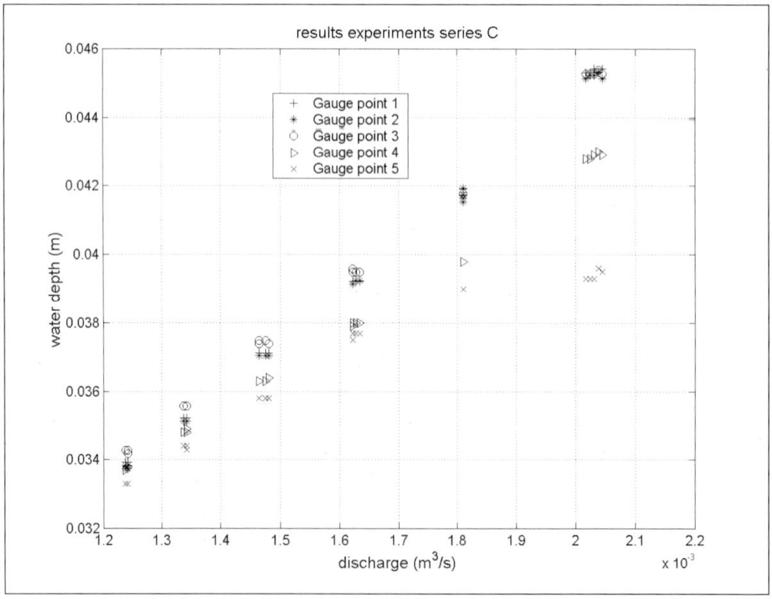

Figure V. 5: Experimental results series C.

V.3 Local loss coefficient.

The local loss coefficient was derived from the experimental data. Since the values for the Boussinesq coefficients are unknown, values of 1.00 were adopted. This implies that any variation in the true values, along with wall friction, is incorporated in the values obtained for the local loss coefficient.

The values for the local loss coefficient were calculated using eq V.3, the definition for the head loss due to friction being:

$$\Delta H_{fr} = \xi \frac{u_1^2}{2g} \qquad \text{(eq. V.10)}$$

The results are shown in Figure V. 6.

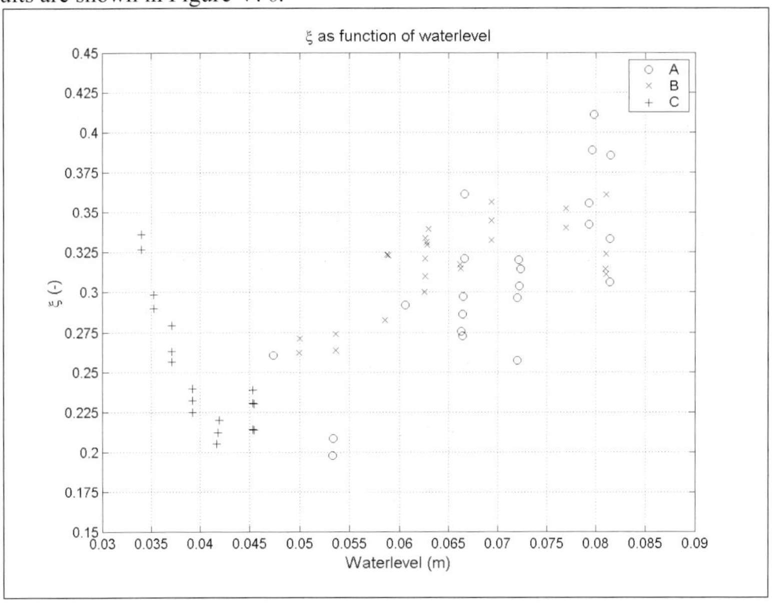

Figure V. 6: Local head loss-coefficient.

As can be seen the results in series A show a relatively large scatter, while the results of series C suffer from backwater effects for water depths below circa 0.04 m.
Also can be seen that the value for the loss deviates from the values found by Wenzel (1975) (circa 0.225), which is explained from the fact that the geometry applied contains sharp edges, while Wenzel used rounded shapes. Since only one pipe diameter was applied, no general relation for the local loss coefficient can be derived.
Therefore, a relation between the water depth at measuring point 1 and the loss coefficients was derived:

$$\xi(h_1) = \alpha h_1^{\beta} \qquad \text{(eq. V.11)}$$

series	α	β	σ_α^2	σ_β^2	ρ
A	1.0838	0.4511	0.0389	0.0044	0.0131
B	1.1224	0.4842	0.3436	0.0387	0.1152
C	Too little information for curve fit				

Table V. 6: Curve fit results for head loss coefficient.

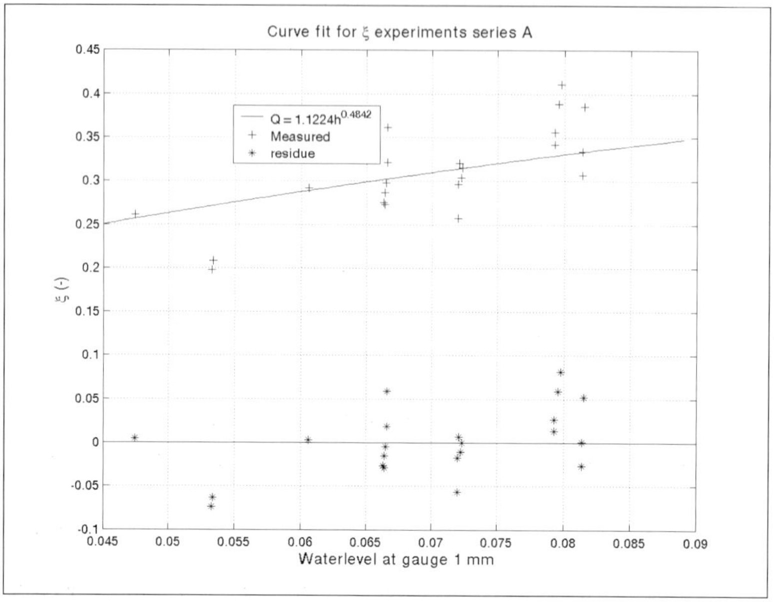

Figure V. 7: Curve fit results head loss coefficient series A.

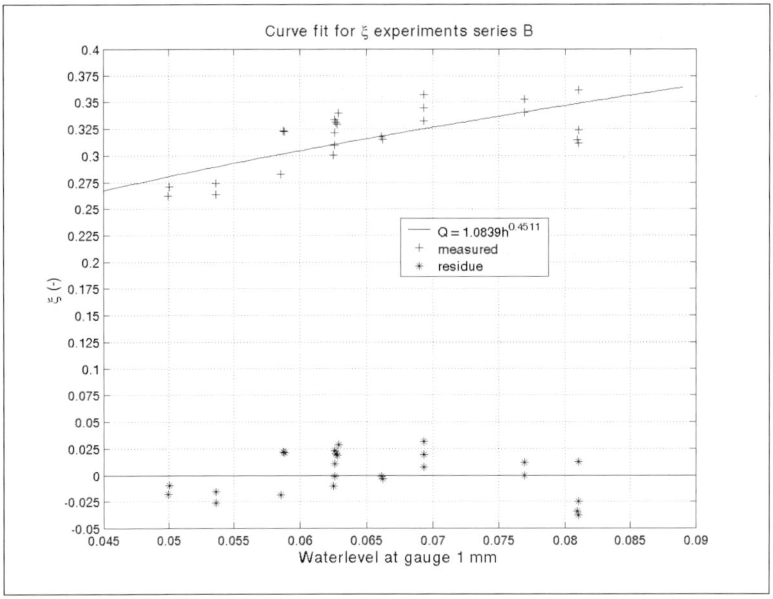

Figure V. 8: Results curve fit series B.

If the relation found for series B are substituted in the equation for the head loss due to friction the following formula is obtained:

$$\Delta H_{fr} = 1.0839 h_1^{0.4511} \frac{u_1^2}{2g} \qquad \text{(eq. V.12)}$$

When this relation is inserted in eq. V.3 the overall accuracy of the measurements with respect to Q-h relation can be quantified([60]). Because of the recursive nature of this formula the 95 % confidence intervals for the Q-h relation given by formula V.3 was calculated using a Monte Carlo simulation with 50.000 function evaluations using three possible measuring accuracies for the water level. The results are shown in Figure V. 9 and Figure V. 10. As can be seen in the graphs, the experimental error in the laboratory experiments (the situation in which $\sigma_h=0$) is significant (up to a relative error of 8 %). When in addition the operational measuring inaccuracy is determined by $\sigma_h=0.001$ m this relative accuracy interval for the flow increases to circa 18 %.

([60]) Here the results for experiment C with water levels less than 0.04 m were left out.

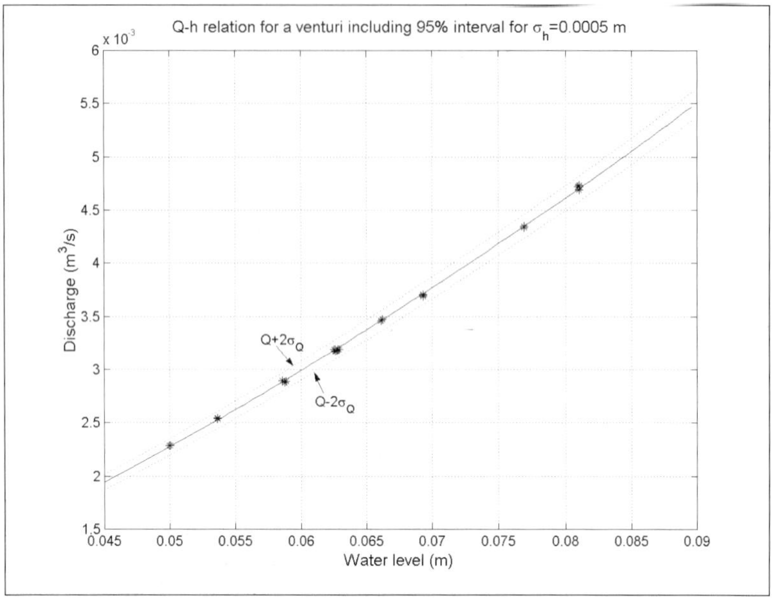

Figure V. 9: Q-h relation including 95% accuracy interval.

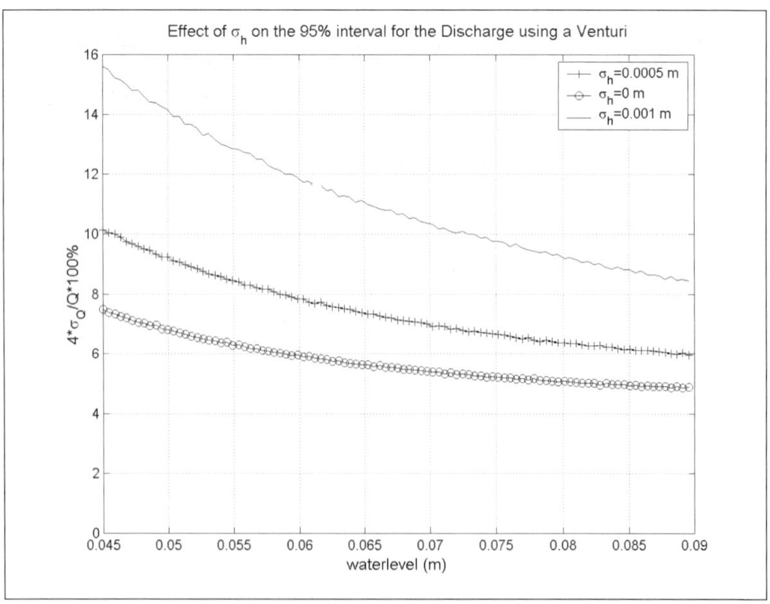

Figure V. 10: 95% accuracy interval as function of discharge and measuring accuracy of the water level measurement.

ANNEXE VI: STORMS APPLIED

VI.1 General

Throughout the thesis 10 artificial and 5 historical storms are applied, in this appendix the characteristics of these storms are shown in tables and graphs.

VI.2 Artificial storm according to the Dutch guidelines

	rain intensity in $l.s^{-1}.ha^{-1}$									
time	return period (years) and storm number									
interval	0.25		0.5		1.0		2.0		5.0	1.0
(minutes)	01	02	03	04	05	06	07	08	09	10
0-5	10	5	10	5	10	5	20	10	50	60
5-10	20	5	20	10	20	10	40	20	90	120
10-15	30	5	30	15	50	15	70	30	160	210
15-20	40	10	50	20	90	20	110	40	160	210
20-25	50	15	70	25	90	25	110	50	140	190
25-30	50	20	70	30	70	30	90	70	110	160
30-35	35	25	50	35	50	35	70	90	90	120
35-40	30	30	40	40	40	40	50	110	70	80
40-45	25	35	35	50	35	50	40	110	50	40
45-50	20	50	30	70	30	70	30	70	30	
50-55	15	50	25	70	25	90	20	40	20	
55-60	10	40	20	50	20	90	10	20	10	
60-65	5	30	15	30	15	50				
65-70	5	20	10	20	10	20				
70-75	5	10	5	10	5	10				

VI.3 Historical storms measured at 'De Hoven'

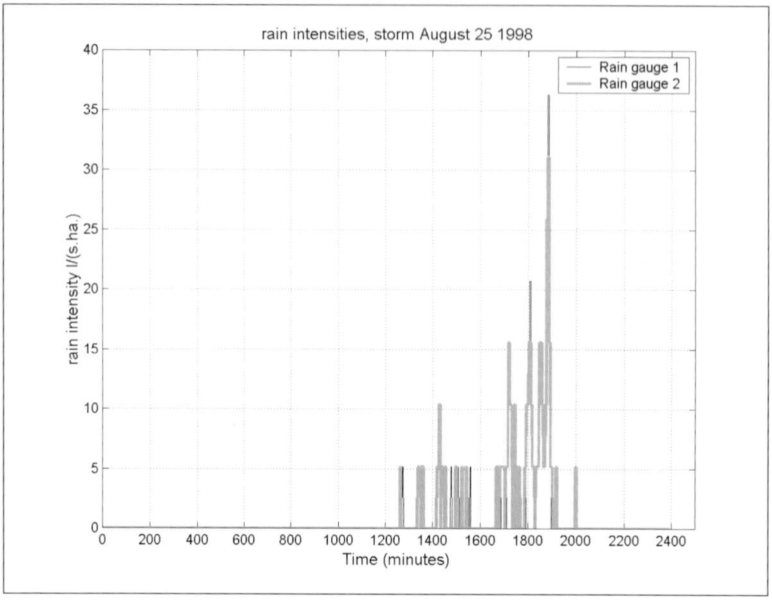

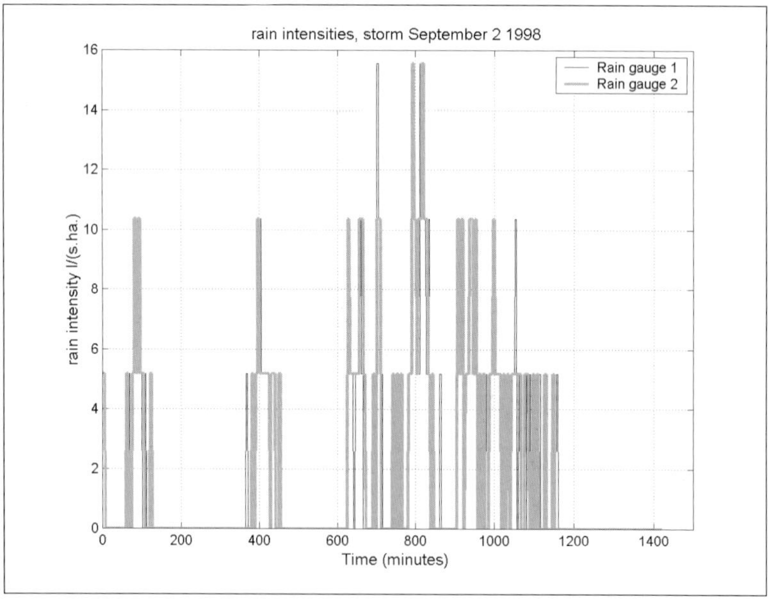

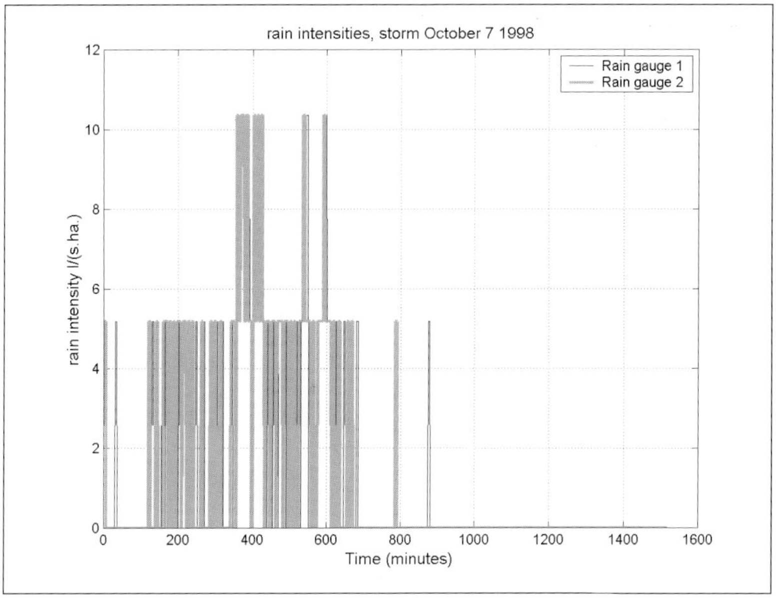
rain intensities, storm October 7 1998
Rain gauge 1
Rain gauge 2
rain intensity l/(s.ha.)
Time (minutes)
0
2
4
6
8
10
12
200
400
600
800
1000
1200
1400
1600

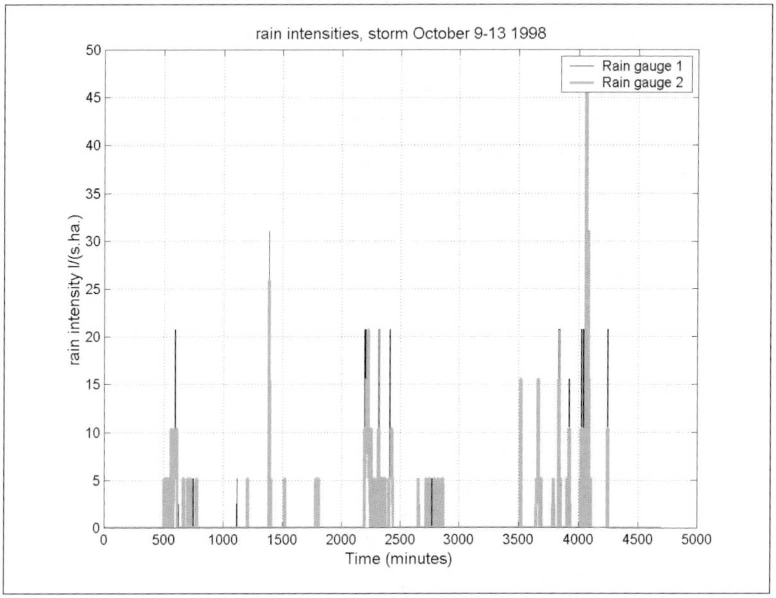
rain intensities, storm October 9-13 1998
Rain gauge 1
Rain gauge 2
rain intensity l/(s.ha.)
Time (minutes)
0
5
10
15
20
25
30
35
40
45
50
500
1000
1500
2000
2500
3000
3500
4000
4500
5000

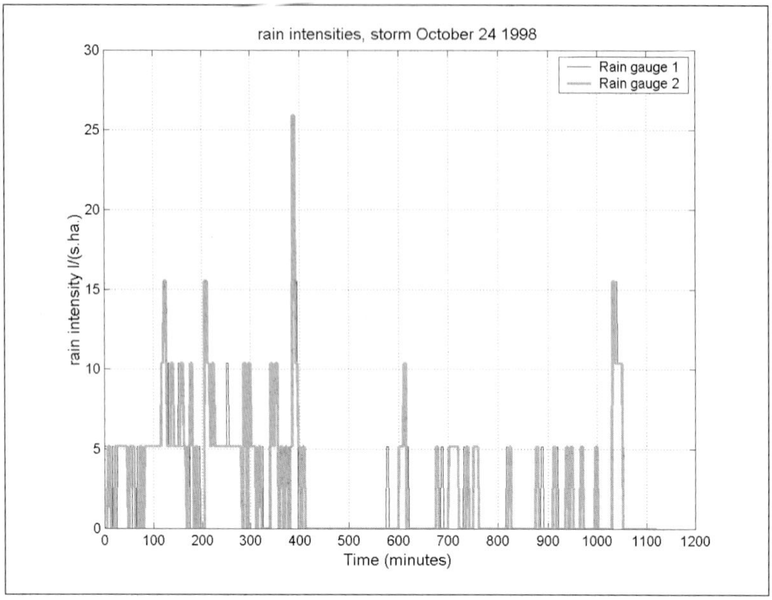
rain intensities, storm October 24 1998
Rain gauge 1
Rain gauge 2
rain intensity l/(s.ha.)
Time (minutes)
0
5
10
15
20
25
30
100
200
300
400
500
600
700
800
900
1000
1100
1200

ANNEXE VII: RESULTS CALIBRATION 'DE HOVEN'

VII.1 Storm event September 2 1998

In Figure VII. 1 the correlation matrix is shown, in this case a noticeable mutual correlation between the DWF (N1) and the values for initial storage loss and infiltration rates is found for the 'slow' surface (the type 2 surface). This is due to the fact the storm event showed only a moderate rise of water level in the system at relatively low rain intensities. This implies that the influence of the DWF was an important factor determining the water level to a relatively large extent throughout the whole storm. Furthermore, a negative correlation between the values for the infiltration rates for both surfaces is present. This is logical, since an increase in the infiltration for one surface leads to a decrease for the other surface while maintaining the same amount of run-off.

In Figure VII. 2 the eigen vectors and singular values are shown, as can be seen every eigen vector has relatively large components for more than one axis in the parameter space. This implies, as was deduced from the correlation matrix also, that none of the parameters can be identified separately

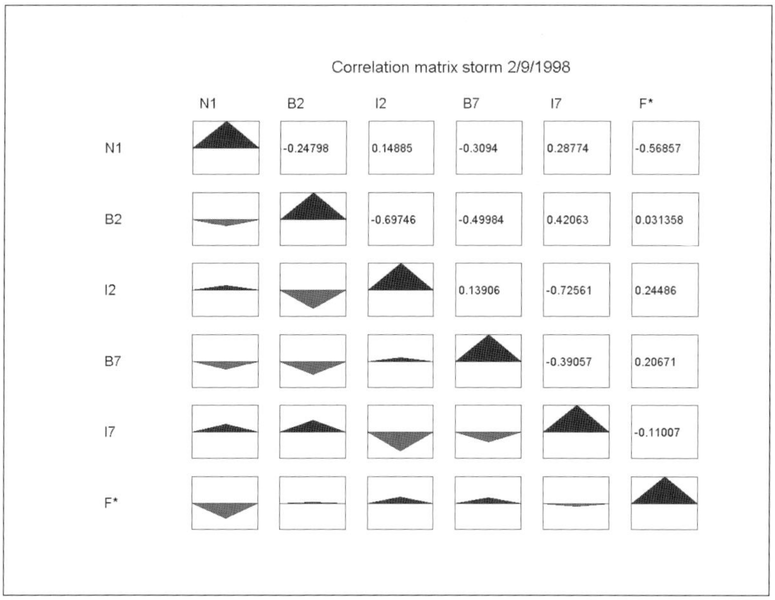

Figure VII. 1: Correlation matrix storm September 2 1998.

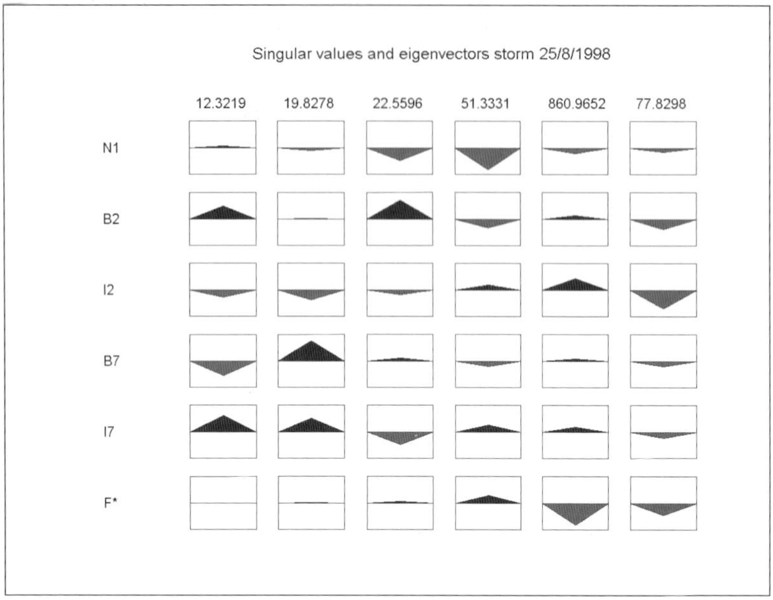

Figure VII. 2: Singular values and eigen vectors for storm on September 2 1998.

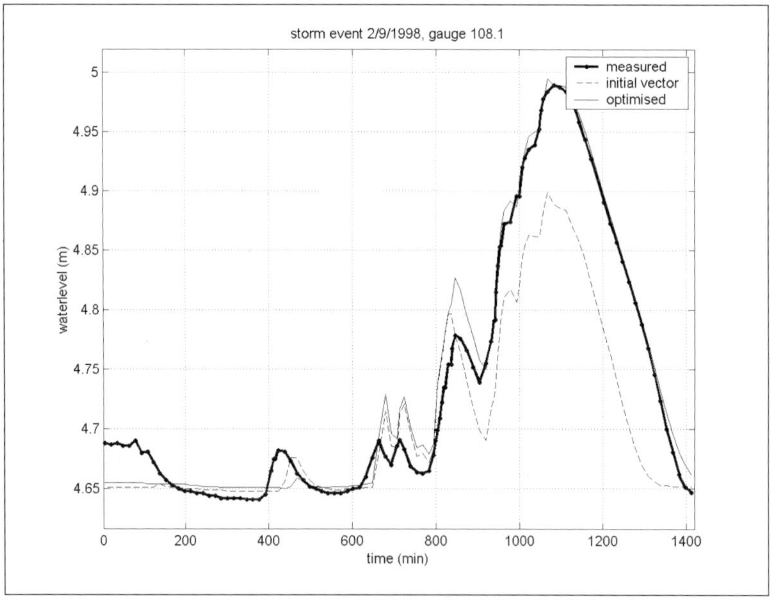

Figure VII. 3 : Measured versus model results, gauge 108.1.

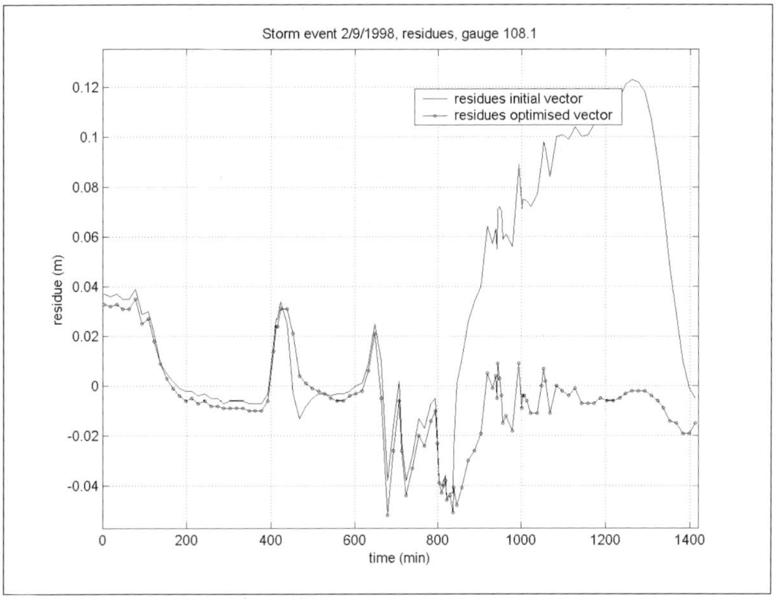

Figure VII. 4: Residues gauge 108.1.

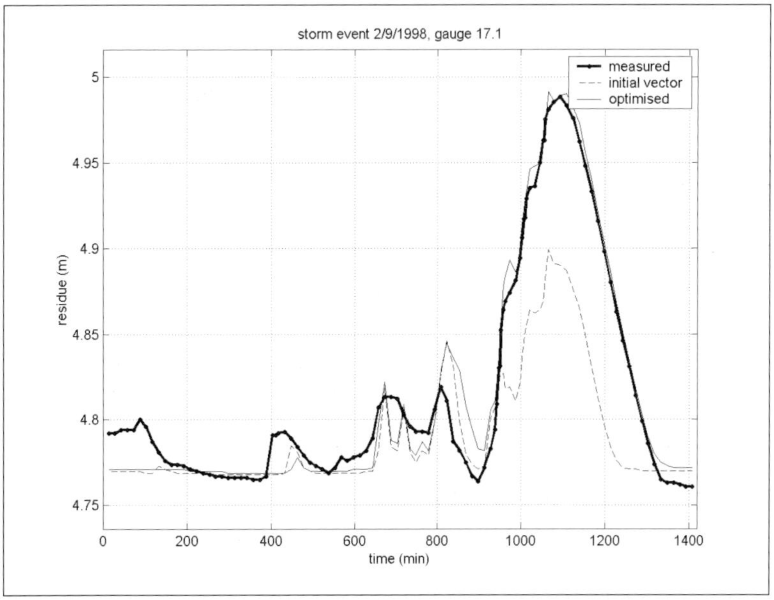

Figure VII. 5 : Measured versus model results, gauge 17.1.

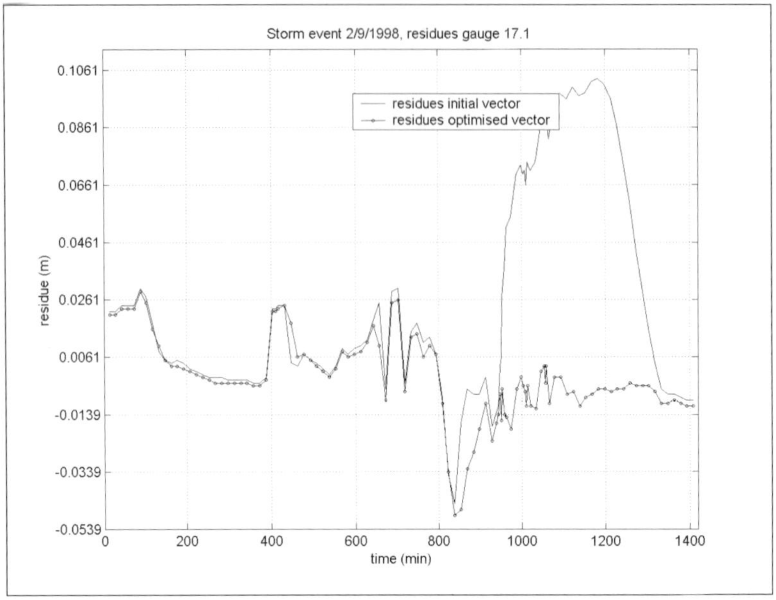

Figure VII. 6: Residues gauge 17.1

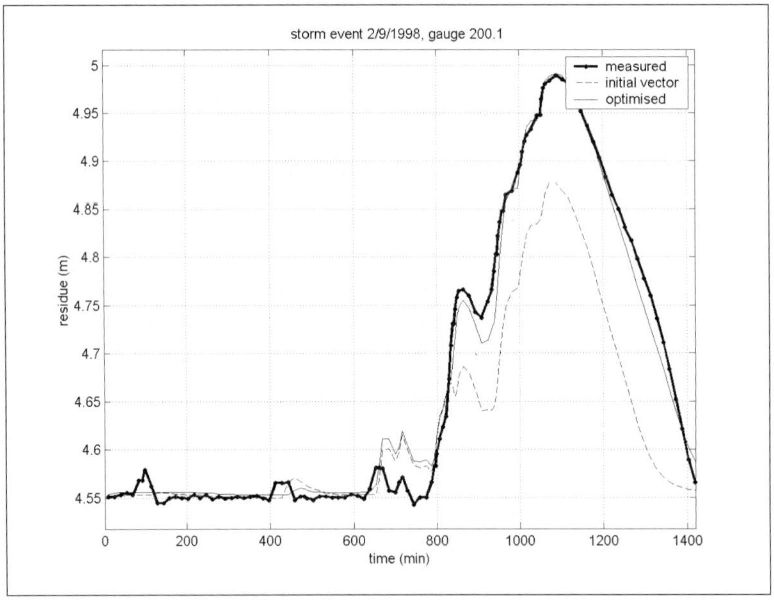

Figure VII. 7 : Measured versus model results, gauge 200.1

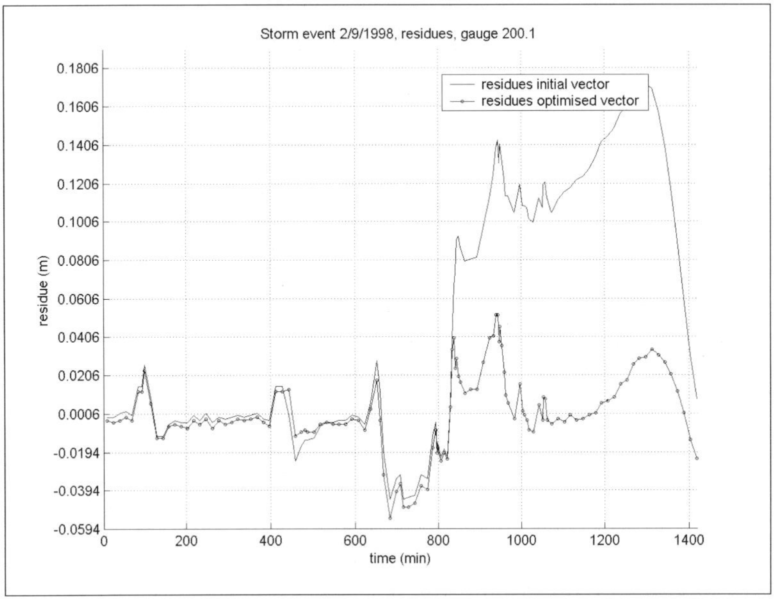

Figure VII. 8: residues for gauge 200.1

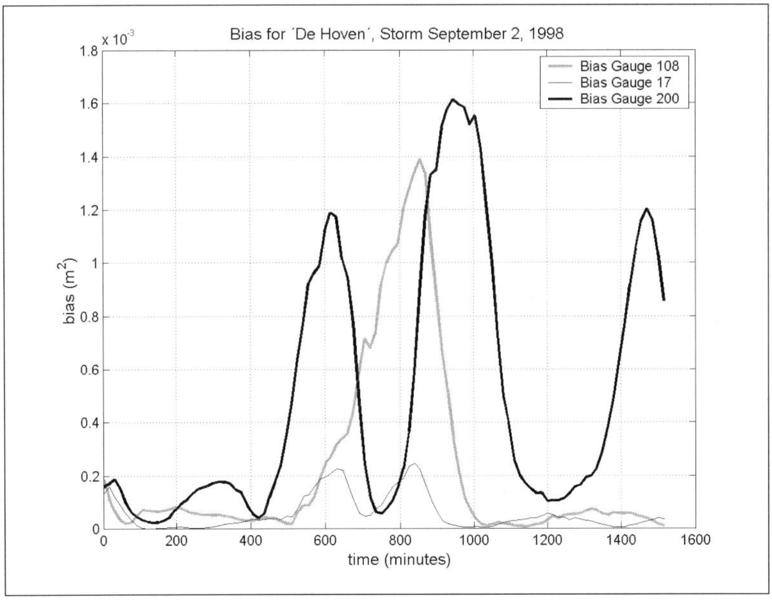

Figure VII. 9: Absolute Bias in a shifting time window.

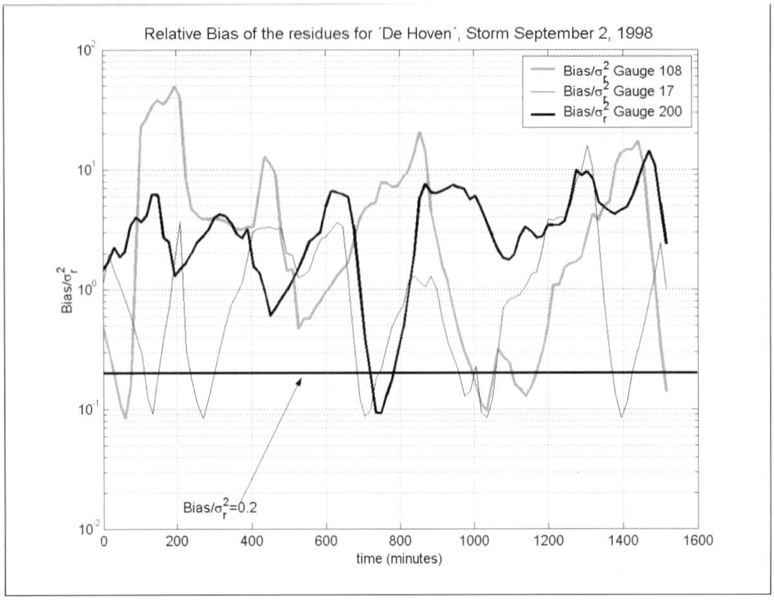

Figure VII. 10: Relative Bias in a shifting time window.

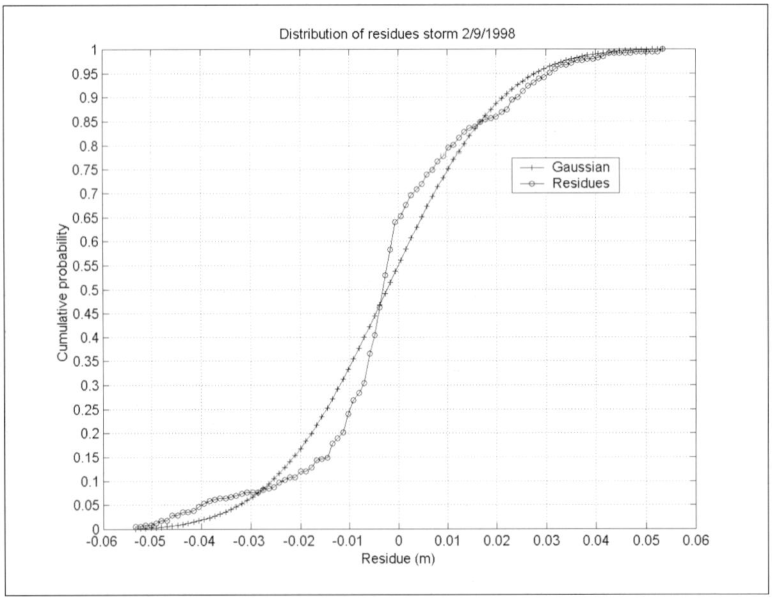

Figure VII. 11: Probability distribution if the residues, storm September 2 1998.

In Figures VII.3 to VII.8 the water levels and residues are shown, as can be seen a fair fit is obtained in the level range above 4.80 m, in the range between 4.55 and 4.80 the overall reaction of the model follows the measurements only in qualitative sense. The reasons for this can be divers; sludge in the system disturbing the database, small DWF fluctuations unaccounted for, run-off effects etc. No physical evidence could be found for either of these causes, therefore the results are accepted as the best obtainable, given the choices made. When studying the residues the presence of bias is immediately clear (see Figure VII.9 and Figure VII.10). Along with the correlation matrix and the eigen vectors this leads to the conclusion that the parameter values found in the calibration process can not be trusted. This is also confirmed when studying the distribution of the residues obtained (see Figure VII.11 they deviate substantially from the assumed Gaussian distribution).

VII.2 Storm event October 7 1998

In Figure VII. 12 the correlation matrix is shown, a strong correlation between DWF (N1) and the linear reservoir constant. As can be seen in the eigen vectors (Figure VII.13) the DWF (N1) and the initial losses on surface type 2 (B2) dominate the vectors with the largest eigen singular values. In Figure VII.14 through Figure VII. 20 the results and the residues per gauging location are shown. At gauge 17 the initial water level in the measurement was circa 4 cm above the invert level (see Figure VII.16), this is either caused by a measuring error or by a coincidental local discharge of DWF. During the months October and November several house owners drained their cellars using small pumps so as to avoid damage due to the high groundwater table.

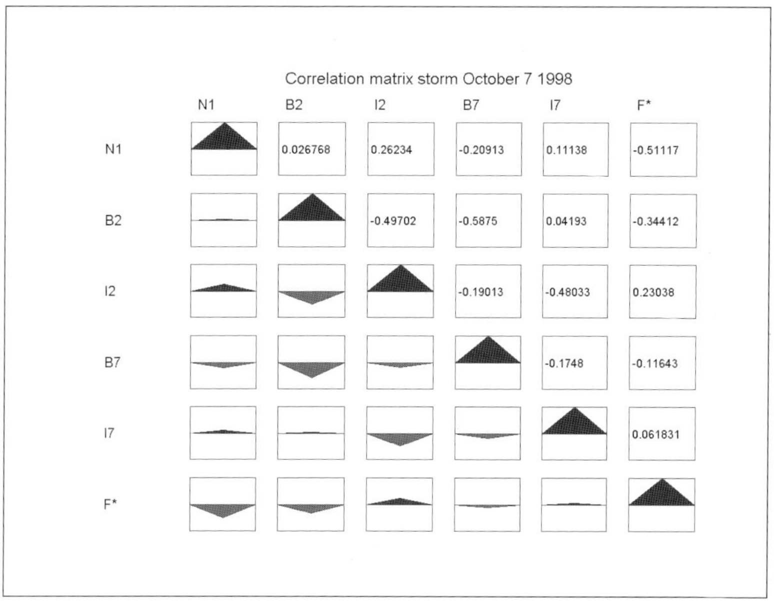

Figure VII.12: Correlation matrix Storm October 7 1998.

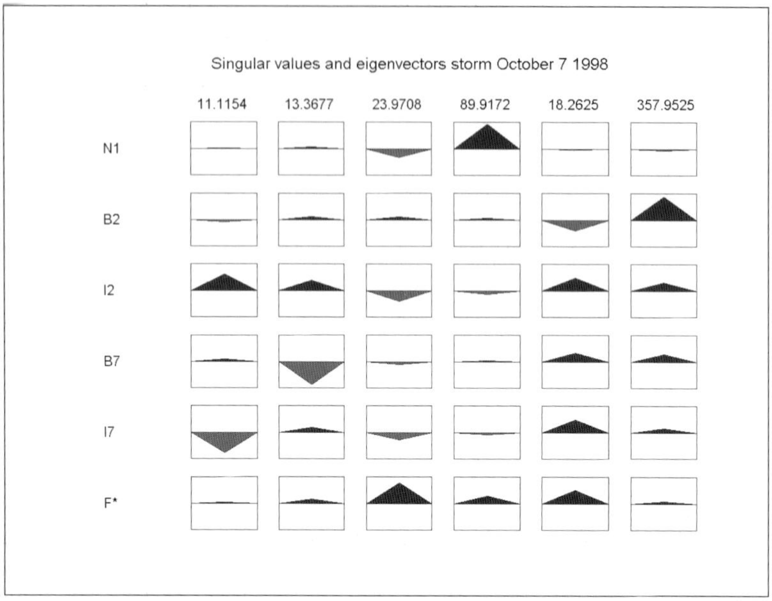

Figure VII. 13: Eigen vectors storm October 7, 1998.

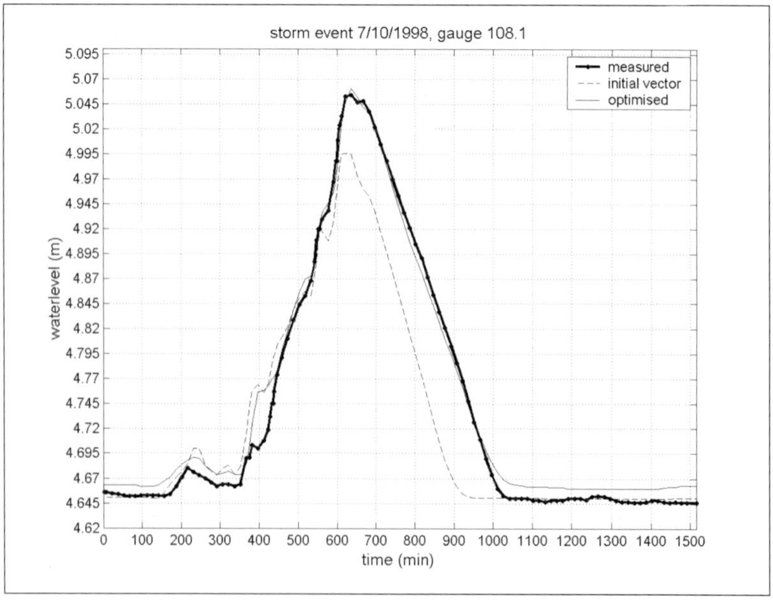

Figure VII. 14

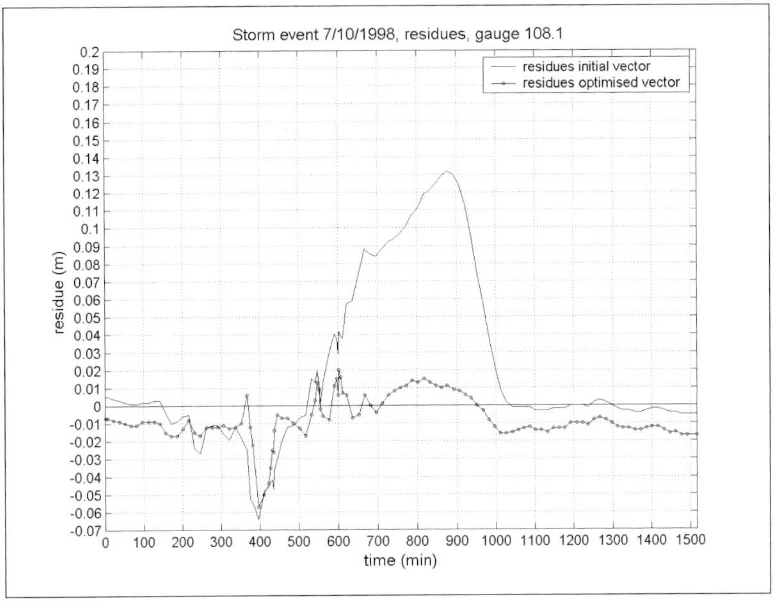

Figure VII. 15

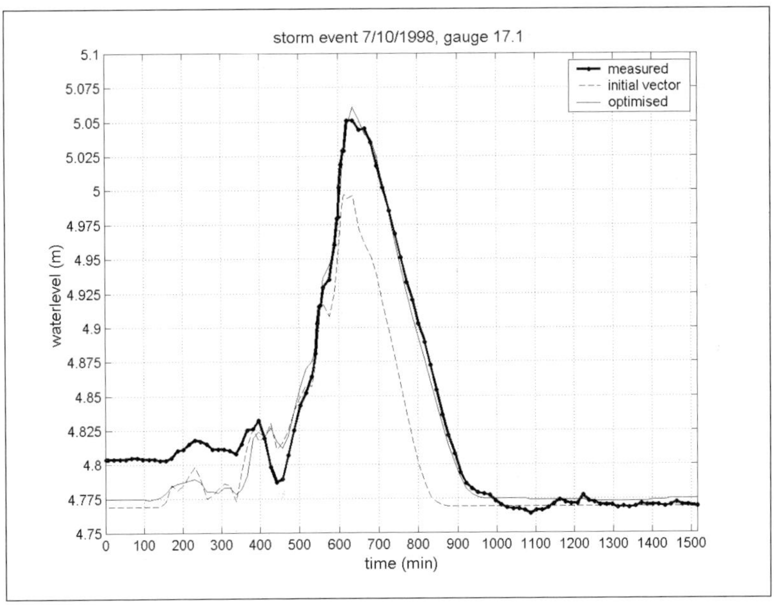

Figure VII. 16

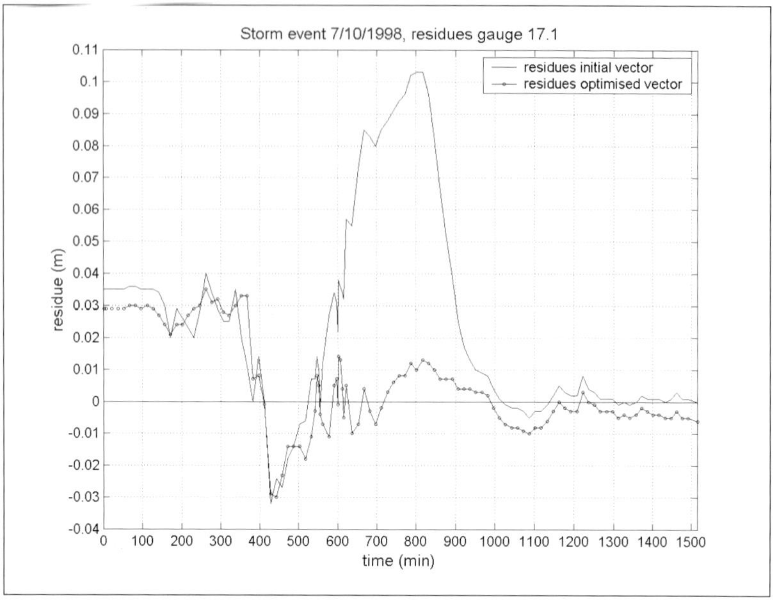

Figure VII. 17

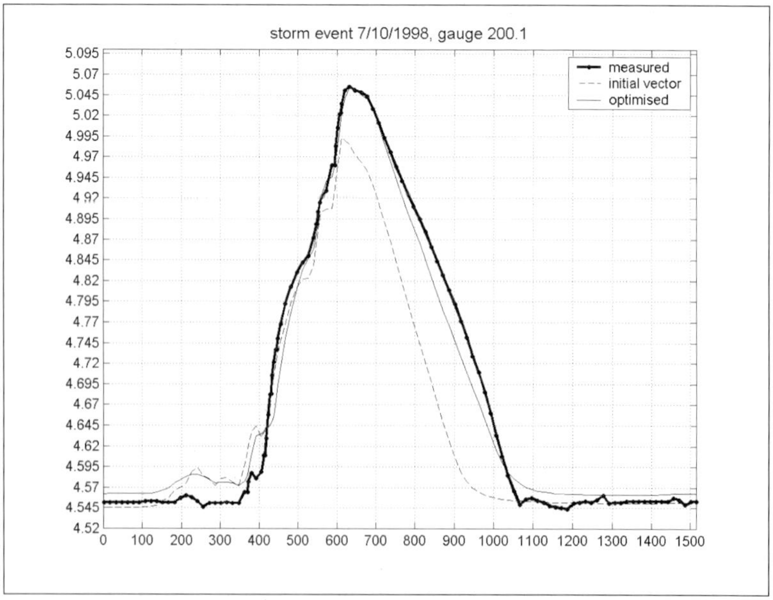

Figure VII. 18

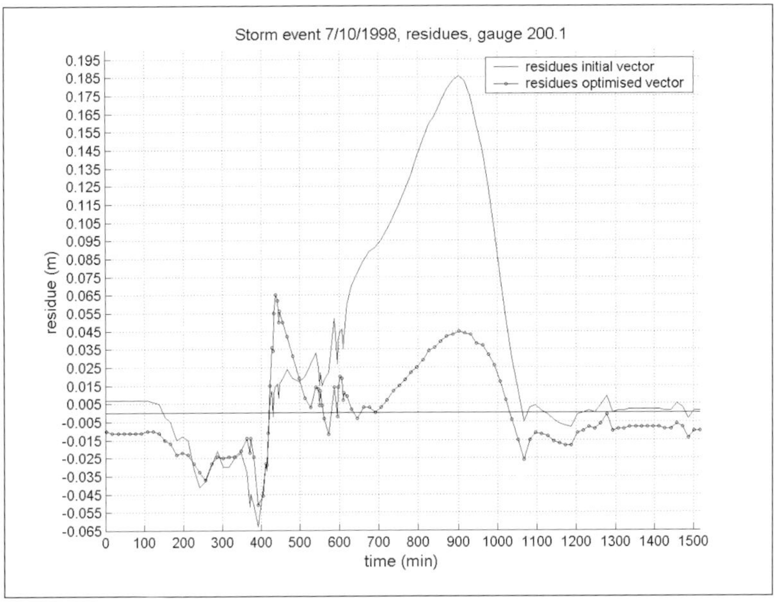

Figure VII. 19

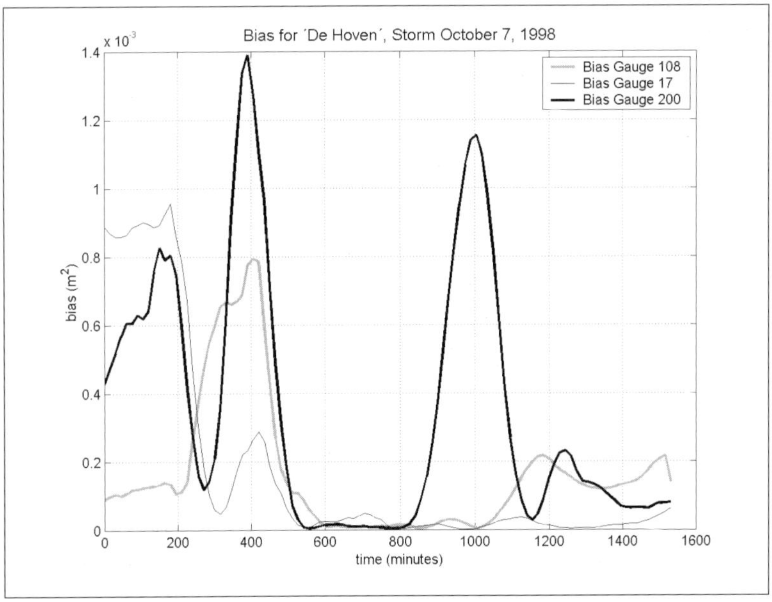

Figure VII. 20: Bias for storm on October 7, 1998.

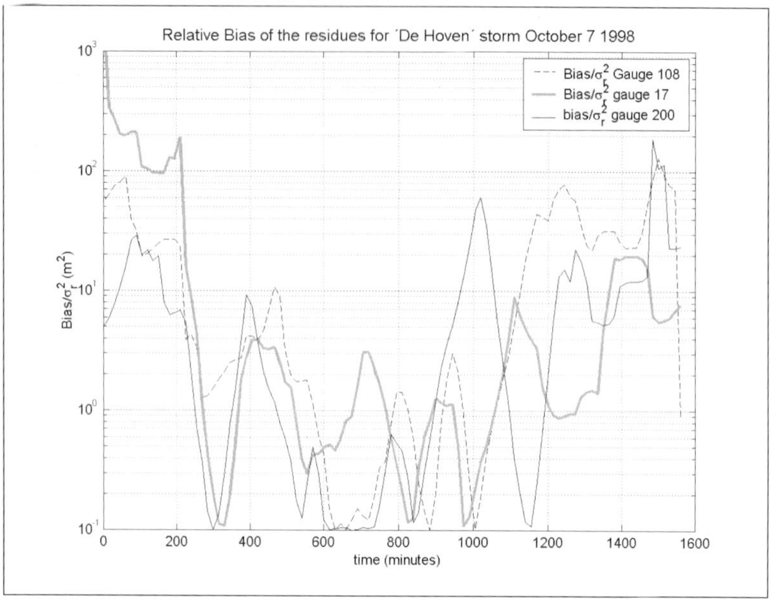

Figure VII. 21: Relative Bias for storm on October 7, 1998.

Bias is present over the whole time range, especially in the period prior to the rain (350 minutes) and during and after emptying of the system(after circa 800 minutes). During the storm itself (between t= circa 350 and t= circa 700 minutes), bias is at an acceptable low level at all three gauges (see Figure VII.20 and VII.21).The distribution of the residues ()Figure VIII.22) shows a deviation from the assumed Gaussian distribution.

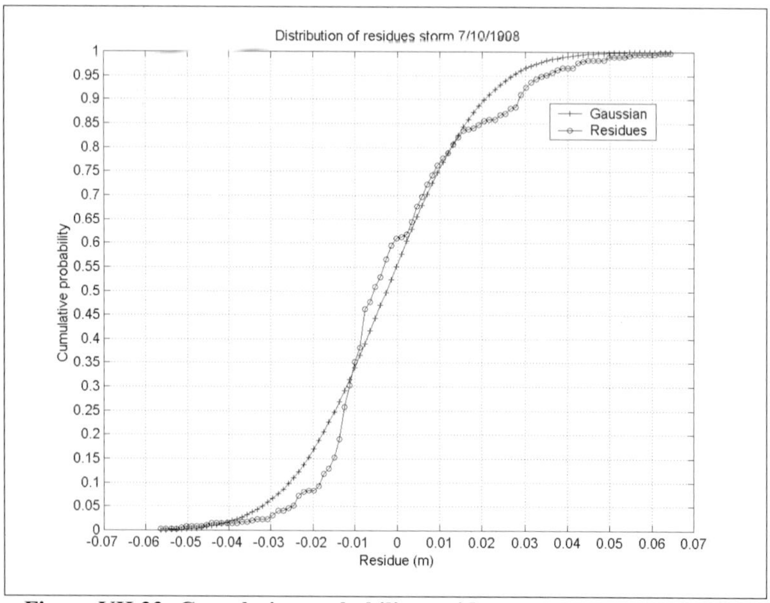

Figure VII.22: Cumulative probability residues storm October 7 1998.

VII.3 Storm event October 9-13 1998

The correlation matrix and eigen vectors (see Figure VII.23 and Figure VII.24) show that there is a significant correlation between almost all parameters. Furthermore it is seen that in non of the eigen vectors one parameter is dominant, implying that the identifiability of the parameters for this storm is rather poor. The calibration results (measured and calculated water levels and residues) are shown in Figure VII.25 through Figure VII.30). As can be seen the main problem is to correctly model the water levels in between the periods during which rain occurs.

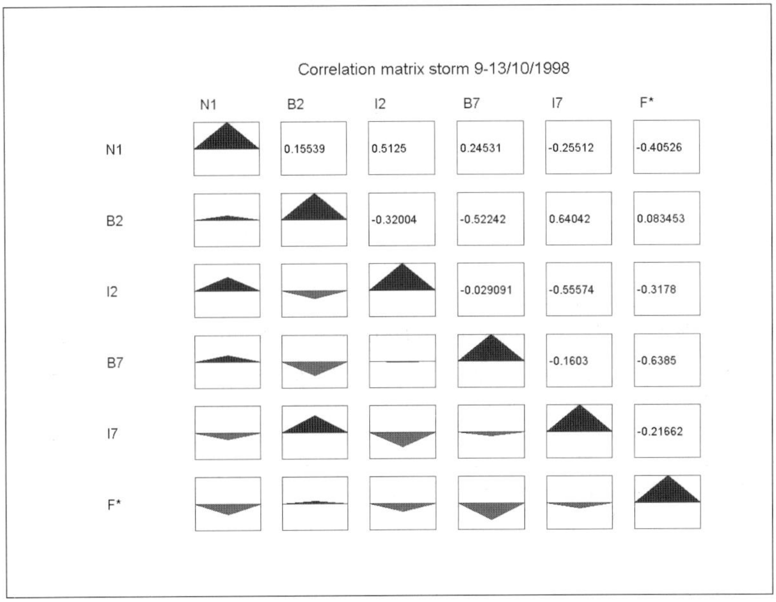

Figure VII.23: Correlation matrix storm October 9-13 1998.

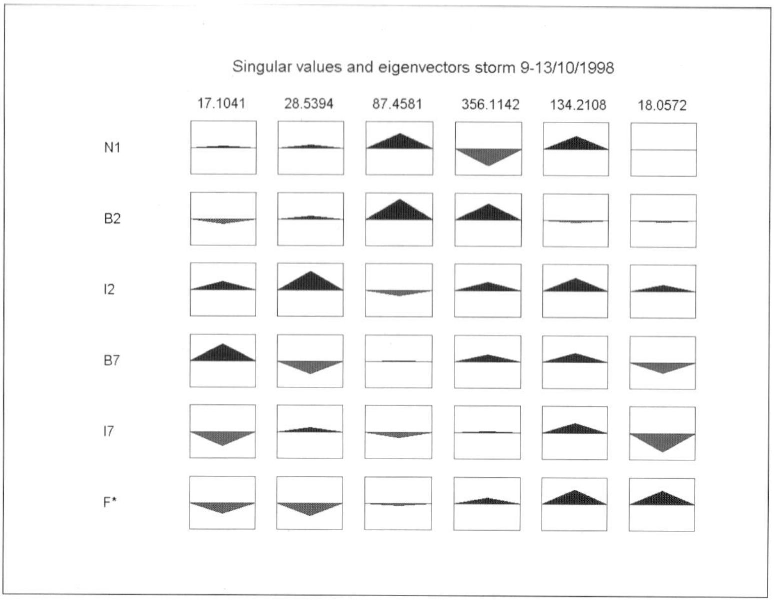

Figure VII.24: Eigen vectors and singular values storm October 9-13 1998.

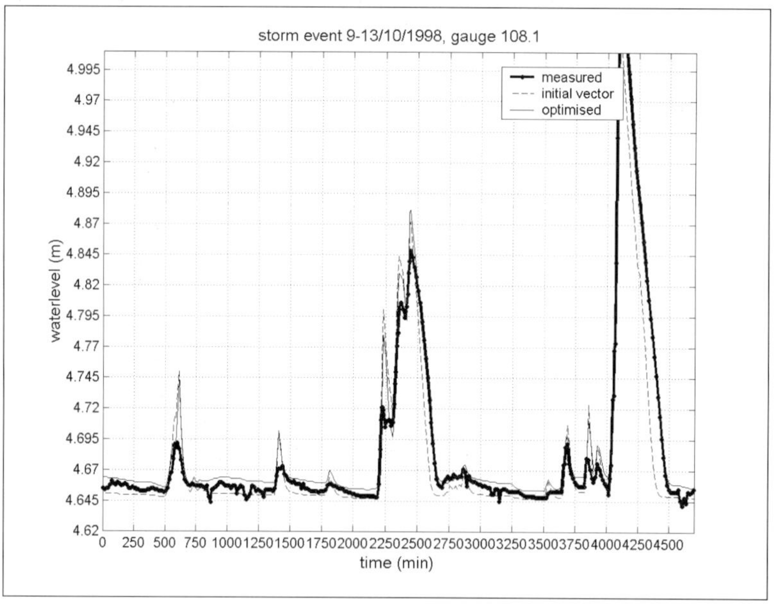

Figure VII. 25

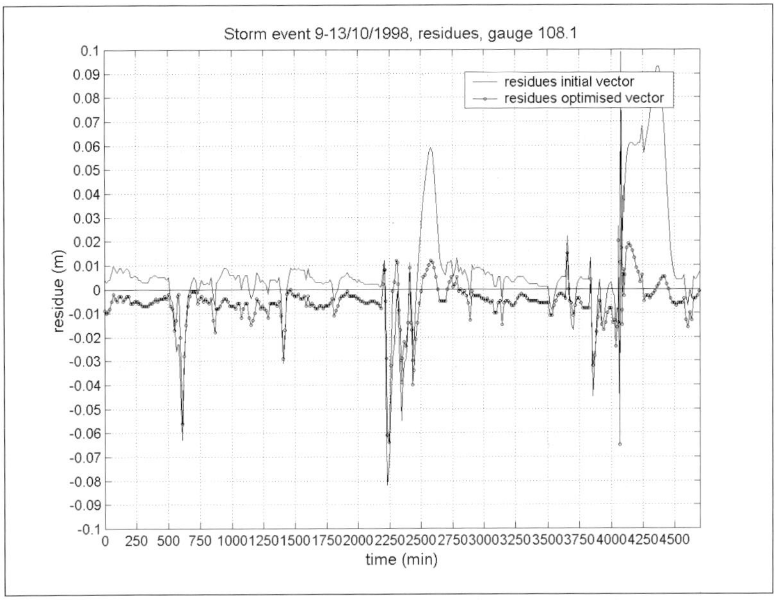

Figure VII. 26

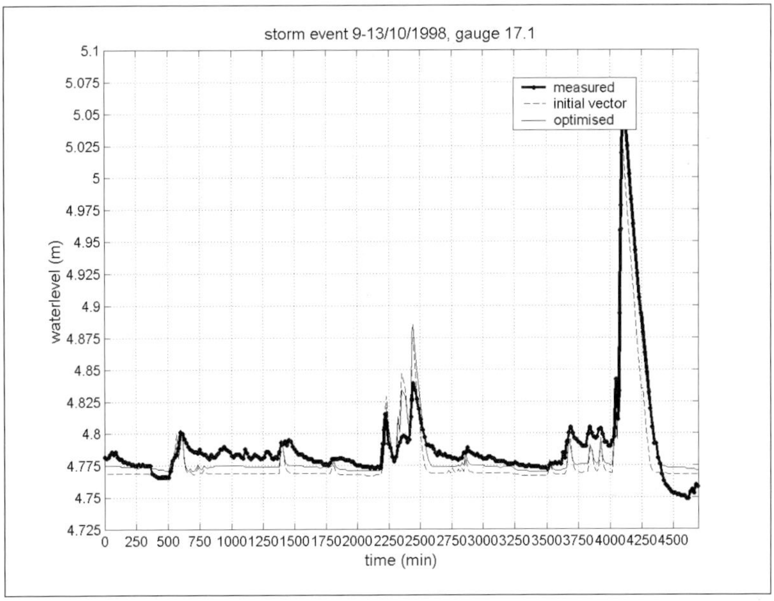

Figure VII. 27

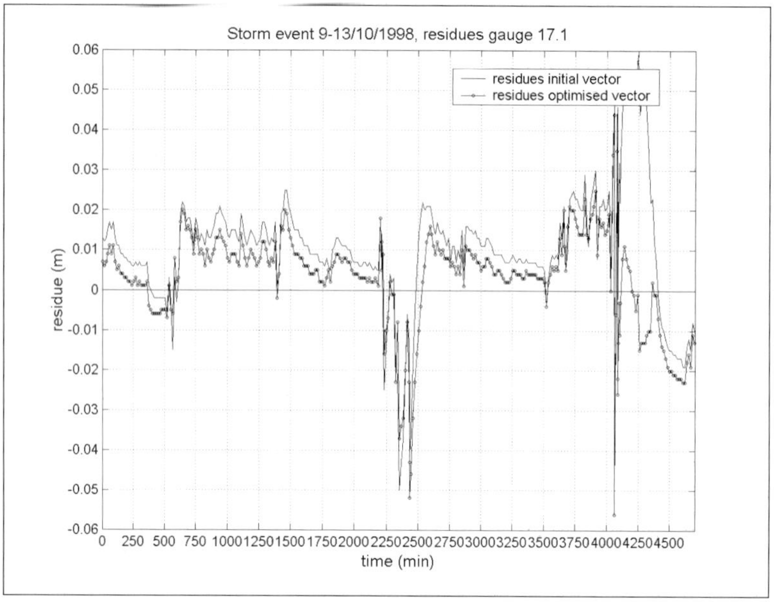

Figure VII. 28

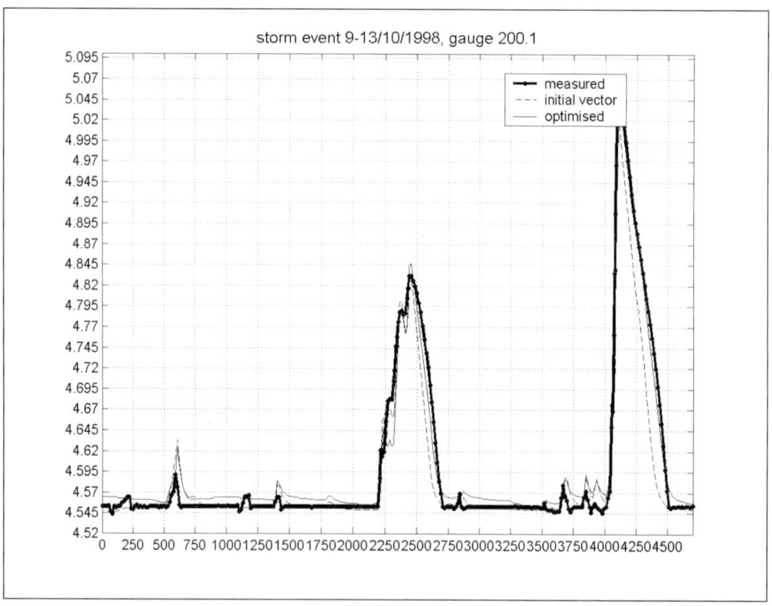

Figure VII. 29

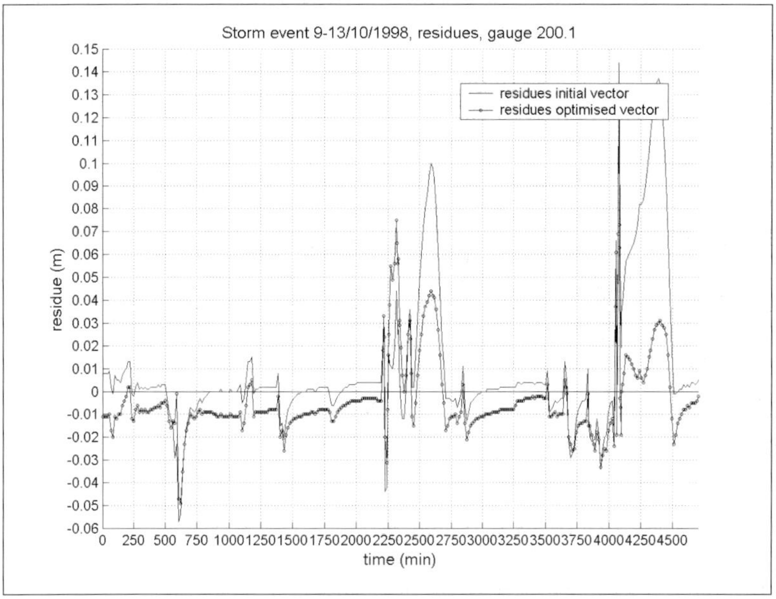

Figure VII. 30

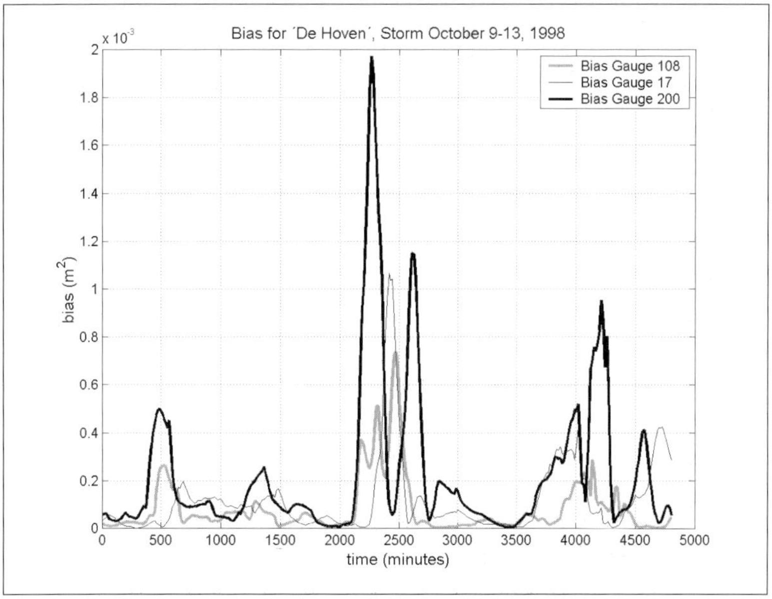

Figure VII.31: Bias during storm on October 9-13 1998.

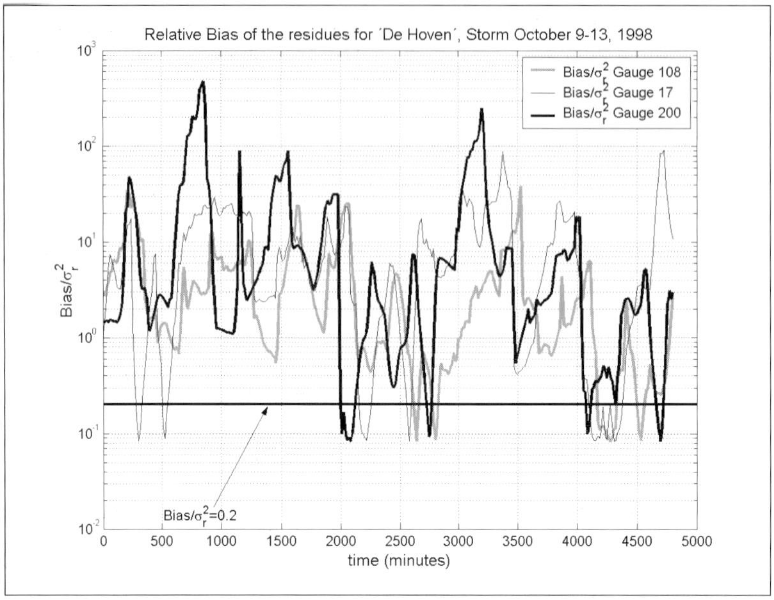

Figure VII.32: Relative Bias during storm on October 9-13 1998.

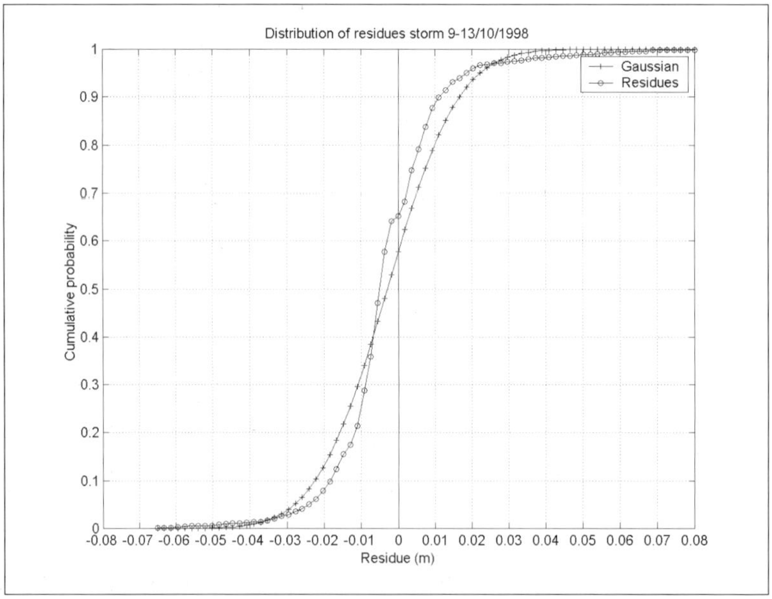

Figure VII. 33

When studying the relative bias (see Figure VII.32) it is seen that bias is significantly lower during wet periods when compared with DWF periods. This indicates that especially during DWF a noticeable systematic deviation between model and measureents is present.

The distribution of the residues (see Figure VII.33), shows a fair match with the assumed Gaussian distribution, this does not imply that the resulting parameter values are correct, the residues show bias implying systematic errors are still present.

VII.4 Storm event October 24 1998

This storm event led to an external overflow, and was therefore fit to be used to calibrate the overflow coefficients of the weirs. Further more the run-off parameters were incorporated in the calibration process. In this storm a similar situation with respect to the parameterisation evolved as was found for the storm on August 25. Therefore the results obtained on the full parameter set is shown also. In Figure VII.34 and Figure VII.35 respectively the correlation matrix and eigen vectors are shown. As can be seen the values for the overflow coefficients show a mutual correlation (the reason for this is explained while discussing storm event 1). Furthermore, relatively strong mutual correlation between DWF (N1) and I2 on one hand and I2 and I7 are present on the other hand. Both dependencies are logical, and increase in infiltration has the same effect on the water level as an increase in DWF. The two infiltration rates are negatively correlated since an increase in one of them is compensated by a decrease of the other. In any case, DWF, I2 and I7 cannot be identified separately. The correlation matrix and eigen vectors for the reduces parameter set are shown in Figure VII.36 and Figure VII.37 respectively. As can be seen the value for the singular values is somewhat increased, indicating an enhancement of the identifiability.

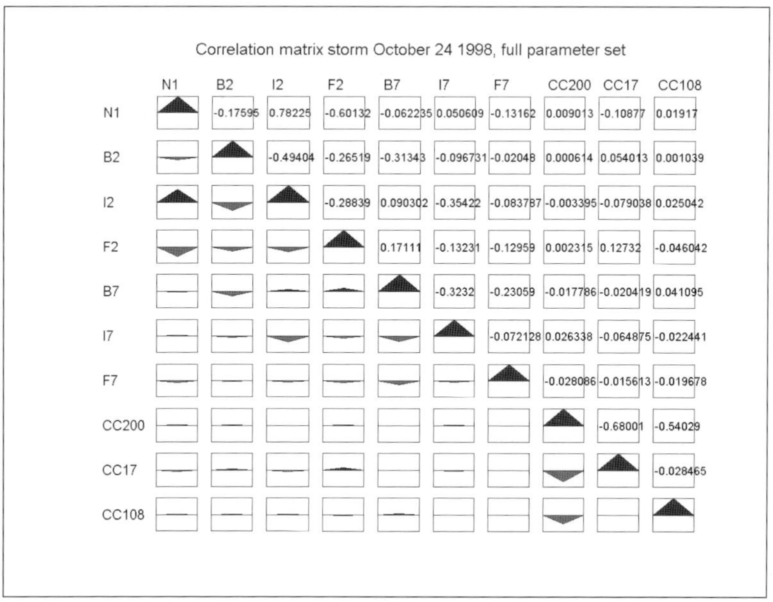

Figure VII. 34 Correlation matrix full parameter set storm October 24 1998.

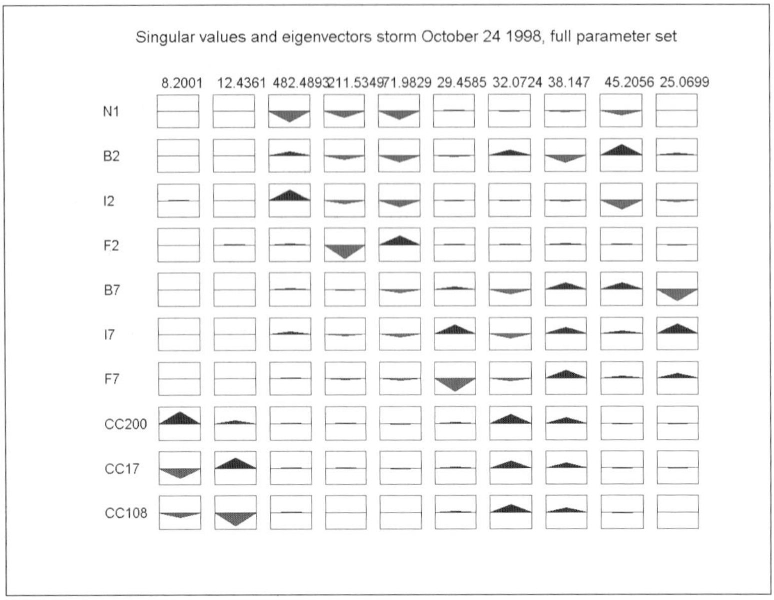

Figure VII.35: Singular values and eigen vectors full parameter set, storm October 24 1998.

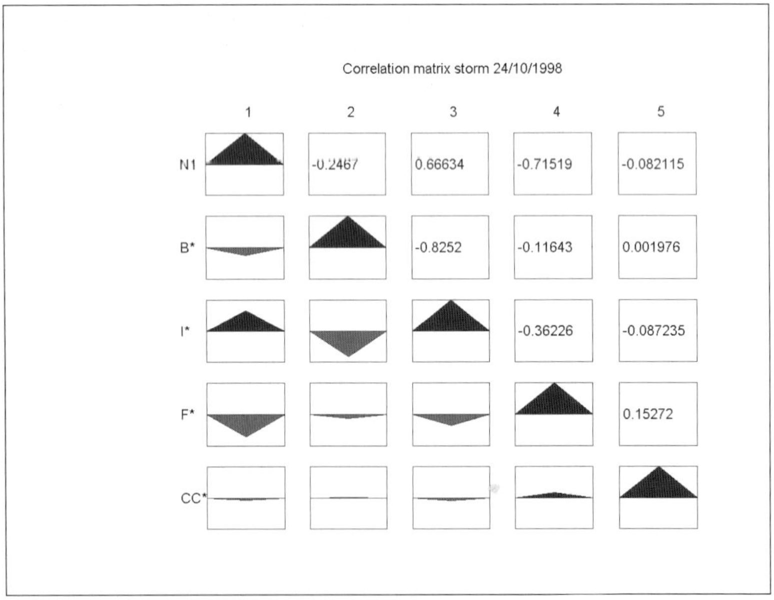

Figure VII.36: Correlation matrix reduced parameter set, storm October 24 1998.

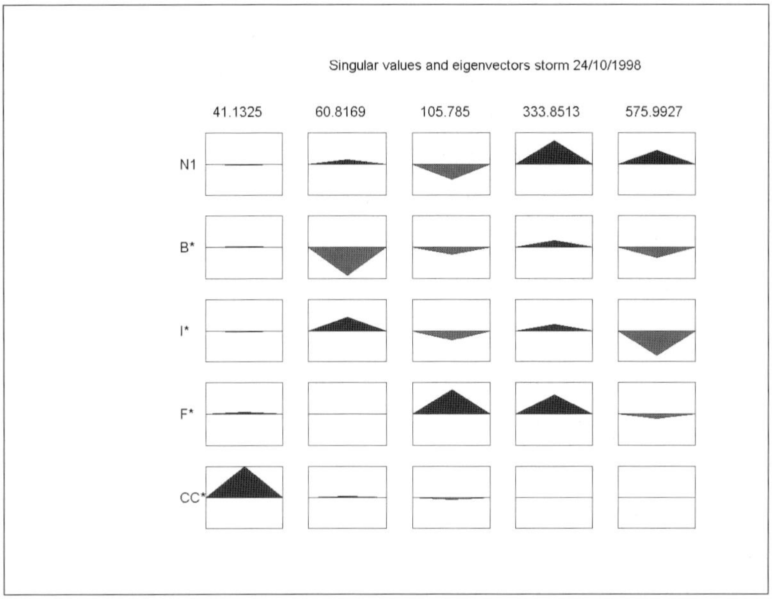

Figure VII.37: Singular values and eigen vectors reduced parameter set, storm October 24 1998.

Figure VII.38 through Figure VII.43 shows the calculation results and the residues. As can be seen at gauge 17 the initial conditions is incorrect, a possible explanation for this is given in the discussion of the storm on 7 October. As can be seen the model responds too quickly (see e.g. period between t=0 and t=250 minutes and the period between 600 and 1100 minutes) on the rain intensity. This might explain the fact that the parameter optimisation algorithm tries to compensate for this by sharply increasing the value for the linear reservoir constant resulting in an unrealistic high value.

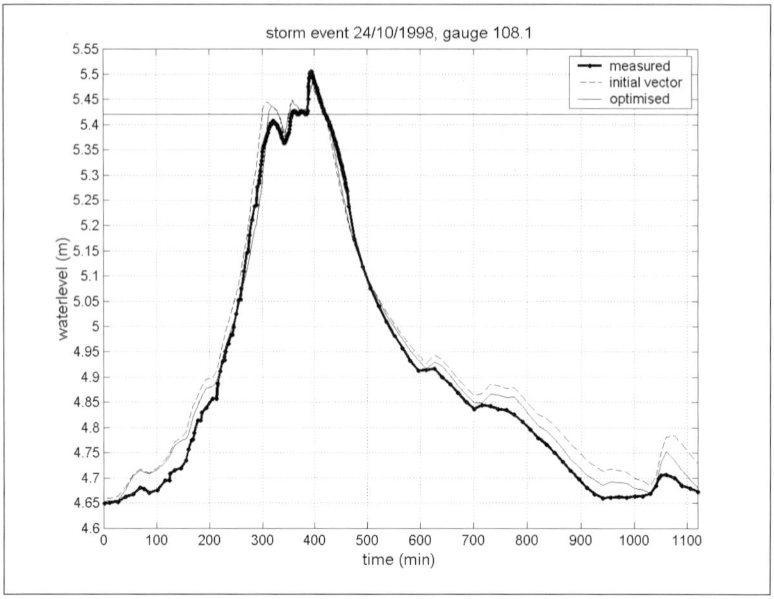

Figure VII.38

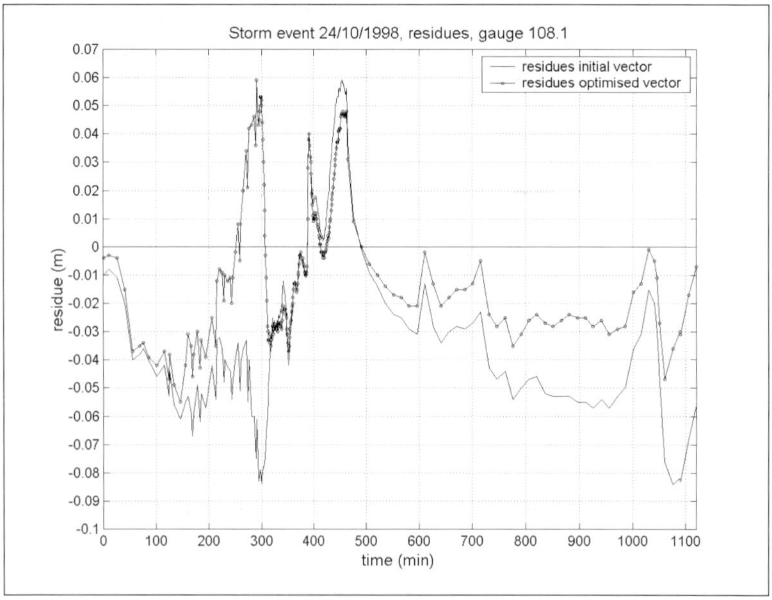

Figure VII.39

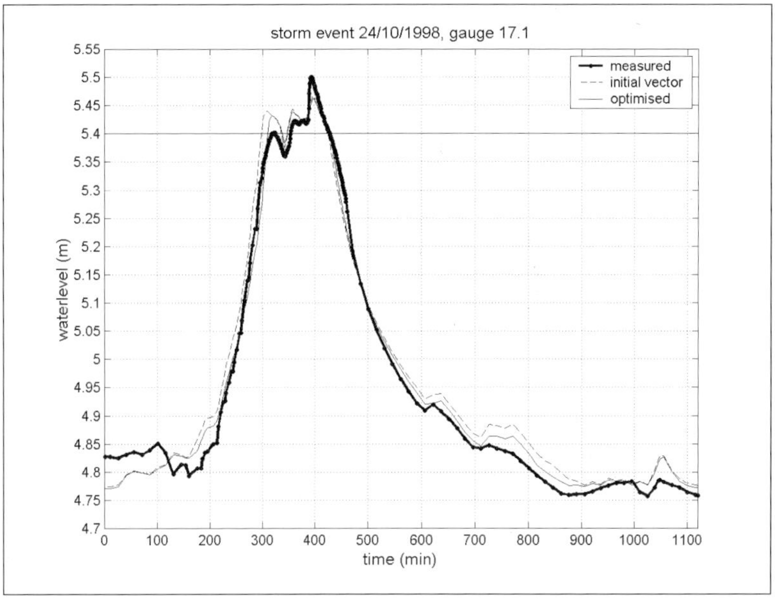

Figure VII.40

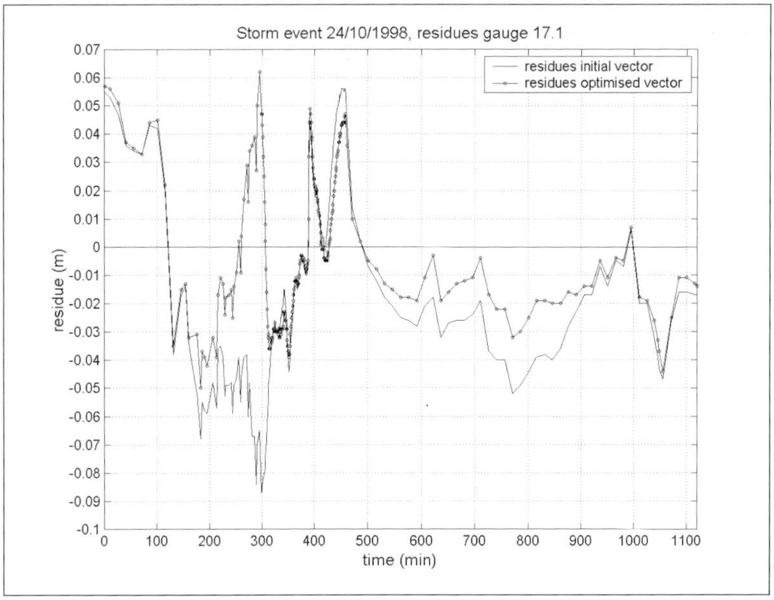

Figure VII.41

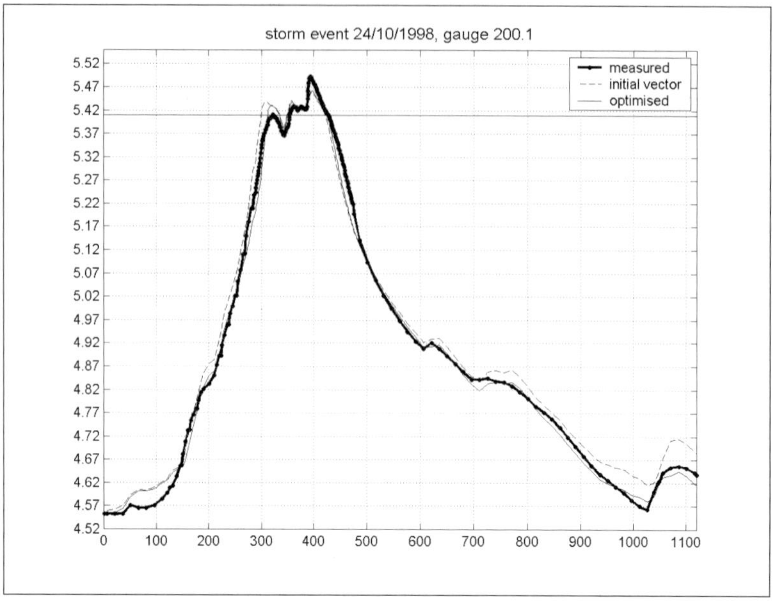

Figure VII.42

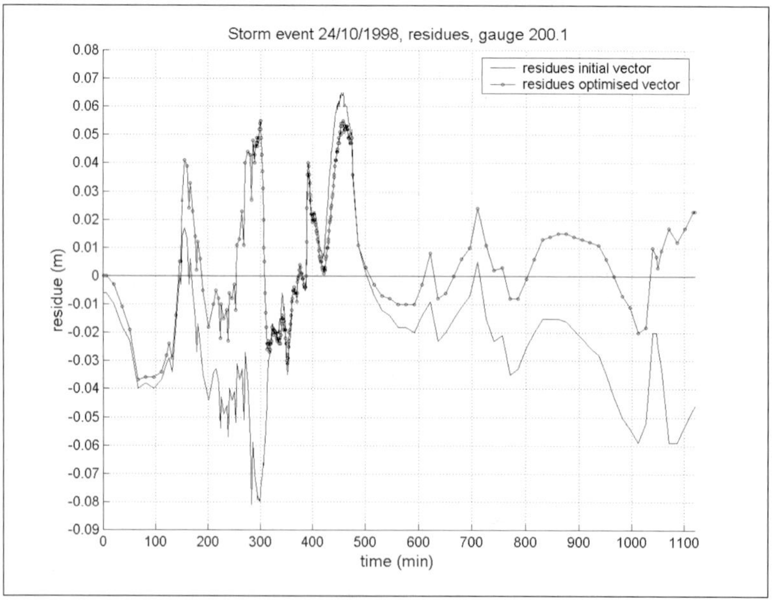

Figure VII.43

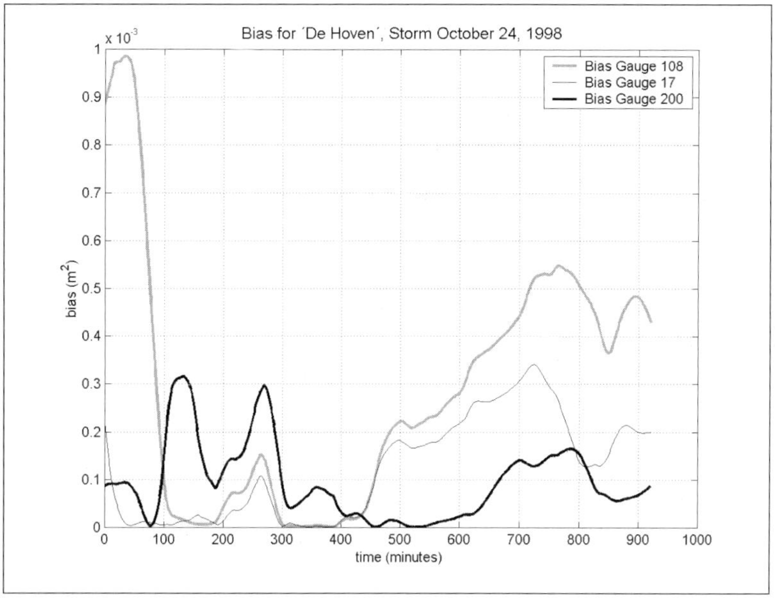

Figure VII.44

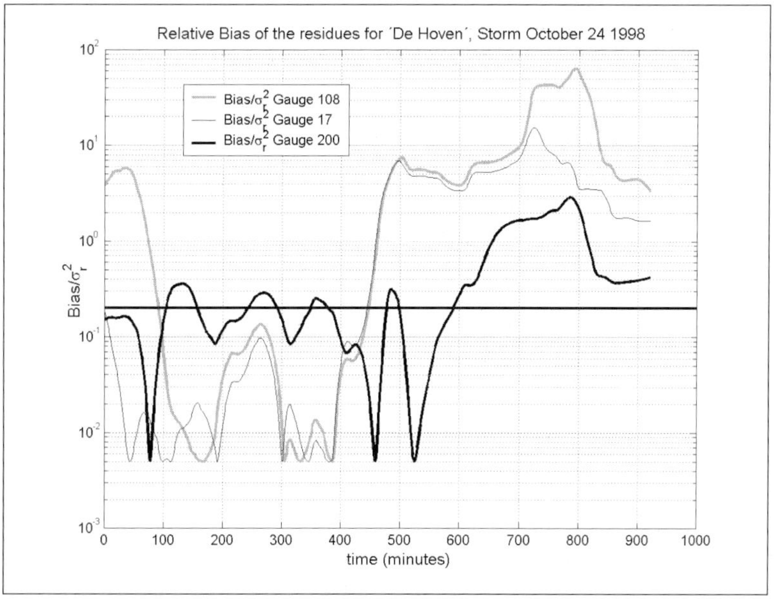

Figure VII.45

When studying the relative bias (see Figure VII.45), it is seen that in the period between circa 100 and 500 minutes bias is at an acceptable low level.

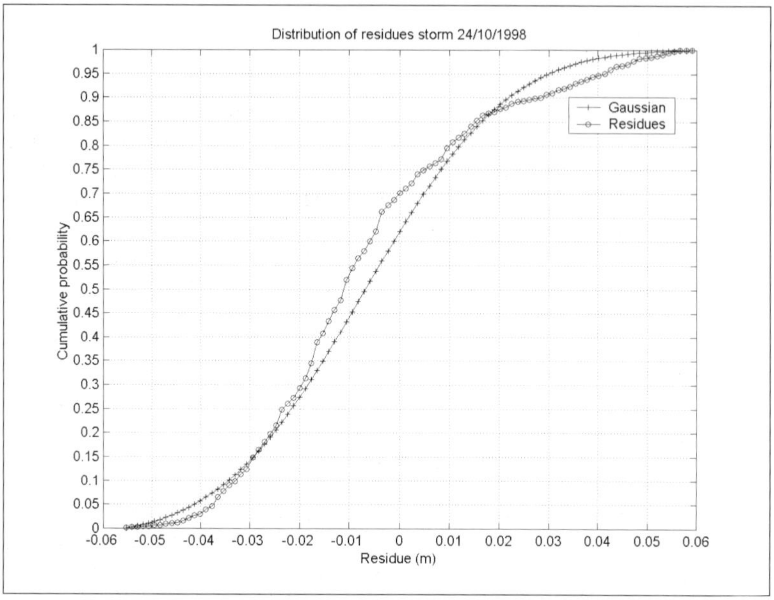

Figure VII.46

The distribution of the residues (Figure VII.46) shows that a noticeable deviation from the assumed Gaussian distribution is present.

CURRICULUM VITEA

François Henri Léon Raymond Clemens, geboren op 9 juni 1962 te Leiden, behaalde in 1980 het VWO diploma aan de RGS Petrus Hondius te Terneuzen. In datzelfde jaar begon hij de opleiding tot civiel ingenieur aan de toenmalige Technische Hogeschool Delft, Afdeling der Civiele Techniek. In 1986 is hij afgestudeerd op de wiskundige beschrijving van het dynamisch suspensietransport van sediment in een ronde buis en een deelonderwerp op het gebied van alternatieve rioleringssystemen. Tijdens zijn studie heeft hij onder meer de functie van secretaris van het Dispuut Gezondheidstechniek en Waterbeheersing vervuld.

Na zijn studie is hij in dienst getreden van Witteveen + Bos raadgevende ingenieurs te Deventer als projectingenieur riolering. In 1988 trad hij in dienst van DHV te Amersfoort als specialist riolering, om in 1996 terug te keren naar Witteveen + Bos te Deventer als afdelingshoofd riolering, waterhuishouding en bouwrijpmaken. Per 1 Augustus 2000 is hij benoemd als bijzonder hoogleraar riolering aan de TU Delft.

In zijn verschillende functies is gewerkt aan een groot aantal projecten waaronder het formuleren van de in Nederland algemeen gehanteerde hydraulische berekeningsmethode voor rioolstelsels neergelegd in de module 'Rioleringsberekeningen; hydraulisch functioneren' in de leidraad riolering en een tweetal rapporten voor de STOWA handelend over de opzet en uitvoering van meetprogramma's aan riolering en oppervlaktewaterkwaliteit. Daarnaast is een bijdrage geleverd aan de ontwikkeling van een tweetal vuiluitworp reducerende randvoorzieningen; het parallelbergbezinkriool en de zogenaamde RIOTUBE, waarvan de laatste is gepatenteerd in 1997. In opdracht van de waterkwaliteitsbeheerders van Noord-Brabant is een ontwerphandleiding voor randvoorzieningen opgesteld die landelijk wordt gehanteerd bij de beoordeling van ontwerpen van diverse soorten randvoorzieningen.

Naast werkzaamheden op het gebied van riolering en waterkwaliteit heeft hij ook gewerkt aan diverse automatiseringsprojecten: de ontwikkeling van een systeem voor rationeel oeverbeheer, de ontwikkeling van een 2-dimensionaal stromingsmodel op basis van de eindige elementen methode en een model voor de berekening van sedimentatie en sediment uitspoeling in een zandvang ten behoeve van een groot irrigatieproject aan de Kosi rivier in Nepal.

Binnen het raamwerk van de NVA, later de stichting wateropleidingen, de stichting PAO en de stichting GEOPLAN heeft hij cursussen verzorgd op het gebied van ontwerpen, hydrodynamisch rekenen, stank- en aantasting problemen van riolering, modellering en planvorming.
Daarnaast is begeleiding verzorgd van verschillende afstudeerders van de TU Delft, het IHE te Delft en diverse HBO-opleidingen in Nederland.

Als rode draad in de werkzaamheden kan het meten en rekenen aan de hydraulica van rioolstelsels worden aangemerkt, het voorliggende proefschrift is het resultaat van diverse werkzaamheden verricht op dit gebied in de periode tussen 1986 en 2000.

'Against the march of time
the child, now man stands alone
and demands just to be; to grow; to live
to love and to do the things that man
must do.
And yet in his soul he questions
and cries out for the simple ways of
yesterday'.

Soliloguy II, C. Simpson

Stellingen

behorende bij het proefschrift

Hydrodynamic Models in Urban Drainage: Application and Calibration.

F.H.L.R. Clemens

(Civiel ingenieur)

Bathmen:26-1-2001

I

'*Hoe slimmer de vis hoe duurder hij is.*' (Observatie van T.A. Clemens, 1995)

II

Het milieueffect van te stellen emissienormen en de vergaderingen daarover is verwaarloosbaar en naar alle waarschijnlijkheid zelfs negatief.

III

Het schilderij 'Le soir qui tombe' van René Magritte dient door elke gebruiker van modellen aandachtig bestudeerd te worden.

IV

Elk model bevat fouten; dit is inherent aan het feit dat de werkelijkheid vaak te gecompliceerd is om in elk detail te beschrijven. Het herkennen, onderkennen en zo mogelijk elimineren van deze fouten is een gezamenlijke verantwoordelijkheid van wetenschapbeoefenaars, software-programmeurs en gebruikers van modellen.

V

Het optreden van '$2\Delta x$' golven in numerieke schema's zoals bijvoorbeeld het Preissmann-schema is inherent aan deze schema's, de suggestie dat door kunstmatige vergroting van de weerstandsterm dit verschijnsel wordt vermeden is niet correct.(Dit proefschrift, hoofdstuk 2.)

VI

Het winnen van de ronde van Frankrijk na het overleven van een kwaadaardige aandoening is een, voor het grote publiek, aansprekender prestatie dan het voltooien van een proefschrift bij identieke initiële condities.

VII

De verificatie van een gecalibreerd model is een korte, confronterende, maar, bij nadere beschouwing, zeer leerzame ervaring voor een modelgebruiker.

VIII

Als het onderwijs aan de TU Delft zich teveel gaat richten op de maatschappelijke en beleidsmatige vaardigheden van aanstaande ingenieurs dan zal in de toekomst de faalkans van technische voortbrengselen dusdanig toenemen dat dit tot zowel maatschappelijk als beleidsmatig onaanvaardbare situaties leidt.

IX

Zowel de ontbossing als de stress bij mensen werkzaam in het onderwijs zullen drastisch afnemen als op het ministerie van onderwijs de helft van de daar nu werkzame beleidsambtenaren wordt ontslagen. Een zelfde effect is wellicht te bereiken door deze ambtenaren allen in dienst te houden en daarbij tevens het gebruik van moderne communicatiemiddelen te verbieden. Welk alternatief is te verkiezen dient nader te worden onderzocht.

X

Het gros van de Nederlandse wetgeving luidt samengevat: 'Alles is verboden, tenzij men vergunning heeft'.

XI

Het gebruik van de term 'optimaal' zonder daarbij een helder en kwantificeerbaar criterium te definiëren moet strafbaar worden gesteld.

XII

De bijna dwangmatige neiging om alles in richtlijnen of normen vast te leggen is wellicht niet uniek, maar wel tekenend voor het vakgebied riolering.

XIII

Na ampele overweging is de term 'riolist' te prefereren boven de term 'rioleur'; immers, men spreekt ook van een violist en niet van een violeur.

XIV

De verantwoordelijkheden van medici en gezondheidstechnici zijn niet wezenlijk anders; de praktische verschillen zitten in de schaalgrootte en de mate van persoonlijke betrokkenheid bij de individuen die afhankelijk zijn van het door beide beroepsgroepen tentoongespreide vakmanschap.